Alan

With the complements
of the author.
Chapter X would
not be the same . . . . .

[illegible]

1/11/01

DYNAMICS OF REGENERATIVE HEAT TRANSFER

Series in Computational and Physical Processes in Mechanics and Thermal Sciences

W. J. Minkowycz and E. M. Sparrow, *Editors*

Anderson, Tannehill, and Pletcher, Computational Fluid Mechanics and Heat Transfer

Aziz and Na, Perturbation Methods in Heat Transfer

Baker, Finite Element Computational Fluid Mechanics

Beck, Cole, Haji-Shiekh, and Litkouhi, Heat Conduction Using Green's Functions

Carey, Computational Grids: Generation, Adaptation, and Solution Strategies

Chung, Editor, Numerical Modeling in Combustion

Comini, Del Giudice, and Nonino, Finite Element Analysis in Heat Transfer: Basic Formulation and Linear Problems

Heinrich and Pepper, Intermediate Finite Element Method: Fluid Flow and Heat Transfer Applications

Jaluria, Computer Methods for Engineering

Jaluria and Torrance, Computational Heat Transfer

Koenig, Modern Computational Mechanics

Patankar, Numerical Heat Transfer and Fluid Flow

Pepper and Heinrich, The Finite Element Method: Basic Concepts and Application

Shih, Numerical Heat Transfer

Shyy, Udaykumar, Rao, and Smith, Computational Fluid Dynamics with Moving Boundaries

Tannehill, Pletcher, and Anderson, Computational Fluid Mechanics and Heat Transfer, Second Edition

DYNAMICS OF REGENERATIVE HEAT TRANSFER

A. John Willmott
Retired
University of York
United Kingdom

TAYLOR & FRANCIS
New York

Denise T. Schanck, *Vice President*
Robert H. Bedford, *Editor*
Catherine M. Caputo, *Assistant Editor*
James A. Wright, *Marketing Director*

Published in 2002 by
Taylor & Francis
29 West 35th Street
New York, NY 10001

Published in Great Britain by
Taylor & Francis
11 New Fetter Lane
London EC4P 4EE

Printed in the United States of America on acid-free paper

Library of Congress Cataloging-in-Publication Data
Willmott, A. John, 1934–
Dynamics of regenerative heat transfer / by A. John Willmott.
p. cm. – (Series in computational & physical processes in mechanics and thermal sciences)
Includes bibliographical references and index.
ISBN 1-56032-369-8 (alk. paper)
1. Heat—Transmission. 2. Thermal regenerators. I. Title. II. Series.
TJ260.W495 2001
621.402′2—dc21 00-050785

CONTENTS

PREFACE

This book is concerned with the mathematical modeling of the processes of regenerative heat transfer. It is directed at industrialists who wish to develop computer simulations for the sizing of conventional process equipment for new, possibly novel applications or in the invention of fresh designs of such equipment for entirely new applications. This area of modeling has advanced considerably over the last twenty years and it is expected that this area of research will not stagnate: it is hoped that this book might prove to be a reference as to the state of the art at the start of the twenty-first century for research workers both in the academic world and in industry.

The text aims to be comprehensive in that it covers the two main themes that should run through any work on regenerative heat transfer: namely, the solution of the descriptive differential equations, on the one hand, and the representation of thermal conductivity effects within the solid phase, on the other. Orthogonal to these two themes are several concerns: at the one extreme, the storage of heat in a solid packing consequent upon the flow of a fluid through the channels of such a packing is considered and, at the other, the operation of regenerative heat exchangers under operating conditions that vary with time is discussed. Between these two concerns lies a most important area, namely, operation at periodic steady state. Overlaying all of this is the development of models that permit the incorporation of the temperature dependence of the relevant physical properties of fluid and solid, as well as time-varying operating conditions. This proves important in relating model performance to practical measurements.

While this book started out as a revision of an earlier text, *Thermal Energy Storage and Regeneration* by Frank W. Schmidt and myself, published in 1981, the title signals that this must be regarded as fresh text embodying most of the results of research in this area that have appeared in the scientific literature since 1980. It is based, of course, on the transformation of the analysis of regenerative heat transfer made possible by the advent of the digital computer. I came to the study of regenerators in 1960, not as an engineer but as a mathematician and computer scientist. Wally Voice and Maurice Ridgion at the British Iron and Steel Research Association (BISRA), who were my head of division and head of laboratory respectively, concluded that a person with my background might contribute to breaking the back of the theory of regenerative heat transfer. Little did they know that this would be the topic of this book, published four decades later.

It would be discourteous not to thank Frank Schmidt, who, having retired, happily gave me a free hand to produce this fresh book that is offered here. I take this opportunity, therefore, to thank Frank for our transatlantic collaboration in bringing together some of this material for a continuing education course at The Pennsylvania State University in 1977, and subsequently in our book mentioned above. Some of the material in the 1981 book has been omitted from this text to make room for the products of new research.

This book has wide origins, in particular the worries about global warming due to the so-called greenhouse gases. These have overtaken and are potentially far more serious than the concerns of the 1970s over energy conservation following the oil crisis.

It has been a great pleasure to work with other people in this field, all of whom have made significant contributions. Professor Ian Pyle, Professor Ian Wand, Professor Keith Mander, and Professor Alan Burns, successive Heads of the Department of Computer Science at the University of York, all gave me considerable encouragement in the pursuit of my research. Dr Ron Thomas and Dr Roger Duggan discovered, in the early 1970s, the inherent weaknesses in the integral equation methods of solving the relevant equations, as they then stood. This was to point to the important work of Brane Baclic and others in this area in which efforts were made to overcome these weaknesses. Professor Alan Burns and Dr Clare Simmons (née Hinchcliffe) later in that decade each made important advances in the understanding of regenerator operation. Alan made large strides toward the understanding of the transient performance of these regenerative heat exchangers. The early work of David Green in this area cannot be forgotten. Clare's main contribution, in my view, was the development of her representation of the inversion of the solid temperature profile at regenerator reversals, although she did important work also on the effect of gas that remains in the channels of a regenerator immediately after a reversal. Ron Thomas had worked in this latter area also.

In the 1980s, Dr Andrew Hill worked with me (or perhaps I should say that I worked with him) on the development of a closed method that could handle nonlinear problems. Dr Sue Brooks looked at the effect of radiative heat transfer in particular. Other D.Phil. students of that period were Dr Manus Henry, who worked on the optimal design of regenerative heat exchangers, and Dr Mark Penney, who examined the effect of the maldistribution of flow of the gases upon thermal regenerator performance.

In 1991, I was fortunate enough to secure a SERC Senior Research Fellowship and a Fitzwilliam College Fellowship, which enabled me to work for five months with Dr David Scott at the Department of Chemical Engineering at the University of Cambridge, UK. David suggested to me a matrix method for handling the transient response of regenerators that was to be the subject of a number of papers subsequently. This method provided a means of unifying regenerator theory that had so long eluded previous workers in this field. Finally, before I retired in 1998, I had the pleasure to work with Dr David Evans, ostensibly on the nonlinear modeling of regenerators. In addition, David facilitated the tidying up of a number of ends that I had left loose over the years.

To all these people and the others I have not mentioned, let me take this opportunity to say thank you for the opportunity to work with them. Let me express my appreciation of the ongoing help and encouragement of Dr Bert Wraith of the University of Newcastle upon Tyne, who has been a patient mentor for almost 40 years. Help, in no small measure, has been forthcoming from the Software Support Group of the Department of Computer Science, University of York, and I am especially grateful to John Murdie who has shown much kindness. My wife, Joan, has exhibited patience beyond the call of duty during my preparation of this text: thank you.

I should like to dedicate this book to Eleanor Beatrice Penny, my granddaughter, who was born about two years before this book was published.

John Willmott

NOMENCLATURE

A	available heating surface area	m^2
Bi	Biot modulus	—
c_p	specific heat of fluid/gas	kJ/(kg K)
C_s	specific heat of packing	kJ/(kg K)
f	fraction of cross-sectional area of a rotary regenerator	—
$F(\xi)$	spatial dimensionless temperature distribution at start of period of regenerator operation	—
i	defined by $i^2 = -1$	—
I	unit (identity) matrix	—
j	Colburn factor	—
H	enthalpy	kJ/kg
J	Bessel function	—
$K(x)$	function defined by equation (5.15)	—
L	length of the packing	m
L_0	beam length	m
M_s, M	mass of the packing	kg
m_f	mass of gas resident in the packing voids	kg
$\dot{m}_f$	mass flow rate of the gas	kg/s
Nu	Nusselt number	—
P	duration of period	s
Pr	Prandtl number	—
Q	heat	kJ
R	"thickness" of wall of hollow cylinder	m
Re	Reynolds number	—
r	radial distance in cylinder or sphere	m
St	Stanton number	—

t	temperature	K
T	dimensionless temperature	—
U	utilization factor	—
U^+	curvature of hollow cylinder	—
v	gas velocity	m/s
w	semithickness of the packing also used as $(z/2)^2$	m
X	dimensionless distance from solid surface	—
x	distance from solid surface	m
y	distance from the gas entrance	m
z	dimensionless distance (x/w)	—
z	dimensionless radial distance in cylinder	—
Z_n	pseudothermal ratio after nth cycle	—

Greek symbols

α	surface heat transfer coefficient	kW/(m^2 K)
α_f	absorptivity of gas	—
$\bar{\alpha}$	lumped heat transfer coefficient	kW/(m^2 K)
β	imbalance factor	—
δ	characteristic width of channel	m
δ	temperature difference between models	K
Δ	characteristic semithickness of packing (equation (2.59))	m
ε	effectiveness $= Q_{act}/Q_{max}$	—
ε_s	emissivity of solid surface	—
ε_f	emissivity of fluid	—
$\varepsilon f_1, \varepsilon f_2$	London's transient responses	—
ε	sometimes used as a dummy variable	—
ξ	dimensionless distance	—
$\bar{\xi}$	dimensionless distance embodying $\bar{\alpha}$	—
$\Delta\xi$	dimensionless step length	—
γ	axial conduction parameter (6.95)	—
γL	Hahnemann's axial conduction factor	—
Γ	zero of function defined by equation (6.56)	—
η	dimensionless time within a period	—
$\bar{\eta}$	dimensionless time within a period embodying $\bar{\alpha}$	—
$\Delta\eta$	dimensionless step length in time	—
η_f	dynamic viscosity	kg/(m s)

η_{REG}	thermal ratio	—
κ	thermal diffusivity	m^2/s
κ	dimensionless time constant (9.32)	—
λ	thermal conductivity	kW/(m K)
Λ	reduced length	—
$\bar{\Lambda}$	reduced length embodying $\bar{\alpha}$	—
ν_f	kinematic viscosity	m^2/s
Π	reduced period	—
$\bar{\Pi}$	reduced period embodying $\bar{\alpha}$	—
Ψ	measure of discrepancy between 2-D and 3-D models	—
ρ	density	kg/m^3
σ	Stefan–Boltzmann constant	—
σ	Λ_1/Λ_2	—
Φ	Hausen phi-factor	—
Φ_H	Hahnemann's heat resistance parameter	—
τ	time	s
Θ	dimensionless time to periodic steady state	—
ω	dimensionless time	—
$\Delta\omega$	step in dimensionless time	—
Ω	dimensionless duration of period	—

Superscripts

prime (′)	refers to the cold period
double prime (″)	refers to the third sector of a rotary regenerator
H	refers to harmonic mean

Subscripts

1	refers to the weaker period
2	refers to the stronger period
act	actual
av	average
f	refers to fluid/gas
f,in	refers to fluid/gas at the entrance
f,x	refers to fluid/gas at the exit
H	refers to Hausen model
H	refers to harmonic mean

i	refers to radius of inner surface of hollow cylinder
max	maximum
min	minimum
o	refers to radius of inner surface of hollow cylinder
s	refers to solid packing
s,x	refers to solid at the exit
s,o	refers to surface of solid packing
s,O	refers to solid packing, at the start
S	refers to Schumann model
w	refers to wetted surface of hollow cylinder

Chapter 1

DYNAMICS OF REGENERATIVE HEAT TRANSFER

AN INTRODUCTION

1.1 Introduction

This book represents a development of an earlier text by Schmidt and Willmott [1] published in 1981. That book emerged out of the perceived interest in thermal energy storage and heat recovery following the oil crisis of the mid-1970s when the price of crude oil quadrupled. Since those days, although the price of oil has dropped (until a recent rise, which may or may not be maintained), the interest in waste heat recovery has been sustained by international concerns about air pollution and, in particular, the effect of CO_2 and other greenhouse gases upon global warming.

Simultaneous improvements in computing power – many personal computers that can be used for scientific calculations are one or two orders of magnitude faster and larger than most of the computers available to scientists and engineers in the 1970s – have made earlier concerns less relevant today. Modern computers enable the demands of engineers for the use of more sophisticated models of the processes of heat transfer involved in waste heat recovery equipment to be satisfied. A notable example is the use of nonlinear models described in Chapter 8. It has been found that better calculated estimates of the temperature performance of thermal regenerators can be obtained if the temperature dependence of the heat transfer coefficients and of the gas specific heats is embodied in the model. The inevitable increase in the demand for computing power arising as a consequence has been accommodated easily.

Since 1981, there have been notable developments in the modeling of regenerative heat exchangers, especially in the area of integral equation methods. These developments are outlined in Chapter 7 and the reader is referred, in addition, to the book by Dragutinovic and Baclic [2] published recently and devoted almost entirely to this topic.

1.2 Underlying Problems

The basic problem discussed in this text is the modeling of the exchange of heat between fluid and solid as the fluid passes through the channels and over the heating surface area of the solid. In its simplest form, this is the *single-blow problem*. In this, it is assumed that an initially isothermal solid is subject to heating/cooling by a fluid, usually a gas, passing through the interparticle channels. The entrance gas temperature is treated as not varying with time and not equal to the initial temperature of the solid. The single blow is assumed to continue for ever, in practice until the solid has become pretty well isothermal again at a temperature equal to that of the incoming gas. The problem is *dynamic* in the sense that its solution consists of the time variations of the solid and the gas down the length of the bed of solid material. This problem and its solution are described in Chapter 2. During a single blow, heat is extracted from or stored in the solid by the passage of the gas.

The somewhat more complicated problem is the determination of the spatial and time variations of the temperature of the solid and the gases for a *thermal regenerator* or *regenerative heat exchanger*. Here, for the contraflow operation of regenerators, the two gases between which thermal energy is exchanged, pass alternately in opposite directions through the channels and thus, over the surface of the solid packing. Heat is recovered from the hotter of the two gases and stored in the packing in the *hot period*. In the subsequent *cold period*, this heat is regenerated from the solid by the colder of the gases, which passes over the same heating surface area. This is the basis of the process of *regenerative heat transfer* between the gases. This can be contrasted with *recuperative* heat transfer, where heat is transferred directly through a partition wall located between the passages for the two gases, which flow simultaneously rather than alternately through the heat exchanger.

The second basic problem also discussed in this text is how the effect of the thermal conductivity of the solid packing can be represented in the model. Dragutinovic and Baclic [2] seem to imply that the major problems associated with simulating regenerative heat exchangers, as mentioned by Jakob [3], are those associated with the integral equation methods for solving the relevant equations. The present author considers that this is only partly the case but that more important are those difficulties associated with the development of reliable lumped heat transfer coefficients. In these coefficients the resistance to heat transfer at the solid surface is combined with the effect of the finite thermal conductivity of the solid in a direction perpendicular to gas flow. This problem is addressed in Chapter 3 as it relates to the single-blow problem and in Chapter 6 for thermal regenerator calculations. Jakob must have been concerned about the scale of the calculations necessary in this area. At the time when Jakob's book was published, computers were not widely used, if at all. Such computers were probably not available to Jakob in the late 1940s and early 1950s: Jakob died some while before his book was published. The idea of solving the Fourier equation within the full 3-D model of a regenerator could only be imagined. Indeed, this full model and the solution of the relevant equations was first published only some twelve years later in 1969 [4]. This model is discussed in detail in Chapter 6. Nowadays, modern computer power in some ways eliminates some of those earlier concerns to be replaced, needless to say, by new problems revealed by the capability to undertake the most complex and extensive of calculations.

The basic concepts, in particular, of the *cycle of operation* and *periodic steady state* for regenerative heat exchanger operation are presented in Chapter 4. At the same time, the differences between fixed bed and rotary regenerators are explained. Descriptions of a variety of different thermal regenerators are offered.

The mathematical models describing the single-blow problem and the operations of a regenerator are the same: each period of regenerator operation can be regarded as an incomplete single blow. The development of such models for regenerators is outlined in Chapters 4 and 5. A key issue is addressed in these chapters and emphasized elsewhere in this book, namely that of the assumptions made in building up a mathematical model of the equipment used for the exploitation of regenerative heat transfer. In particular, the difference between *linear* and *nonlinear* models is explained, starting with an introduction to a linear model with the possible dimensionless parameters for temperature, length, and time. This appears initially in Chapter 4 for regenerators and is developed as the text proceeds. A thorough discussion of nonlinear models is presented in Chapter 8. In nonlinear models, the temperature dependence of any of the thermophysical properties can be accommodated, for example.

The physical details of particular packings for different configurations of regenerators, applicable also to single-blow beds, are described. These details are set in the context of the operating conditions required for efficient counterflow regenerator operation, best understood in terms of the dimensionless parameters that emerge from the linear model. Briefly, counterflow regenerators with a low heating surface area to mass ratio can be operated with long cycle times as long as 1 hr, or possibly even longer, as, for example, in Cowper stoves used for iron making. Regenerative burners must be reversed every 30–180 s because their ceramic sphere packing has a relatively high area to mass ratio. The situation can be even more acute in the case of rotary regenerators. On the other hand, stoves are massive and very expensive, whereas regenerative burners and rotary regenerators can be considerably smaller.

There is a whole raft of methods that have been published over the last 80 years or so for the solution of the differential equations describing the process of regenerative heat transfer. The author has elected, therefore, to introduce an *open* method (in which the model of the regenerator is cycled to equilibrium) and a *closed* method (where periodic steady state is computed directly) of regenerator calculation, just to enable the reader to grasp the essential principles without, hopefully, being overwhelmed.

On the other hand, a thorough discussion of integral equation closed methods is presented in Chapter 7 in which important issues like computational stability, or lack of it, and the means whereby nonsymmetric models can be handled efficiently, are raised. The state of the art in this area is presented: this has been advanced since the publication of the texts by Schmidt and Willmott [1] and by Hausen [5]. Another method of the closed variety, originally devised by Razelos [6] is described in Chapter 8 because it is amenable to extension to deal with nonlinear problems.

What are not covered in this book are the early methods based on the idea that a regenerator can be approximated by an equivalent recuperator. The reader is referred to the text by Hausen [5] where these methods are discussed in detail. A historical overview of some of these techniques is presented by Willmott [7].

The thermal capacity of the packing, a fundamental requirement of regenerative heat transfer, means that a regenerative heat exchanger does not respond immediately to any step change in operation, for example, in hot-side inlet temperature, but instead a number of *transient cycles* must be endured until periodic steady state is restored. A key development in this area has been a matrix theory whereby the transient response is described mathematically in terms of the eigenvalues of an appropriate matrix. This development is described in Chapter 9 where it is noted how earlier work on the transient behavior of regenerators can be explained in terms of this matrix theory. At the same time, this theory enables the open and closed methods for solving the equations that model counterflow regenerators to be unified in a most interesting manner. This provides a unity in the methodology not previously apparent.

This text focuses in general upon the counterflow of the gases in regenerative heat transfer. This is the most common and proves to be the most efficient. Nevertheless, Chapter 10 briefly covers parallel-flow regenerators in which the gases pass through the exchanger in the same direction. Although this is seemingly a far simpler method of operating a regenerator, it turns out that parallel-flow operation is less efficient and far more complicated than is the case for counterflow.

This is a suitable juncture at which to take an overview of the material in this book. Some of the material offered here is new and represents some of the developments to regenerative heat transfer realized since the books of Schmidt and Willmott [1] and of Hausen [5] were published about 20 years ago. The remainder is that which, in many respects, provides the foundation for these later developments.

It is speculated that the work on nonlinear models will probably be extended and advanced in the years to come. This is inevitable as experimental work, so vital in this area, is carried out and better correlations between theory and practice are sought.

What is less certain is the future of the extensive work on integral equations. Baclic and Dragutinovic [8] rightly refer to the "specialists and connoisseurs of the state of [the] open literature" in this area. It could well be that this work will greatly facilitate the mathematical understanding of other modeling problems involving integral equations. On the other hand, if these integral equations can be extended to nonlinear problems – a formidable task – economic methods might be evolved for optimization procedures where many simulations of thermal regenerators must be undertaken as optimal designs are sought. The extension of these integral equation methods to cope with axial thermal conduction in the solid packing would represent a significant advance. Both such developments could prove to be invaluable.

In the years to come, the author considers that the importance of the work done on the *transient performance* of regenerators will become more widely realized. Regenerators frequently are required to operate in time-varying conditions, rarely at steady state. If bounds can be placed on the variation of any operating conditions, then the corresponding bounds should be able to be calculated on the consequent variation in temperature performance. Equally important will be the relationship between the size of the regenerative heat exchanger and the possible damping out of *relatively* high-frequency changes in operation. There is a good deal of mileage to be made in this area which could well be of practical importance.

1.3 Concluding Remarks

The response of the oil-consuming nations to the oil crisis of 1973 was first to realize their dependence upon the world's oil producers and, secondly, to reduce their demand for oil by, in part, exploiting waste heat recovery methodology using, again in part, regenerative heat transfer technology. It seems that, for the time being, the oil-producing nations are cautious about excessive increases in the price of oil for fear of forcing the industrial nations to make themselves permanently less dependent on oil. In the longer term, as the earth's stocks of oil are depleted and, equally important, as steps are taken to stem the greenhouse effect, the use of heat exchangers for thermal energy saving will widen and perhaps new forms of the thermal regenerator will be invented to meet new requirements.

References

1. F. W. Schmidt, A. J. Willmott, *Thermal Energy Storage and Regeneration*, McGraw-Hill, New York (1981).
2. G. D. Dragutinovic, B. S. Baclic, *Operation of Counterflow Regenerators*, WIT Press, London (Computational Mechanics Publications) (1998).
3. M Jakob, *Heat Transfer*, Vol. 2, John Wiley, New York (1957).
4. A. J. Willmott, "The Regenerative Heat Exchanger Computer Representation," *Int. J. Heat Mass Transfer* **12**, 997–1014 (1969).
5. H. Hausen, *Heat Transfer in Counterflow, Parallel Flow and Crossflow*, English translation (1983) edited by A. J. Willmott, McGraw-Hill, New York (originally published 1976).
6. P. Razelos, "An Analytic Solution to the Electric Analog Simulation of the Regenerative Heat Exchanger with Time-Varying Fluid Inlet Temperatures," *Warme- und Stoffubertragung* **12**, 59–71 (1979).
7. A. J. Willmott, "The Development of Thermal Regenerator Theory from 1931 to the Present," *J. Inst. Energy* **66**, 54–70 (June 1993).
8. B. S. Baclic, G. D. Dragutinovic, private communication, 1997.

Chapter 2

THE STORAGE OF HEAT IN A PACKING

THE SINGLE-BLOW PROBLEM

2.1 Introduction

The storage of thermal energy in, or extraction from, a solid material is commonly realized by the passage of a fluid, usually a gas, over the surface of the material. At this juncture, it is unnecessary to specify the geometrical configuration of the storage material except to mention that discrete passages in the material must be available for the flow of the gas, that is, the material must be *porous* in some way. Typically, the heat storing unit has rectangular or circular flow passages or consists of a bed of relatively small particles, when it is called a *packed bed*.

From a theoretical point of view, the simplest conditions arise when the solid material is initially isothermal and the gas is at the same temperature. The gas is then subject to a single step change in temperature and the flow rate of the gas remains unaltered. The storage unit is said to be operating under *single-blow* conditions. An analysis of this problem is offered shortly. A practical application of this might be a test unit for electronic components that are required to be subject to a thermal shock as part of a trial of such components. Heat is stored slowly in a packed bed, for example, perhaps by means of electrical heating until the bed has become isothermal at the required temperature. The components are then exposed to a stream of air at a high mass flow rate for a short period when the thermal shock is imposed. The thermal energy is extracted quickly from the packed bed as the air is heated from ambient by passing it through the bed. Such a unit is sometimes called a blowdown-type wind tunnel.

In general, the inlet temperature and the flow rate of the fluid may remain constant after an initial disturbance, may experience arbitrary timewise changes, or may vary in a period fashion. Units subject to arbitrary time changes in temperature can be treated as a succession of single blows that can be superimposed provided the model used is linear, that is, no allowance is made for the temperature dependence of the thermophysical properties of the packing and the fluid. Units subject to operating conditions that vary periodically are called *regenerators*. These will be dealt with in detail in later chapters.

We begin by focusing upon the single-blow problem, a first solution to which was published in 1926. It has been subject to a good deal of analysis and the present author has elected to present the analysis afresh. Means were sought by several authors to simplify this analysis so that the required temperatures could be evaluated quickly, even without the aid of computing machines. This is no longer necessary because the integrals needed can be computed quite quickly with modern, fast, desktop computers.

2.2 The Single-Blow Problem

This problem is concerned with a porous bed of solid material whose temperature is initially uniform. Through this bed is blown a gas whose entry temperature is held constant but not equal to the initial solid temperature. Should this flow of gas through the bed continue long enough, the temperature of the packed bed will become uniform again and equal to the inlet temperature of the gas. The *single-blow problem* addresses the issue of the calculation of the spatial and chronological variations of the gas and solid temperatures during the time until the packed bed temperature becomes uniform once more.

This calculation was first attempted in 1926 by Anzelius [1] in Sweden and independently, shortly thereafter, by Schumann [2] in the USA. The model that both employed was the same. Its linear form was based on the simplifying assumptions set out below.

1. The thermophysical properties of the gas and solid are constant and are not regarded as temperature dependent.
2. The entrance gas temperature and the flow rate of the gas do not vary with time.
3. The cross-sectional area of the bed does not vary down its length. Further, the flow of gas, as also the size of the solid particles in the

bed and their interspacing, is uniform in the cross-section of the packed bed at all positions down its length. The model ignores any possible maldistribution of flow that might arise in practice.
4. The longitudinal thermal conductivities of the gas and the solid, that is, in the direction of gas flow, are ignored.
5. Both Anzelius and Schumann considered that the latitudinal thermal conductivity of the solid was infinitely large, that is, in any direction perpendicualr to gas flow. The implication of this assumption is that the resistance to heat transfer lies solely at the surface of the solid (the interface between gas and solid) and is represented solely by the surface heat transfer coefficient.

The differential equations that describe heat transfer in a packed bed can be written as

$$\frac{\partial t_f}{\partial y} = \frac{\alpha A}{\dot{m}_f c_p L}(t_s - t_f) - \frac{m_f}{\dot{m}_f L}\frac{\partial t_f}{\partial \tau} \quad (2.1)$$

$$\frac{\partial t_s}{\partial \tau} = \frac{\alpha A}{M C_s}(t_f - t_s) \quad (2.2)$$

Equation (2.1) represents the gain/loss of heat by the gas across the heating surface of the packing, as the gas passes through the packed bed. On the other hand, equation (2.2) represents the loss/gain of heat by the solid packing as thermal energy is transferred through the surface of the packing to/from the gas.

The equations are simplified by the introduction of dimensionless parameters. The dimensionless temperature, T, is defined by

$$T = \frac{t - t_{s,0}}{t_{f,in} - t_{s,0}} \quad (2.3)$$

where $t_{s,0}$ is the initial temperature of the packing and where $t_{f,in}$ is the unchanging temperature of the gas as it enters the packed bed. Note that T and t in equation (2.3) can refer to either gas or solid temperatures.

From equation (2.3) it follows that

$$T_{s,0} = 0 \quad \text{and} \quad T_{f,in} = 1$$

on this dimensionless temperature scale.

On the basis of the first assumption given above, it is possible to define the parameters ξ and η in the following way:

$$\xi = \frac{\alpha A}{\dot{m}_f c_p L} y \tag{2.4}$$

$$\eta = \frac{\alpha A}{M C_s}\left(\tau - \frac{m_f y}{\dot{m}_f L}\right) \tag{2.5}$$

The parameter ξ can be viewed as dimensionless distance down the bed from the entrance of the gas and the dimensionless length or *reduced length*, Λ, of the packed bed is given by setting $y = L$ in equation (2.4), in which case

$$\Lambda = \frac{\alpha A}{\dot{m}_f c_p} \tag{2.6}$$

The parameter η can be regarded as dimensionless time from the start of the heating up of the bed.

Equations (2.1) and (2.2) now take the simplified form†

$$\frac{\partial T_f}{\partial \xi} = T_s - T_f \tag{2.7}$$

$$\frac{\partial T_s}{\partial \eta} = T_f - T_s \tag{2.8}$$

which constitutes the so-called *Schumann model*. When $\xi = 0$ at the entrance to the bed, then

$$T_{s,in}(\eta) = 1 - e^{-\eta} \tag{2.9}$$

---

†Equation (2.7) can be deduced in the following way. We note that

$$\frac{\partial T_f}{\partial y} = \frac{\partial T_f}{\partial \xi}\frac{\partial \xi}{\partial y} + \frac{\partial T_f}{\partial \eta}\frac{\partial \eta}{\partial y} = \frac{\partial T_f}{\partial \xi}\frac{\alpha A}{\dot{m}_f c_p L} - \frac{\alpha A}{M C_s}\frac{m_f}{\dot{m}_f L}\frac{\partial T_f}{\partial \eta}$$

From this, it follows that

$$\frac{\partial T_f}{\partial \xi} = T_s - T_f + \frac{\partial T_f}{\partial \eta}\frac{m_f c_p}{M C_s} - \frac{m_f c_p}{\alpha A}\frac{\partial T_f}{\partial \tau} \tag{A}$$

Now

$$\frac{\partial T_f}{\partial \tau} = \frac{\partial T_f}{\partial \eta}\frac{\partial \eta}{\partial \tau} = \frac{\partial T_f}{\partial \eta}\frac{\alpha A}{M C_s}$$

From this and equation (A), it follows that

$$\frac{\partial T_f}{\partial \xi} = T_s - T_f + \frac{\partial T_f}{\partial \eta}\left(\frac{m_f c_p}{M C_s} - \frac{m_f c_p}{M C_s}\right)$$

and therefore

$$\frac{\partial T_f}{\partial \xi} = T_s - T_f$$

Jakob [3] introduces two dimensionless variables, Φ and Ψ by

$$T_f = (\Phi + \Psi)e^{-\xi-\eta} \tag{2.10}$$

$$T_s = (\Phi - \Psi)e^{-\xi-\eta} \tag{2.11}$$

Upon substitution into equations (2.7) and (2.8) we obtain

$$\frac{\partial\Phi}{\partial\xi} + \frac{\partial\Psi}{\partial\xi} = \Phi - \Psi \tag{2.12}$$

$$\frac{\partial\Phi}{\partial\eta} - \frac{\partial\Psi}{\partial\eta} = \Phi + \Psi \tag{2.13}$$

Differentiating equation (2.12) with respect to η and differentiating equation (2.13) with respect to ξ we obtain, after then substituting from the same equations,

$$\frac{\partial^2\Phi}{\partial\xi\partial\eta} + \frac{\partial^2\Psi}{\partial\xi\partial\eta} = \frac{\partial\Phi}{\partial\eta} - \frac{\partial\Phi}{\partial\eta} = \Phi + \Psi \tag{2.14}$$

$$\frac{\partial^2\Phi}{\partial\xi\partial\eta} - \frac{\partial^2\Psi}{\partial\xi\partial\eta} = \frac{\partial\Phi}{\partial\xi} + \frac{\partial\Phi}{\partial\xi} = \Phi - \Psi \tag{2.15}$$

Subtracting equation (2.15) from equation (2.14) yields the same equation as was suggested by Anzelius [1],

$$\frac{\partial^2\Psi}{\partial\xi\partial\eta} = \Psi \tag{2.16}$$

Jakob sets $\zeta^2 = -4\xi\eta$ and reduces equation (2.16) to the Bessel differential equation:

$$\frac{d^2\Psi}{d\zeta^2} + \frac{1}{\zeta}\frac{d\Psi}{d\zeta} + \Psi = 0 \tag{2.17}$$

It is necessary to consider the boundary conditions for Φ and Ψ. When $\xi = 0$, then

$$T_{f,in} = 1$$

from which it follows from equation (2.10) that, at $\xi = 0$,

$$(\Phi + \Psi)e^{-\eta} = 1 \tag{2.18}$$

Recall that at the entrance to the bed,

$$T_{s,in}(\eta) = 1 - e^{-\eta} \tag{2.9}$$

in which case, at $\xi = 0$, substituting into equation (2.11),

$$1 - e^{-\eta} = (\Phi - \Psi)e^{-\eta} \tag{2.19}$$

Upon adding equations (2.18) and (2.19), we obtain

$$2\Phi e^{-\eta} = 2 - e^{-\eta} \quad \text{which yields} \quad \Phi = e^{\eta} - \tfrac{1}{2} \tag{2.20}$$

Upon substituting (2.20) into equation (2.18), we obtain for $\xi = 0$

$$\Psi = \tfrac{1}{2} \tag{2.21}$$

We now consider the boundary condition where $\eta = 0$. We recall that

$$T_s = (\Phi - \Psi)e^{-\xi-\eta} \tag{2.11}$$

and note that at $\eta = 0$, $T_s = 0$, from which it follows that

$$e^{-\xi}(\Phi - \Psi) = 0$$

which implies that, because $e^{-\xi} \neq 0$ for finite values of ξ,

$$\Phi - \Psi = 0 \tag{2.22}$$

It also follows that the differential equation

$$\frac{\partial T_f}{\partial \xi} = T_s - T_f \tag{2.7}$$

takes the form at $\eta = 0$

$$\frac{\partial T_f}{\partial \xi} = -T_f \tag{2.23}$$

It can easily be seen that

$$\log_e T_f = -\xi + \log_e T_{f,in} \quad \text{and that} \quad \frac{T_f}{T_{f,in}} = e^{-\xi} \tag{2.24}$$

where $T_{f,in} = 1$ on the dimensionless temperature scale. From this we can deduce that

$$T_f = e^{-\xi} \quad \text{at} \quad \eta = 0 \tag{2.25}$$

Substituting into equation (2.10) for $\eta = 0$, we obtain

$$e^{-\xi} = (\Phi + \Psi)e^{-\xi} \quad \text{or} \quad \Phi + \Psi = 1 \tag{2.26}$$

Upon solving equations (2.22) and (2.26), we deduce that, at $\eta = 0$,

$$\Phi = \Psi = \tfrac{1}{2} \tag{2.27}$$

McLachlan [4] shows that the general solution to Bessel's equation

$$\frac{d^2\Psi}{d\zeta^2} + \frac{1}{\zeta}\frac{d\Psi}{d\zeta} + \Psi = 0 \tag{2.17}$$

takes the form

$$\Psi(\zeta) = K_1 J_0(\zeta) + K_2 Y_0(\zeta) \tag{2.28}$$

where $J_0(\zeta)$ and $Y_0(\zeta)$ are the zero-order Bessel functions of the *first* and *second* kind respectively. Application of the boundary conditions given by equations (2.21) and (2.27) requires $K_2 = 0$ (because Ψ is finite in size at $\zeta = 0$, whereas $Y_0(\zeta) \to -\infty$ as $\zeta \to 0$) and that

$$K_1 = \tfrac{1}{2}$$

This yields a final solution of

$$\Psi = \tfrac{1}{2} J_0\left(2i\sqrt{\xi\eta}\right) \quad \text{where} \quad i^2 = -1 \tag{2.29}$$

Subtracting equation (2.11) from equation (2.12), and noting equations (2.8) and (2.29), we obtain

$$\frac{\partial T_s}{\partial \eta} = T_f - T_s = 2\Psi e^{-\xi-\eta} = e^{-\xi-\eta} J_0\left(2i\sqrt{\xi\eta}\right) \tag{2.30}$$

The integration of equation (2.30) is undertaken in the following manner:

$$\int_0^{\eta} \frac{\partial T_s}{\partial \eta} d\eta = T_s - T_{s,0}$$

where $T_{s,0}$, the initial solid temperature, is zero on the dimensionless temperature scale. It follows from equation (2.30) that

$$T_s(\xi, \eta) = e^{-\xi} \int_0^{\eta} e^{-\eta} J_0\left(2i\sqrt{\xi\eta}\right) d\eta \tag{2.31}$$

Equation (2.30) can be rearranged using equation (2.7) instead of equation (2.8) to obtain

$$\frac{\partial T_f}{\partial \xi} = T_s - T_f = -2\Psi e^{-\xi-\eta} = -e^{-\xi-\eta} J_0\left(2i\sqrt{\xi\eta}\right) \tag{2.32}$$

The integration of equation (2.32) is undertaken in a similar manner:

$$\int_0^{\xi} \frac{\partial T_f}{\partial \xi} = T_f - T_{f,in} = T_f - 1$$

where $T_{f,in} = 1$ on the dimensionless temperature scale. It follows that

$$1 - T_f(\xi, \eta) = e^{-\eta} \int_0^{\xi} e^{-\xi} J_0\left(2i\sqrt{\xi\eta}\right) d\xi \tag{2.33}$$

Upon comparing equations (2.31) and (2.33), it can readily be seen that

$$T_s(\xi, \eta) = 1 - T_f(\eta, \xi) \tag{2.34}$$

In other words, it is sufficient to compute one table of values of $T_s(\xi, \eta)$ and to exploit the symmetry indicated by equation (2.34) to obtain values of T_f (ξ, η) from the same table. This result was pointed out by Larsen [5].

The present author has recomputed a table of values for $T_s(\xi, \eta)$ by evaluating the integral (2.31), simply using Simpson's rule, taking as many data points as required to achieve the accuracy required in the table. The Bessel function with complex imaginary argument using the standard convergent series is

$$J_0(iz) = 1 + \frac{z^2}{2^2} + \frac{z^4}{2^2 \cdot 4^2} + \frac{z^6}{2^2 \cdot 4^2 \cdot 6^2} + \cdots \tag{2.35}$$

which can be written as

$$J_0(iz) = 1 + \sum_{r=1}^{\infty} \frac{z^{2r}}{2^2 \cdot 4^2 \cdots (2r)^2} = \sum_{r=0}^{\infty} c_r$$

where

$$c_{r+1} = \frac{z^2}{[2(r+1)]^2} c_r \quad \text{and} \quad c_0 = 1 \tag{2.36}$$

It is noted that for integral (2.31) for example, where $z = 2\sqrt{(\xi\eta)}$, a possible explosion might occur in the values in c_r as r increases, for large values in either ξ or η. We avoid this by calculating

$$e^{-\xi-\eta} J_0\left(2i\sqrt{\xi\eta}\right) = \sum_{r=0}^{\infty} d_r \tag{2.37}$$

where

$$d_{r+1} = \frac{z^2}{[2(r+1)]^2} d_r \quad \text{and} \quad d_0 = e^{-\xi-\eta} \tag{2.38}$$

In other words, possibly large values in c_r (equation (2.36)) arising from large values in ξ, for example, are scaled down in size by multiplying by $e^{-\xi}$.

Table 2.1 presents the solid temperatures, $T_s(\xi, \eta)$ and, for completeness, Table 2.2 shows the fluid temperatures, $T_f(\xi, \eta)$. The values of $T_s(\xi, \eta)$ obtained replicate those offered by Larsen [5] with the exception of a very few entries where, possibly, rounding to four significant figures appears to have been applied in a different manner. Larsen, however, computes these values using the finite difference method described later in this chapter.

Klinkenberg [6] complained that authors previous to himself fell "short in mentioning the accuracy of the solution, so that it may not be so good as an older one or it may be better than they seem to think." The present author, therefore, specifies that an approximation of the function in equation (2.37) is employed here, that is

$$e^{-\xi-\eta}J_0\left(2i\sqrt{\xi\eta}\right) \approx B_n = \sum_{r=0}^{n} d_r \tag{2.39}$$

where the $\{d_r\}$ are given by equations (2.38) and n is determined such that

$$|B_n - B_{n-1}| < \tfrac{1}{2} \times 10^{-8}$$

The Simpson's rule integration of equation (2.31) is performed to the same accuracy by using a small enough step length. All arithmetic was performed in double-precision floating point on an IBM RS6000 computer. The results offered are rounded to four decimal places after the decimal point.

The maximum amount of heat, Q_{max}, that can be accumulated in a bed of solid material is simply

$$Q_{max} = MC_s(t_{f,in} - t_{s,0}) \tag{2.40}$$

when the solid has become isothermal again, having reached the inlet temperature, $t_{f,in}$, of the fluid.

The energy stored in the bed, $Q(\tau)$, at time τ is given by

$$Q(\tau) = MC_s(t_{s,m}(\tau) - t_{s,0}) \tag{2.41}$$

where $t_{s,m}(\tau)$ is the *average* solid temperature in the bed at time τ, and

$$t_{s,m}(\tau) = \frac{1}{L}\int_0^L t_s(y, \tau)\, dy \tag{2.42}$$

Placing this on the dimensionless temperature scale (equation (2.3)), we note that

$$t_{s,m}(\tau) - t_{s,0} = T_{s,m}(\tau)(t_{f,in} - t_{s,0})$$

and that equation (2.41) takes the form

$$Q(\tau) = MC_s(t_{s,in} - t_{s,0})T_{s,m}(\tau) \tag{2.43}$$

The *dimensionless heat storage, $Q^+(\tau)$* at time τ is defined by the relation

$$Q^+(\tau) = \frac{Q(\tau)}{Q_{max}} \tag{2.44}$$

It follows that, for the single-blow problem,

Table 2.1: Dimensionless solid temperatures, $T_s(\xi, \eta)$

	Dimensionless time, η										
ξ	0	1	2	3	4	5	6	7	8	9	10
0.0	0.0000	0.6321	0.8647	0.9502	0.9817	0.9933	0.9975	0.9991	0.9997	0.9999	1.0000
1.0	0.0000	0.3457	0.6057	0.7750	0.8766	0.9344	0.9659	0.9827	0.9913	0.9957	0.9979
2.0	0.0000	0.1826	0.3965	0.5853	0.7300	0.8314	0.8983	0.9404	0.9659	0.9809	0.9895
3.0	0.0000	0.0939	0.2470	0.4167	0.5731	0.7018	0.7997	0.8698	0.9177	0.9492	0.9694
4.0	0.0000	0.0472	0.1481	0.2830	0.4283	0.5649	0.6820	0.7756	0.8465	0.8978	0.9335
5.0	0.0000	0.0233	0.0861	0.1851	0.3070	0.4361	0.5590	0.6672	0.7567	0.8271	0.8802
6.0	0.0000	0.0114	0.0488	0.1172	0.2124	0.3245	0.4418	0.5544	0.6554	0.7411	0.8106
7.0	0.0000	0.0055	0.0270	0.0722	0.1425	0.2337	0.3379	0.4462	0.5508	0.6459	0.7281
8.0	0.0000	0.0026	0.0147	0.0435	0.0931	0.1635	0.2509	0.3486	0.4497	0.5478	0.6379
9.0	0.0000	0.0012	0.0079	0.0256	0.0594	0.1116	0.1813	0.2651	0.3575	0.4526	0.5453
10.0	0.0000	0.0006	0.0042	0.0149	0.0371	0.0744	0.1279	0.1966	0.2771	0.3649	0.4551
11.0	0.0000	0.0003	0.0022	0.0085	0.0228	0.0486	0.0883	0.1425	0.2099	0.2875	0.3713
12.0	0.0000	0.0001	0.0011	0.0048	0.0137	0.0311	0.0597	0.1012	0.1556	0.2216	0.2965
13.0	0.0000	0.0001	0.0006	0.0026	0.0082	0.0196	0.0397	0.0705	0.1131	0.1674	0.2320
14.0	0.0000	0.0000	0.0003	0.0014	0.0048	0.0122	0.0259	0.0482	0.0806	0.1240	0.1781
15.0	0.0000	0.0000	0.0001	0.0008	0.0028	0.0074	0.0167	0.0324	0.0565	0.0903	0.1342
16.0	0.0000	0.0000	0.0001	0.0004	0.0016	0.0045	0.0106	0.0215	0.0390	0.0647	0.0995
17.0	0.0000	0.0000	0.0000	0.0002	0.0009	0.0027	0.0066	0.0140	0.0265	0.0456	0.0725
18.0	0.0000	0.0000	0.0000	0.0001	0.0005	0.0016	0.0041	0.0090	0.0178	0.0317	0.0521
19.0	0.0000	0.0000	0.0000	0.0001	0.0003	0.0009	0.0025	0.0058	0.0118	0.0217	0.0369
20.0	0.0000	0.0000	0.0000	0.0000	0.0002	0.0005	0.0015	0.0036	0.0077	0.0147	0.0258

Table 2.1 *(contd.)*

	Dimensionless time, η										
ξ	11	12	13	14	15	16	17	18	19	20	21
0.0	1.0000	1.0000	1.0000	1.0000	1.0000	1.0000	1.0000	1.0000	1.0000	1.0000	1.0000
1.0	0.9990	0.9995	0.9998	0.9999	1.0000	1.0000	1.0000	1.0000	1.0000	1.0000	1.0000
2.0	0.9943	0.9969	0.9984	0.9992	0.9996	0.9998	0.9999	0.9999	1.0000	1.0000	1.0000
3.0	0.9819	0.9894	0.9940	0.9966	0.9981	0.9989	0.9994	0.9997	0.9998	0.9999	1.0000
4.0	0.9577	0.9736	0.9838	0.9902	0.9942	0.9966	0.9980	0.9989	0.9994	0.9996	0.9998
5.0	0.9189	0.9462	0.9650	0.9776	0.9859	0.9913	0.9947	0.9968	0.9981	0.9989	0.9993
6.0	0.8647	0.9054	0.9352	0.9564	0.9712	0.9812	0.9880	0.9924	0.9952	0.9971	0.9982
7.0	0.7963	0.8508	0.8929	0.9247	0.9479	0.9646	0.9763	0.9843	0.9898	0.9934	0.9958
8.0	0.7170	0.7839	0.8383	0.8815	0.9147	0.9396	0.9579	0.9711	0.9805	0.9870	0.9914
9.0	0.6311	0.7074	0.7728	0.8271	0.8708	0.9052	0.9316	0.9514	0.9659	0.9765	0.9839
10.0	0.5431	0.6252	0.6990	0.7630	0.8169	0.8610	0.8963	0.9238	0.9449	0.9607	0.9723
11.0	0.4572	0.5412	0.6201	0.6915	0.7542	0.8076	0.8519	0.8878	0.9163	0.9385	0.9554
12.0	0.3769	0.4591	0.5396	0.6155	0.6848	0.7461	0.7990	0.8434	0.8798	0.9091	0.9323
13.0	0.3045	0.3818	0.4607	0.5381	0.6114	0.6788	0.7388	0.7911	0.8354	0.8723	0.9023
14.0	0.2413	0.3116	0.3862	0.4621	0.5368	0.6078	0.6733	0.7322	0.7838	0.8280	0.8651
15.0	0.1878	0.2497	0.3180	0.3901	0.4634	0.5356	0.6044	0.6683	0.7260	0.7770	0.8210
16.0	0.1437	0.1967	0.2574	0.3237	0.3936	0.4646	0.5345	0.6014	0.6637	0.7203	0.7707
17.0	0.1081	0.1525	0.2049	0.2644	0.3290	0.3968	0.4657	0.5335	0.5986	0.6594	0.7150
18.0	0.0801	0.1163	0.1607	0.2125	0.2708	0.3338	0.3998	0.4666	0.4326	0.5960	0.6555
19.0	0.0585	0.0874	0.1241	0.1683	0.2196	0.2767	0.3382	0.4025	0.4675	0.5318	0.5936
20.0	0.0422	0.0648	0.0944	0.1314	0.1755	0.2261	0.2822	0.3423	0.4049	0.4684	0.5310

Table 2.2: Dimensionless fluid temperatures, $T_f(\xi, \eta)$

	Dimensionless time, ξ										
ξ	0	1	2	3	4	5	6	7	8	9	10
0.0	1.0000	0.3679	0.1353	0.0498	0.0183	0.0067	0.0025	0.0009	0.0003	0.0001	0.0000
1.0	1.0000	0.6543	0.3943	0.2250	0.1234	0.0656	0.0341	0.0173	0.0087	0.0043	0.0021
2.0	1.0000	0.8174	0.6035	0.4147	0.2700	0.1686	0.1017	0.0596	0.0341	0.0191	0.0105
3.0	1.0000	0.9061	0.7530	0.5833	0.4269	0.2982	0.2003	0.1302	0.0823	0.0508	0.0306
4.0	1.0000	0.9528	0.8519	0.7170	0.5717	0.4351	0.3180	0.2244	0.1535	0.1022	0.0665
5.0	1.0000	0.9767	0.9139	0.8149	0.6930	0.5639	0.4410	0.3328	0.2433	0.1729	0.1198
6.0	1.0000	0.9886	0.9512	0.8828	0.7876	0.6755	0.5582	0.4456	0.3446	0.2589	0.1894
7.0	1.0000	0.9945	0.9730	0.9278	0.8575	0.7663	0.6621	0.5538	0.4492	0.3541	0.2719
8.0	1.0000	0.9974	0.9853	0.9565	0.9069	0.8365	0.7491	0.6514	0.5503	0.4522	0.3621
9.0	1.0000	0.9988	0.9921	0.9744	0.9406	0.8884	0.8187	0.7349	0.6425	0.5474	0.4547
10.0	1.0000	0.9994	0.9958	0.9851	0.9629	0.9256	0.8721	0.8034	0.7229	0.6351	0.5449
11.0	1.0000	0.9997	0.9978	0.9915	0.9772	0.9514	0.9117	0.8575	0.7901	0.7125	0.6287
12.0	1.0000	0.9999	0.9989	0.9952	0.9863	0.9689	0.9403	0.8988	0.8444	0.7784	0.7035
13.0	1.0000	0.9999	0.9994	0.9974	0.9918	0.9804	0.9603	0.9295	0.8869	0.8326	0.7680
14.0	1.0000	1.0000	0.9997	0.9986	0.9952	0.9878	0.9741	0.9518	0.9194	0.8760	0.8219
15.0	1.0000	1.0000	0.9999	0.9992	0.9972	0.9926	0.9833	0.9676	0.9435	0.9097	0.8658
16.0	1.0000	1.0000	0.9999	0.9996	0.9984	0.9955	0.9894	0.9785	0.9610	0.9353	0.9005
17.0	1.0000	1.0000	1.0000	0.9998	0.9991	0.9973	0.9934	0.9860	0.9735	0.9544	0.9275
18.0	1.0000	1.0000	1.0000	0.9999	0.9995	0.9984	0.9959	0.9910	0.9822	0.9683	0.9479
19.0	1.0000	1.0000	1.0000	0.9999	0.9997	0.9991	0.9975	0.9942	0.9882	0.9783	0.9631
20.0	1.0000	1.0000	1.0000	1.0000	0.9998	0.9995	0.9985	0.9964	0.9923	0.9853	0.9742

Table 2.2 *(contd.)*

ξ	Dimensionless time, η										
	11	12	13	14	15	16	17	18	19	20	21
0.0	0.0000	0.0000	0.0000	0.0000	0.0000	0.0000	0.0000	0.0000	0.0000	0.0000	0.0000
1.0	0.0010	0.0005	0.0002	0.0001	0.0000	0.0000	0.0000	0.0000	0.0000	0.0000	0.0000
2.0	0.0057	0.0031	0.0016	0.0008	0.0004	0.0002	0.0001	0.0000	0.0000	0.0000	0.0000
3.0	0.0181	0.0106	0.0060	0.0034	0.0019	0.0011	0.0006	0.0003	0.0002	0.0001	0.0000
4.0	0.0423	0.0264	0.0162	0.0098	0.0058	0.0034	0.0020	0.0011	0.0006	0.0004	0.0002
5.0	0.0811	0.0538	0.0350	0.0224	0.0141	0.0087	0.0053	0.0032	0.0019	0.0011	0.0007
6.0	0.1353	0.0946	0.0648	0.0436	0.0288	0.0188	0.0120	0.0076	0.0048	0.0029	0.0018
7.0	0.2037	0.1492	0.1071	0.0753	0.0521	0.0354	0.0237	0.0157	0.0102	0.0066	0.0042
8.0	0.2830	0.2161	0.1617	0.1185	0.0853	0.0604	0.0421	0.0289	0.0195	0.0130	0.0086
9.0	0.3689	0.2926	0.2272	0.1729	0.1292	0.0948	0.0684	0.0486	0.0341	0.0235	0.0161
10.0	0.4569	0.3748	0.3010	0.2370	0.1831	0.1390	0.1037	0.0762	0.0551	0.0393	0.0277
11.0	0.5428	0.4588	0.3799	0.3085	0.2458	0.1924	0.1481	0.1122	0.0837	0.0615	0.0446
12.0	0.6231	0.5409	0.4604	0.3845	0.3152	0.2539	0.2010	0.1566	0.1202	0.0909	0.0677
13.0	0.6955	0.6182	0.5393	0.4619	0.3886	0.3212	0.2612	0.2089	0.1646	0.1277	0.0977
14.0	0.7587	0.6884	0.6138	0.5379	0.4632	0.3922	0.3267	0.2678	0.2162	0.1720	0.1349
15.0	0.8122	0.7503	0.6820	0.6099	0.5366	0.4644	0.3956	0.3317	0.2740	0.2230	0.1790
16.0	0.8563	0.8033	0.7426	0.6763	0.6064	0.5354	0.4655	0.3986	0.3363	0.2797	0.2293
17.0	0.8919	0.8475	0.7951	0.7356	0.6710	0.6032	0.5343	0.4665	0.4014	0.3406	0.2850
18.0	0.9199	0.8837	0.8393	0.7875	0.7292	0.6662	0.6002	0.5334	0.4674	0.4040	0.3445
19.0	0.9415	0.9126	0.8759	0.8317	0.7804	0.7233	0.6618	0.5975	0.5325	0.4682	0.4064
20.0	0.9578	0.9352	0.9056	0.8686	0.8245	0.7739	0.7178	0.6577	0.5951	0.5316	0.4690

$$Q^+(\tau) = \frac{MC_s(t_{s,in} - t_{s,0})T_{s,m}(\tau)}{MC_s(t_{s,in} - t_{s,0})} = T_{s,m}(\tau) \tag{2.45}$$

Employing the dimensionless parameters, ξ and η, we deduce that

$$Q^+(\eta) = T_{s,m}(\eta) = \frac{1}{\Lambda}\int_0^{\Lambda} T_s(\xi, \eta)\, d\xi \tag{2.46}$$

from which it follows from equation (2.31) that

$$Q^+(\eta) = \frac{1}{\Lambda}\int_0^{\Lambda} e^{-\xi} \int_0^{\eta} e^{-\eta} J_0\left(2i\sqrt{\xi\eta}\right) d\eta\, d\xi \tag{2.47}$$

This double integral has been evaluated, again applying Simpson's rule, taking a sufficient number of points both in the ξ and η directions to realize the accuracy needed. Table 2.3 presents the values of the dimensionless heat storage, $Q^+(\xi, \eta)$, so computed. The Bessel function is calculated in the manner described previously.

Example 2.1 Consider a Feolite thermal storage unit whose available heating surface area, A, is 5.8 m^2 and whose mass, M, is 905 kg. The specifc heat of Feolite is considered to be 920.0 J/(kg °C). The entire unit is well insulated. Air flows through the packing at a rate, $\dot{m}_f$ equal to 0.156 kg/s. The initial temperature of the packing is 10°C and the warm air enters the packing at a temperature of 80°C. The convective heat transfer coefficient, α is assumed to have a value of 54.39 W/(m^2 °C). Determine the exit solid and gas temperatures and the heat stored in the packing after 3.665 hr and 7.331 hr.

The dimensionless length, Λ, of the storage unit is given by

$$\Lambda = \frac{\alpha A}{\dot{m}_f c_f} = \frac{(54.39)(5.8)}{(0.156)(1011.0)} = 2.0$$

where it is assumed that the specific heat of air at 50°C is equal to 1011 J/(kg °C).

The dimensionless time, η, is given by

$$\eta = \frac{\alpha A}{MC_s}\tau$$

where it is assumed that the term $m_f y/(\dot{m}_f L)$ can be neglected. It follows that

$$\eta = \frac{(54.39)(5.8)}{(905.0)(920.0)}\tau = 0.0003789\tau$$

After 3.665 hr (= 13 196 s), $\eta = 0.0003789 \times 13\,196$ and thus $\eta \approx 5.0$. Similarly, after 7.331hr (= 26 392 s), $\eta \approx 10.0$. From Table 2.1, we note that the dimensionless solid temperatures are given by $T_s(2.0, 5.0) = 0.8314$ and $T_s(2.0, 10.0) = 0.9895$. Similarly, from Table 2.2, we note that the dimensionless fluid temperatures are given by $T_f(2.0, 5.0) = 0.9139$ and $T_f(2.0, 10.0) = 0.9958$.

The dimensionless temperatures are given by

$$T = \frac{t - t_{s,0}}{t_{f,in} - t_{s,0}} \tag{2.3}$$

Here, $t_{s,0} = 10°\text{C}$ and $t_{f,in} = 80°\text{C}$, thus

$$T = \frac{t - 10.0}{80.0 - 10.0} \quad \text{and} \quad t = (70.0)T + 10.0$$

At the exit, $\Lambda = 2.0$. The solid temperatures are given at $\eta = 5$ by

$$t_s(2, 5) = (70.0)(0.8314) + 10.0 = 68.2°\text{C}$$

whereas at $\eta = 10$

$$t_s(2, 10) = (70.0)(0.9895) + 10.0 = 79.3°\text{C}$$

The fluid temperatures are given at $\eta = 5$ by

$$t_f(2, 5) = (70.0)(0.9139) + 10.0 = 74.0°\text{C}$$

whereas at $\eta = 10$

$$t_f(2, 10) = (70.0)(0.9958) + 10.0 = 79.7°\text{C}$$

The maximum heat that can be stored in the packing, relative to the initial isothermal temperature, $t_{s,0} = 10°\text{C}$, is given by

$$Q_{max} = MC_s(t_{f,in} - t_{s,0}) \tag{2.40}$$

Here, $M = 905\,\text{kg}$ and $C_s = 920.0\,\text{J/(kg °C)}$. The inlet temperature of the gas is $t_{f,in} = 80°\text{C}$. It follows that

$$\begin{aligned} Q_{max} &= (905.0)(920.0)(80.0 - 10.0)\,\text{J} \\ &= 58\,282\,\text{kJ} \end{aligned}$$

The dimensionless heat storage at $\Lambda = 2$ for $\eta = 5$ is extracted from Table 2.3 where it is found that $Q^+(2, 5) = 0.9270$. Similarly, for $\eta = 10$, it is found that $Q^+(2, 10) = 0.9969$.

Table 2.3: Dimensionless heat storage, $Q^+(\xi, \eta)$

	Dimensionless length, ξ									
η	1	2	3	4	5	6	7	8	9	10
1.0	0.4762	0.3662	0.2887	0.2335	0.1936	0.1641	0.1418	0.1246	0.1109	0.0999
2.0	0.7324	0.6142	0.5152	0.4348	0.3708	0.3200	0.2795	0.2471	0.2209	0.1994
3.0	0.8660	0.7727	0.6813	0.5977	0.5244	0.4618	0.4091	0.3651	0.3283	0.2974
4.0	0.9340	0.8697	0.7969	0.7224	0.6510	0.5854	0.5269	0.4756	0.4311	0.3927
5.0	0.9680	0.9270	0.8740	0.8138	0.7509	0.6889	0.6301	0.5759	0.5271	0.4835
6.0	0.9846	0.9599	0.9236	0.8782	0.8266	0.7721	0.7173	0.6643	0.6143	0.5682
7.0	0.9927	0.9783	0.9546	0.9221	0.8821	0.8369	0.7887	0.7396	0.6914	0.6452
8.0	0.9966	0.9885	0.9736	0.9511	0.9215	0.8857	0.8453	0.8021	0.7577	0.7136
9.0	0.9984	0.9939	0.9848	0.9699	0.9487	0.9215	0.8890	0.8524	0.8133	0.7727
10.0	0.9993	0.9969	0.9914	0.9818	0.9671	0.9470	0.9218	0.8920	0.8586	0.8227
11.0	0.9997	0.9984	0.9952	0.9891	0.9792	0.9649	0.9458	0.9223	0.8947	0.8640
12.0	0.9998	0.9992	0.9974	0.9936	0.9871	0.9771	0.9631	0.9450	0.9229	0.8973
13.0	0.9999	0.9996	0.9986	0.9963	0.9921	0.9853	0.9752	0.9617	0.9444	0.9236
14.0	1.0000	0.9998	0.9992	0.9979	0.9952	0.9906	0.9836	0.9737	0.9605	0.9440
15.0	1.0000	0.9999	0.9996	0.9988	0.9971	0.9941	0.9893	0.9822	0.9723	0.9596
16.0	1.0000	0.9999	0.9998	0.9993	0.9983	0.9964	0.9931	0.9881	0.9809	0.9712
17.0	1.0000	1.0000	0.9999	0.9996	0.9990	0.9978	0.9956	0.9921	0.9869	0.9797
18.0	1.0000	1.0000	0.9999	0.9998	0.9994	0.9987	0.9972	0.9949	0.9912	0.9859
19.0	1.0000	1.0000	1.0000	0.9999	0.9997	0.9992	0.9983	0.9967	0.9941	0.9903
20.0	1.0000	1.0000	1.0000	0.9999	0.9998	0.9995	0.9989	0.9979	0.9961	0.9934

Table 2.3 *(contd.)*

η	Dimensionless length, ξ									
	11	12	13	14	15	16	17	18	19	20
1.0	0.0909	0.0833	0.0769	0.0714	0.0667	0.0625	0.0588	0.0556	0.0526	0.0500
2.0	0.1815	0.1665	0.1538	0.1428	0.1333	0.1250	0.1176	0.1111	0.1053	0.1000
3.0	0.2714	0.2493	0.2304	0.2141	0.1999	0.1875	0.1765	0.1667	0.1579	0.1500
4.0	0.3597	0.3312	0.3066	0.2851	0.2663	0.2498	0.2352	0.2222	0.2105	0.2000
5.0	0.4451	0.4113	0.3816	0.3554	0.3324	0.3120	0.2938	0.2776	0.2631	0.2500
6.0	0.5263	0.4885	0.4547	0.4246	0.3977	0.3736	0.3522	0.3329	0.3155	0.2999
7.0	0.6019	0.5618	0.5251	0.4918	0.4617	0.4345	0.4100	0.3878	0.3678	0.3496
8.0	0.6707	0.6300	0.5918	0.5564	0.5238	0.4940	0.4669	0.4422	0.4197	0.3992
9.0	0.7321	0.6922	0.6538	0.6175	0.5834	0.5517	0.5225	0.4956	0.4709	0.4483
10.0	0.7854	0.7477	0.7105	0.6743	0.6397	0.6070	0.5763	0.5477	0.5212	0.4967
11.0	0.8309	0.7963	0.7612	0.7263	0.6921	0.6592	0.6278	0.5981	0.5702	0.5442
12.0	0.8687	0.8380	0.8058	0.7730	0.7401	0.7078	0.6764	0.6462	0.6175	0.5904
13.0	0.8996	0.8730	0.8443	0.8142	0.7833	0.7523	0.7216	0.6916	0.6627	0.6350
14.0	0.9244	0.9018	0.8768	0.8499	0.8216	0.7925	0.7632	0.7340	0.7053	0.6775
15.0	0.9438	0.9251	0.9039	0.8803	0.8549	0.8283	0.8008	0.7729	0.7451	0.7177
16.0	0.9588	0.9437	0.9259	0.9058	0.8835	0.8595	0.8343	0.8082	0.7817	0.7551
17.0	0.9702	0.9582	0.9437	0.9267	0.9075	0.8864	0.8637	0.8397	0.8149	0.7897
18.0	0.9787	0.9693	0.9577	0.9437	0.9275	0.9092	0.8891	0.8675	0.8847	0.8210
19.0	0.9849	0.9778	0.9686	0.9573	0.9438	0.9283	0.9108	0.8916	0.8710	0.8492
20.0	0.9895	0.9841	0.9769	0.9679	0.9569	0.9439	0.9290	0.9123	0.8939	0.8742

The heat stored in the packing after dimensionless time, η is given by

$$Q(\Lambda, \eta) = Q^+(\Lambda, \eta)Q_{max}$$

from which it follows that

$$Q(2, 5) = 0.9270 \times 58\,282 = 54\,027.4\,\text{kJ}$$
$$Q(2, 10) = 0.9969 \times 58\,282 = 58\,101.3\,\text{kJ}$$

Example 2.2 Consider the same Feolite thermal storage unit. Calculate the time variation of the exit gas and solid temperatures over the first 6 hr of heating the unit.

These temperatures can be computed using equation (2.31) and exploiting equation (2.34). A graphical display of the exit solid, $t_{s,x}$ and the exit gas, $t_{f,x}$, temperatures as they vary with time is displayed in Figure 2.1.

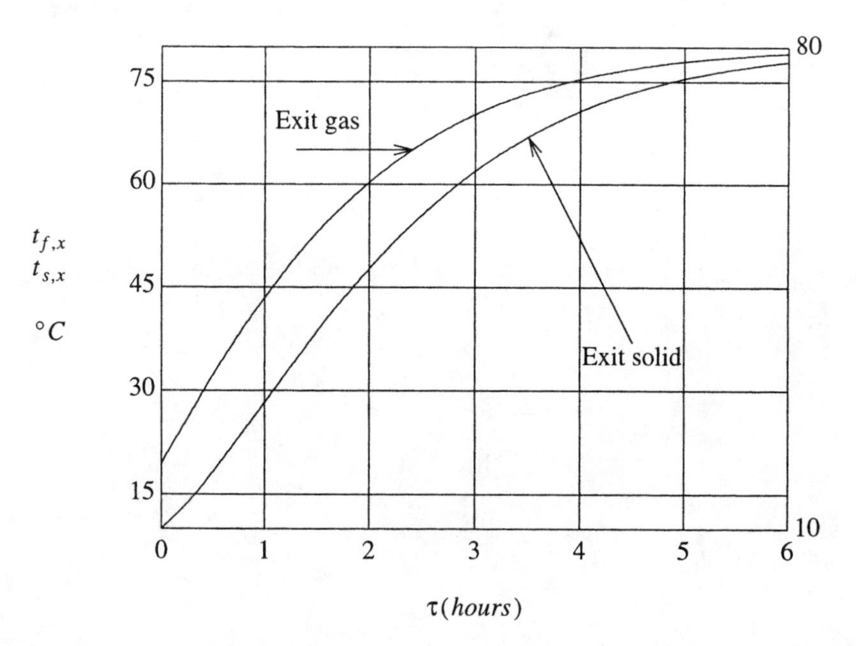

Figure 2.1: Transient response of a heat storage unit. Variation of $t_{f,x}$ and $t_{s,x}$ with time τ.

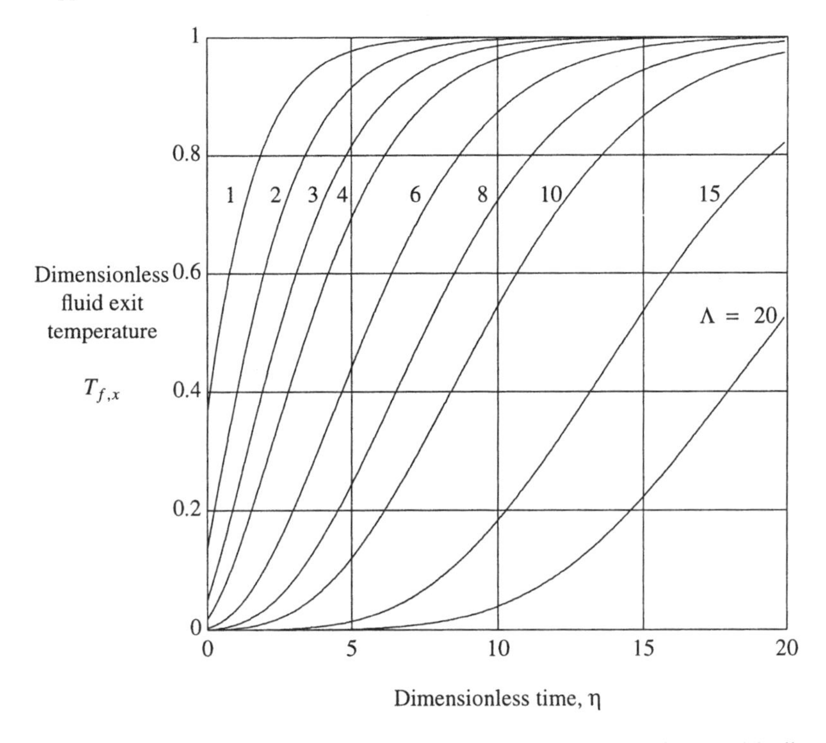

Figure 2.2: Dimensionless fluid exit temperature. Variation of $T_{f,x}$ with dimensionless time η, for different values of dimensionless length, Λ.

A complete evaluation of the dimensionless outlet fluid temperatures, again using equation (2.31) and making use of equation (2.34), has been undertaken and the results are presented in Figure 2.2.

2.3 Approximations

Numerous approximations to the solution of the single-blow problem have been developed. Some of these enabled the calculations to be undertaken at a time when digital computers were not readily available. The finite-difference techniques emerged, it is likely, when computers started to be used to solve the periodic flow regenerator problem – this will be discussed later.

Klinkenberg [6] undertook a complete evaluation of the methods of approximation available at that time. He suggested that for $2 \leq \xi \geq 4$ and $2 \leq \eta \geq 4$, the following formulae might be used:

$$T_f(\xi, \eta) = \frac{1}{2}\left[1 + \operatorname{erf}\left(\sqrt{\eta} - \sqrt{\xi} + \frac{1}{8\sqrt{\eta}} + \frac{1}{8\sqrt{\xi}}\right)\right] \tag{2.48}$$

$$T_s(\xi, \eta) = \frac{1}{2}\left[1 + \operatorname{erf}\left(\sqrt{\eta} - \sqrt{\xi} - \frac{1}{8\sqrt{\eta}} - \frac{1}{8\sqrt{\xi}}\right)\right] \tag{2.49}$$

For higher values of ξ and η, Klinkenberg suggests the even simpler formulation,

$$T_f(\xi, \eta) = \tfrac{1}{2}\left[1 + \operatorname{erf}\left(\left(\eta + \tfrac{1}{4}\right)^{1/2} - \left(\xi - \tfrac{1}{4}\right)^{1/2}\right)\right] \tag{2.50}$$

$$T_s(\xi, \eta) = \tfrac{1}{2}\left[1 + \operatorname{erf}\left(\left(\eta - \tfrac{1}{4}\right)^{1/2} - \left(\xi + \tfrac{1}{4}\right)^{1/2}\right)\right] \tag{2.51}$$

For $\xi < 2$ and $\eta < 2$, Klinkenberg suggests

$$T_f(\xi, \eta) = \frac{1}{2}\left[1 + \operatorname{erf}\left(\sqrt{\eta} - \sqrt{\xi}\right)\right] + \frac{\xi^{1/4}}{\xi^{1/4} + \eta^{1/4}} e^{-\xi-\eta} J_0\left(2i\sqrt{\xi\eta}\right) \tag{2.52}$$

$$T_s(\xi, \eta) = \frac{1}{2}\left[1 + \operatorname{erf}\left(\sqrt{\eta} - \sqrt{\xi}\right)\right] - \frac{\eta^{1/4}}{\xi^{1/4} + \eta^{1/4}} e^{-\xi-\eta} J_0\left(2i\sqrt{\xi\eta}\right) \tag{2.53}$$

Unless it is required to undertake calculations on a very slow desk-laptop computer that has only software floating-point arithmetic facilities, for example, the present author suggests that equation (2.31) with (2.34) be employed. Libraries of scientific and mathematical software will provide the means to obtain the integrals required as well as values of $J_0(ix)$ (where $i^2 = -1$).

On the other hand, the integral in equation (2.46) to obtain $Q^+(\xi, \eta)$ might well be found using the Klinkenberg approximations offered above, thereby avoiding the need to compute the double integral of equation (2.47). This double integral can be computationally expensive to find if many values are required.

Klinkenberg and Harmens [7] extended this work to the case where the initial solid temperature was arbitrary and the inlet gas temperature varied with time.

2.4 Finite-Difference Approximations

Direct use is made of equations (2.7) and (2.8), which are represented in finite difference form. Willmott [8] proposed the application of the trapezium rule; the fluid and solid temperatures are computed at positions (i, j) on a space–time mesh (Figure 2.3), that is, at positions $i\Delta\xi$ from the entrance of the gas, with $i = 0, 1, 2, \ldots, N$, where $N\Delta\xi = \Lambda$, the dimensionless length of the bed, and at times $j\Delta\eta$ from the beginning of the single-blow operation. We call $\Delta\xi$ the *step length* and $\Delta\eta$ the *step length in time*.

The fluid temperature at $(i+1, j+1)$ is estimated using

$$T_{f,i+1,j+1} = T_{f,i,j+1} + \frac{\Delta\xi}{2}\left(\left.\frac{\partial T_f}{\partial \xi}\right|_{i+1} + \left.\frac{\partial T_f}{\partial \xi}\right|_{i}\right)$$
$$= T_{f,i,j+1} + \frac{\Delta\xi}{2}(T_{s,i+1,j+1} - T_{f,i+1,j+1} + T_{s,i,j+1} - T_{f,i,j+1}) \tag{2.54}$$

The solid temperature at $(i+1, j+1)$ is found using

$$T_{s,i+1,j+1} = T_{s,i,j+1} + \frac{\Delta\eta}{2}\left(\left.\frac{\partial T_s}{\partial \eta}\right|_{j+1} + \left.\frac{\partial T_s}{\partial \eta}\right|_{j}\right)$$
$$= T_{s,i+1,s} + \frac{\Delta\eta}{2}(T_{f,i+1,j+1} - T_{s,i+1,j+1} + T_{f,i+1,j} - T_{s,i+1,j}) \tag{2.55}$$

Equations (2.54) and (2.55) can be rearranged as

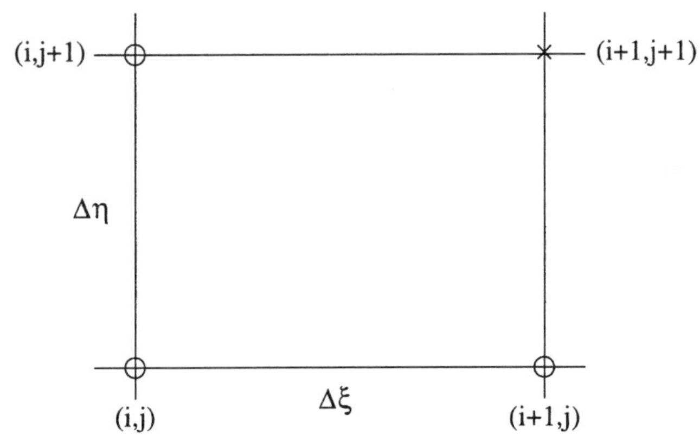

Figure 2.3: Space-time mesh. In general, fluid and solid temperatures are known at (i, j), $(i+1, j)$ and $(i, j+1)$. It is required to compute these at $(i+1, j+1)$.

$$T_{f,i+1,j+1} = A_1 T_{f,i,j+1} + A_2[T_{s,i,j+1} + T_{s,i+1,j+1}] \tag{2.56}$$

$$T_{s,i+1,j+1} = B_1 T_{s,i,j+1} + B_2[T_{f,i,j+1} + T_{f,i+1,j+1}] \tag{2.57}$$

where

$$A_1 = \frac{2 - \Delta\xi}{2 + \Delta\xi} \qquad B_1 = \frac{2 - \Delta\eta}{2 + \Delta\eta}$$

$$A_2 = \frac{\Delta\xi}{2 + \Delta\xi} \qquad B_2 = \frac{\Delta\eta}{2 + \Delta\eta}$$

Except at the entrance to the bed where $T_{f,0,j} = 1$, for $j = 0, 1, 2, \ldots, T_{f,i+1,j+1}$ in equation (2.57) is unknown. However, $T_{f,i+1,j+1}$ can be computed by substituting equation (2.56) into (2.57) to yield

$$T_{s,i+1,j+1} = K_1 T_{s,i+1,j} + K_2 T_{f,i+1,j} + K_3 T_{s,i,j+1} + K_4 T_{f,i,j+1} \tag{2.58}$$

where

$$K_1 = \frac{B_1}{X} \qquad K_2 = \frac{B_2}{X}$$

$$K_3 = \frac{A_2 B_2}{X} \qquad K_4 = \frac{A_1 B_2}{X}$$

$$X = 1 - A_2 B_2$$

This method of integration is then as follows. At $j = 0$, the initial condition holds, namely $T_{s,0} = T_{s,i,0} = 0$ for $i = 0, 1, 2, \ldots, N$, down the length of the bed. The boundary condition representing the inlet fluid temperature is $T_{f,in} = T_{f,0,j} = 1$ for $j = 0, 1, 2, \ldots$. We begin by computing the fluid temperature down the length of the bed, using equation (2.56) in the form

$$T_{f,i+1,0} = A_1 T_{f,i,0} + A_2[T_{s,i,0} + T_{s,i+1,0}]$$

or more simply, using the solution to equation (2.7) for $\eta = 0$, namely

$$T_{f,i,0} = e^{i\Delta\xi}$$

The entrance solid temperature, $T_{s,0,j}$, for $j = 0, 1, 2, \ldots$, can be computed in a similar manner using the solution of equation (2.8) for $\xi = 0$, namely

$$T_{s,0,j} = 1 - e^{j\Delta\eta}$$

Thereafter, the solid temperature at $(i + 1, j + 1)$ is computed using equation (2.58) and the fluid temperature at the same position using equation (2.56).

The same method was developed by Larsen [5] although the coefficients, K_1, K_2, K_3, and K_4 presented by Larsen were in a more restricted form, where $\Delta\xi = \Delta\eta$. Thus Larsen suggests that

$$K_1 = \frac{4 - \Delta\xi^2}{4 + 4\Delta\xi}$$

with similar forms for K_2, K_3, and K_4.

2.5 Use of a Modified Heat Transfer Coefficient in the Simplified Model

It will be recalled that both Anzelius and Schumann considered that the latitudinal thermal conductivity of the solid was infinitely large, that is, in any direction perpendicular to gas flow. Further, the implication of this assumption is that the resistance to heat transfer lies solely at the surface of the solid (the interface between gas and solid) and is represented solely by the surface heat transfer coefficient. The rate of heat transfer, $\delta\dot{Q}$, through an element of heating surface area, δA, is given by

$$\delta\dot{Q} = \alpha\,\delta A(t_s - t_f)$$

In many practical cases, this assumption is not valid and there is a transverse temperature gradient within the solid material. In order to represent this, we introduce a *bulk* heat transfer coefficient $\bar{\alpha}$ and treat the rate of heat transfer to take the form $\delta\dot{Q} = \bar{\alpha}\,\delta A(t_{s,av} - t_f)$. Note that $t_{s,av}$ is the mean solid temperature, which in the case of slabs is

$$t_{s,av} = \frac{1}{w}\int_0^w t_s(x)\,dx$$

where w is the semithickness of the slab and x is the direction perpendicular to gas flow into the solid slab. The key problem is to know the relationship between α, the surface heat transfer coefficient, and $\bar{\alpha}$, the bulk heat transfer coefficient.

Schmidt and Willmott [9] present the necessary relationships in terms of Δ, the characteristic semithickness of the packing material and λ_s, the latitudinal thermal conductivity of this material. The equation

$$\frac{1}{\bar{\alpha}} = \frac{1}{\alpha} + \frac{\Delta}{\lambda_s} \tag{2.59}$$

is based on the simplifying assumption that latitudinal conductivity is finite in value and that the mean solid temperature, $t_{s,av}$ varies linearly with time. This assumption is approximately valid except at the start and the end (when the solid temperature is approaching closely the inlet gas temperature) of the single-blow operation. Certainly, the transient performance can be estimated using the bulk heat transfer coefficient, $\bar{\alpha}$.

Schmidt and Willmott offer the following relationships:

$$\text{Slabs:} \quad \Delta = \frac{w}{3} \qquad \text{Solid cylinders:} \quad \Delta = \frac{r_o}{4}$$

$$\text{Solid spheres:} \quad \Delta = \frac{r_o}{5}$$

$$\text{Hollow cylinders:} \quad \Delta = \frac{r_i}{2(r_o^2 - r_i^2)} \frac{2r_o^4}{r_o^2 - r_i^2} \ln\left(\frac{r_o}{r_i}\right) - 2r_o^2 + r_i^2$$

Here, r_o is the radius of the solid cylinder, the radius of the sphere, and the outer radius of the hollow cylinder, and r_i is the inner radius of the hollow cylinder. Again, w is the semithickness of the slab.

It will be recalled that the parameter ξ is defined in the following way:

$$\xi = \frac{\alpha A}{\dot{m}_f c_p L} y \tag{2.4}$$

Similarly, we can define $\bar{\xi}$, where

$$\frac{1}{\bar{\xi}} = \frac{\dot{m}_f c_p L}{\bar{\alpha} A_y}$$

from which it follows that

$$\frac{1}{\bar{\xi}} = \frac{\dot{m}_f c_p L}{Ay}\left(\frac{1}{\alpha} + \frac{\Delta}{\lambda_s}\right) = \frac{\dot{m}_f c_p L}{\alpha A_y}\left(1 + \frac{\alpha\Delta}{\lambda_s}\right)$$

The quantity $\alpha\Delta/\lambda_s$ is called the Biot modulus relative to the *thickness* Δ and is denoted by

$$\text{Bi}_\Delta = \frac{\alpha\Delta}{\lambda_s}$$

It follows that the dimensionless distance, $\bar{\xi}$, and similarly the dimensionless time, $\bar{\eta}$, are given by

$$\bar{\xi} = \frac{\xi}{1 + \text{Bi}_\Delta} \qquad \bar{\eta} = \frac{\eta}{1 + \text{Bi}_\Delta} \tag{2.60}$$

Example 2.3 Consider again the Feolite thermal storage unit whose available heating surface area, A, is 5.8 m^2 and whose mass, M, is 905 kg. The specific heat of Feolite is considered to be 920.0 J/(kg °C) and the thermal conductivity to be 2.1 W/(m°C). We consider the unit to be made up of slabs of semithickness 0.04 m so that $\Delta = 0.04/3 = 0.01333$ m. The entire unit is well insulated. Air flows through the packing at a rate $\dot{m}_f$, equal to 0.156 kg/s. The initial temperature of the packing is 10°C and the warm air enters the packing at a temperature of 80°C. The convective heat transfer coefficient, α, is assumed to have a value of 54.39 W/(m^2 °C). Determine the exit solid and gas temperatures and the heat stored in the packing after 6.0948 hr.

The Biot modulus takes the value

$$\mathrm{Bi}_\Delta = \frac{\alpha\Delta}{\lambda_s} = \frac{54.39 \times 0.01333}{2.1} = 0.3409$$

The modified dimensionless length, $\bar{\Lambda}$ is calculated to be

$$\bar{\Lambda} = \frac{\Lambda}{1 + \mathrm{Bi}_\Delta} = \frac{2.0}{1.3409} = 1.4915$$

and the dimensionless time becomes

$$\bar{\eta} = \frac{\eta}{1 + \mathrm{Bi}_\Delta} = \frac{0.0003789\tau}{1.3409} = 0.0002826\tau$$

For the case where $\tau = 6.0948\ \mathrm{hr} = 21\,941\ \mathrm{s}$,

$$\bar{\eta} = 0.0002826 \times 21\,941 = 6.2$$

By evaluation of the integral in equation (2.31) and use of symmetry given in equation (2.34), we find that $T_{s,av}(1.4915, 6.2) = 0.9441$ and $T_f(1.4915, 6.2) = 0.9772$. We know from Example 2.1 that

$$t = 70.0T + 10.0$$

from which it follows that

$$t_{s,av}(1.4915, 6.2) = (70.0)(0.9441) + 10.0 = 76.1°\mathrm{C}$$
$$t_f(1.4915, 6.2) = (70.0)(0.9772) + 10.0 = 78.4°\mathrm{C}$$

References

1. A. Anzelius, "Heating by Means of a Flowing Medium," *Z.A.M.M.* **6**(4), 291–294 (1926)
2. T. E. W. Schumann, "Heat Transfer: A Liquid Flowing Through a Porous Prism," *J. Franklin Inst.* **208**, 405–416 (1929).
3. M. Jakob, *Heat Transfer*, Vol. 2, John Wiley, New York (1957).
4. N. W. McLachlan, *Bessel Functions for Engineers*, 2nd edn, Oxford University Press, Oxford (1955).

5. F. W. Larsen, "Rapid Calculation of Temperature in a Regenerative Heat Exchanger Having Arbitrary Initial Solid and Entering Fluid Temperatures," *Int. J. Heat Mass Transfer* **10**, 149–168 (1967).
6. A. Klinkenberg, "Heat Transfer in Cross-Flow Heat Exchangers and Packed Beds," *Indust. Eng. Chem. (Eng. Des. Process Dev.)* **46**, 2285–2289 (1956).
7. A. Klinkenberg, A. Harmens, "Unsteady State Heat Transfer in Stationary Packed Beds," *Chem. Eng. Sci.* **11**, 260–266 (1960).
8. A. J. Willmott, "Digital Computer Simulation of a Thermal Regenerator," *Int. J. Heat Mass Transfer* **7**, 1291–1302 (May 1964).
9. F. W. Schmidt, A. J. Willmott, *Thermal Energy Storage and Regeneration*, McGraw-Hill, New York (1981).

Chapter 3

THE SINGLE-BLOW PROBLEM

EFFECT OF SOLID FINITE CONDUCTIVITY

3.1 Introduction

This chapter is concerned with the effect of finite thermal conductivity in the packing of a storage unit upon the unit's temperature behavior under single-blow conditions. It will be recalled that, in Chapter 2, the assumptions below were embodied in the model:

1. The longitudinal thermal conductivities of the gas and the solid, that is in the direction of gas flow, are ignored.
2. Both Anzelius [1] and Schumann [2] considered that the latitudinal thermal conductivity of the solid was infinitely large, that is in any direction perpendicular to gas flow. The implication of this assumption is that the resistance to heat transfer lies solely at the surface of the solid (the interface between gas and solid) and is represented solely by the surface heat transfer coefficient.

At the end of that chapter, the use of a modified heat transfer coefficient was introduced, whereby the effect of latitudinal conductivity could be approximated using the simplified model embodied in equations (2.7) and (2.8).

Schmidt and Willmott [3] suggest that there are two main resistances to heat transfer between the gas and the interior of the solid, namely a resistance at the surface of the solid (the interface between gas and solid), and a resistance inside the solid, namely that caused by a finite latitudinal or *transverse* thermal conductivity of the solid in a direction perpendicular to gas flow. In addition, the performance of the unit is modified by the longitudinal or *axial* conduction of heat in the solid, in a direction parallel to gas flow. In general, however, the longitudinal conduction effect is small relative to the resistances to heat transfer in the transverse direction.

In terms of equation (2.59),

$$\frac{1}{\bar{\alpha}} = \frac{1}{\alpha} + \frac{\Delta}{\lambda_s} \tag{2.59}$$

it will be seen that if the surface heat transfer coefficient, α, is very large relative to the thermal conduction effect within the solid, then $\bar{\alpha} \to \alpha$ and the temperature gradients will be negligible within the solid material in the latitudinal direction. Under these conditions, the simplified model set out in Chapter 2 should be able to be employed to predict the transient response of the unit. Where the two resistances are approximately equal in magnitude, that is when

$$\frac{1}{\alpha} \approx \frac{\Delta}{\lambda_s}$$

or where the internal resistance to heat transfer is larger than the resistance at the surface of the solid, significant transverse temperature gradients will be present. In this case, a thorough approach is that of Schmidt and Szego [4] in which the latitudinal and longitudinal conduction effects are represented explicitly within the model using the Fourier equation. They refer to this as the *finite-conductivity model.*

The relative influence of the surface and interior resistances to heat transfer is measured by the size of the dimensionless parameter, *Biot modulus*, Bi, where for the case of the plane wall

$$\mathrm{Bi} = \frac{\alpha w}{\lambda_{\mathrm{s}}}$$

where w is the semithickness of the wall. This can be extended to other geometries so that w becomes the radius of the spheres or solid cylinders that might compose the packing, for example. In the case of the hollow cylinder, $w = r_o - r_i$, where r_o is the outer radius and r_i is the inner radius. Here the gas flows through the middle of the cylinder. For spheres, $w = r_i$, where r_i is the sphere radius.

For the plane wall and hollow cylinder, we consider the flow of gas through a single channel. The complete storage unit is then considered to be a bundle of identical channels. In the case of a bed of spherical particles, we consider the temperature transients within a single sphere and then assume that all the spheres at the same height in the storage material behave identically. The same applies to solid cylinders, where it is assumed that the cylinders all lie in the same direction relative to that of gas flow.

Where the Biot modulus is small, the simplified model described in Chapter 2 can be used. If the Biot modulus is large, a condition encountered when the energy-carrying fluid is a liquid or when the thermal conductivity, λ_s, is small, the single-blow problem can only be solved accurately using some form of the finite conductivity model.

3.2 The Finite Conductivity Model

The transient behavior of the storage unit is governed by the energy conservation equation for the moving fluid and the diffusion equation for the solid. We consider the case of the plain wall, in which case the equations are

$$\frac{\partial t_f}{\partial y} = \frac{\alpha A}{\dot{m}_f c_p L}(t_{s,o} - t_f) - \frac{m_f}{\dot{m}_f L}\frac{\partial t_f}{\partial \tau} \tag{3.1}$$

$$\frac{1}{\kappa_s}\frac{\partial t_s}{\partial \tau} = \frac{\partial^2 t_s}{\partial x^2} + \frac{\partial^2 t_s}{\partial y^2} \tag{3.2}$$

respectively. The equations are connected through the boundary conditions. In particular,

$$\frac{\partial t_s}{\partial y} = 0 \quad \text{at} \quad y = 0 \quad \text{and at} \quad y = L \tag{3.3}$$

where $y = 0$ refers to the gas entrance and $y = L$ refers to the gas exit of the storage unit. These boundary conditions embody the assumption that there are no heat losses from the top or bottom of the packing of the storage unit, where we assume that the gas is flowing vertically through the packing. These were called the *adiabatic* boundary conditions by Handley and Heggs [5] in contrast to the situation where it is assumed that the packing gains/loses heat from the top and bottom surfaces of the packing from/to the gas passing through the storage unit. Further, there is the boundary condition

$$\frac{\partial t_s}{\partial x} = 0 \quad \text{at} \quad x = w \tag{3.4}$$

which represents a symmetry of temperature behavior about the semithickness of the packing material. At the surface of the solid,

$$\lambda_s \frac{\partial t_s}{\partial x} = \alpha(t_{s,o} - t_f) \quad \text{at} \quad x = 0 \quad \text{and at} \quad x = 2w \tag{3.5}$$

where $t_{s,o}$ in equations (3.1) and (3.5) is the surface solid temperature.

3.3 Simplification of the Finite Conductivity Model

We consider first the case where the effect of longitudinal (axial) conductivity is ignored, in which equation (3.2) takes the form

$$\frac{1}{\kappa_s}\frac{\partial t_s}{\partial \tau} = \frac{\partial^2 t_s}{\partial x^2} \tag{3.6}$$

We adopt the linearizing assumptions set out in Chapter 2:

1. The thermophysical properties of the gas and solid are constant and are not regarded as temperature dependent.
2. The entrance gas temperature and the flow rate of the gas do not vary with time.
3. The cross-sectional area of the bed does not vary down its length. Further, the flow of gas, as well as the size of the solid particles in the bed and their interspacing, is uniform in the cross-section of the packed bed at all positions down its length. The model ignores any possible maldistribution of flow that might arise in practice.

Under these assumptions, the system of equations (3.1) and (3.6) can be simplified using the substitutions

$$\xi = \frac{\alpha A}{\dot{m}_f c_p L} y \qquad \mathrm{Fo} = \frac{\kappa_s}{w^2}\left(\tau - \frac{m_f y}{\dot{m}_f L}\right) \qquad X = \frac{x}{w} \tag{3.7}$$

where Fo is the Fourier number. The dimensionless temperature, T, defined by equation (2.3) is also introduced, whereupon the equations take the form

$$\frac{\partial T_f}{\partial \xi} = T_{s,o} - T_f \tag{3.8}$$

where $T_{s,o}$ is the dimensionless surface solid temperature, and

$$\frac{\partial T_s}{\partial \mathrm{Fo}} = \frac{\partial^2 T_s}{\partial X^2} \tag{3.9}$$

On the other hand, if axial conductivity is not ignored, then application of substitutions (3.7) to equation (3.2) yields

$$\frac{\partial T_s}{\partial \mathrm{Fo}} = \frac{\partial^2 T_s}{\partial X^2} + \left(\frac{\alpha A w}{\dot{m}_f c_p L}\right)^2 \frac{\partial^2 T_s}{\partial \xi^2} \tag{3.10}$$

The boundary conditions are similarly modified. In particular, equation (3.5) takes the form

$$\frac{\partial T_s}{\partial X} = \mathrm{Bi}(T_{s,o} - T_f) \quad \text{at} \quad X = 0 \quad \text{and at} \quad X = 2 \tag{3.11}$$

where again, Bi is the Biot modulus. At the semithickness of the packing, equation (3.4) takes the form

$$\frac{\partial T_s}{\partial X} = 0 \quad \text{at} \quad X = 1 \tag{3.12}$$

and at the entrance and exit to the packing, equation (3.3) becomes

$$\frac{\partial T_s}{\partial \xi} = 0 \quad \text{at} \quad \xi = 0 \quad \text{and at} \quad \xi = \Lambda \tag{3.13}$$

where Λ is dimensionless length or *reduced length* of the packed bed. It is given by setting $y = L$ in equation (2.4), in which case

$$\Lambda = \frac{\alpha A}{\dot{m}_f c_p} \tag{3.14}$$

3.4 Measure of Longitudinal Conduction

A possible measure, proposed by Tipler [6], of the potential effect of axial conduction down the length of the packing of the storage unit is the ratio γ, where

$$\gamma = \frac{\text{Longitudinal heat conduction}}{\text{Thermal energy input/extracted by the gas}}$$

The temperature gradient down the length of the packing at any instant can be represented very approximately by $(t_{f,in} - t_{f,x})/L$, assuming that the solid temperature gradient is roughly parallel to the gas temperature gradient. Note that $t_{f,in}$ and $t_{f,x}$ are, respectively, the inlet and exit gas temperatures to and from the packing at any instant.

The volume of the plain wall packing is equal to the product of the heating surface area, A and the semithickness, w, of the walls surrounding the channels. It follows that the effective cross-sectional area of the solid packing is Aw/L. It follows that the rate of axial heat conduction down the bed is $Aw\lambda_s(t_{f,in} - t_{f,x})/L^2$.

The rate of supply of heat to the packing by the gas at any instant is given by $\dot{m}_f c_p(t_{f,in} - t_{f,x})$. The factor γ is therefore given by

$$\gamma = \frac{Aw\lambda_s(t_{f,in} - t_{f,x})}{\dot{m}_f c_p(t_{f,in} - t_{f,x})L^2} = \frac{Aw\lambda_s}{\dot{m}_f c_p L^2} \tag{3.15}$$

As we shall discuss in a later chapter, Tipler [6] mentioned a value of $\gamma = 10^{-2}$ for regenerators used in gas turbines and considered that axial conduction effects could be ignored in this case.

We return to a consideration of equation (3.10) and note that

$$\left(\frac{\alpha A w}{\dot{m}_f c_p L}\right)^2 = \Lambda\left(\frac{\alpha A w^2}{\dot{m}_f c_p L^2}\right) = \Lambda\left(\frac{A w \lambda_s}{\dot{m}_f c_p L^2}\right)\left(\frac{\alpha w}{\lambda_s}\right) = \Lambda \gamma \, \mathrm{Bi}$$

Equation (3.10) can now be simplified into the form

$$\frac{\partial T_s}{\partial \mathrm{Fo}} = \frac{\partial^2 T_s}{\partial X^2} + \Lambda \gamma \, \mathrm{Bi} \frac{\partial^2 T_s}{\partial \xi^2} \tag{3.16a}$$

Equation (3.16a) can be rearranged in an illuminating fashion. Now

$$\left(\frac{\alpha A w}{\dot{m}_f c_p L}\right)^2 = \Lambda^2 \left(\frac{w}{L}\right)^2$$

and it follows that equation (3.16a) now becomes

$$\frac{\partial T_s}{\partial \mathrm{Fo}} = \frac{\partial^2 T_s}{\partial X^2} + \Lambda^2 \left(\frac{w}{L}\right)^2 \frac{\partial^2 T_s}{\partial \xi^2} \tag{3.16b}$$

From this it is clear that the effect of axial conduction is strongly influenced by the ratio of the semithickness to the length of the packing, w/L (or V^+ in the notation of Schmidt and Szego), and by the dimensionless length.

Example 3.1 Consider again the Feolite thermal storage unit whose available heating surface area, A, is $5.8\,\mathrm{m}^2$. The thermal conductivity is considered to be $2.1\,\mathrm{W/(m\,°C)}$. The unit is regarded as being made up of slabs of semithickness $w = 0.04\,\mathrm{m}$. The entire unit is well insulated. Air flows through the packing at a rate $\dot{m}_f$ equal to $0.156\,\mathrm{kg/s}$. The convective heat transfer coefficient, α, has the value $54.39\,\mathrm{W/(m^2\,°C)}$. The length of the unit is $1.5\,\mathrm{m}$.

The dimensionless length, Λ, of the storage unit is given by

$$\Lambda = \frac{\alpha A}{\dot{m}_f c_f} = \frac{(54.39)(5.8)}{(0.156)(1011.0)} = 2.0$$

where it is assumed that the specific heat of air at 50°C is equal to 1011 J/(kg °C). The Biot modulus, Bi, takes the value

$$\mathrm{Bi} = \frac{\alpha w}{\lambda_s} = \frac{(54.39)(0.04)}{2.1} = 1.036$$

Finally, the Tipler measure, γ is calculated:

$$\gamma = \frac{Aw\lambda_s}{\dot{m}_f c_p L^2} = \frac{(5.8)(0.04)(2.1)}{(0.156)(1011)(1.5^2)} = 1.373 \times 10^{-3}$$

It follows that

$$\Lambda\gamma \text{Bi} = (2.0)(1.373 \times 10^{-3})(1.036) = 2.844 \times 10^{-3}$$

Similarly

$$\left[\Lambda\left(\frac{w}{L}\right)\right]^2 = \left[2.0\left(\frac{0.04}{1.5}\right)\right]^2 = 2.844 \times 10^{-3}$$

On the basis of Tipler's observations, one should expect the effect of longitudinal conductivity to be negligible in this case and it should be sufficient to employ equations (3.8) and (3.9).

3.5 Simplified Approach to Axial Conduction

Handley and Heggs [5] examined this problem by separating the latitudinal and longitudinal effects. We present this approach although formulated in a slightly different fashion.

The differential equations that describe heat transfer in a packed bed can be written in the following way if longitudinal conduction effects are included but transverse conduction effects are ignored.

$$\frac{\partial t_f}{\partial y} = \frac{\alpha A}{\dot{m}_f c_p L}(t_s - t_f) - \frac{m_f}{\dot{m}_f L}\frac{\partial t_f}{\partial \tau} \tag{3.17}$$

$$\frac{\partial t_s}{\partial \tau} = \frac{\alpha A}{MC_s}(t_f - t_s) + \kappa_s \frac{\partial^2 t_s}{\partial y^2} \tag{3.18}$$

The dimensionless temperatures T_s and T_f given by equation (2.3) are introduced. We again use the dimensionless parameters ξ and η given by

$$\xi = \frac{\alpha A}{\dot{m}_f c_p L} y \qquad \eta = \frac{\alpha A}{MC_s}\left(\tau - \frac{m_f y}{\dot{m}_f L}\right) \tag{3.19}$$

The key thing to note immediately is that

$$\frac{\partial^2 T_s}{\partial y^2} = \left(\frac{\alpha A}{\dot{m}_f c_p L}\right)^2 \frac{\partial^2 T_s}{\partial \xi^2}$$

We consider the introduction of these parameters to equation (3.18):

$$\begin{aligned}\frac{\partial T_s}{\partial \eta} &= \frac{MC_s}{\alpha A}\frac{\partial T_s}{\partial \tau} \\ &= \frac{MC_s}{\alpha A}\left[\frac{\alpha A}{MC_s}(T_f - T_s) + \kappa_s\left(\frac{\alpha A}{\dot{m}_f c_p L}\right)^2 \frac{\partial^2 T_s}{\partial \xi^2}\right] \\ &= T_f - T_s + \psi\frac{\partial^2 T_s}{\partial \xi^2}\end{aligned} \tag{3.20}$$

where

$$\psi = \frac{MC_s\kappa_s\alpha A}{(\dot{m}_f c_p L)^2}$$

It turns out that the factor ψ can be rearranged. We note that the diffusivity of the packing material $\kappa_s = \lambda_s/(\rho_s C_s)$ where ρ_s is the density of the packing. It will be observed that the mass, M, of the packing is equal to the density times the packing volume, that is $M = Aw\rho_s$. It follows that

$$\begin{aligned}\psi &= \frac{\alpha A}{\dot{m}_f c_p}\frac{\lambda_s}{\rho_s C_s}\frac{Aw\rho_s C_s}{L^2\dot{m}_f c_p} \\ &= \Lambda\frac{\lambda_s Aw}{L^2\dot{m}_f c_p} = \gamma\Lambda\end{aligned} \tag{3.21a}$$

Alternatively,

$$\begin{aligned}\psi &= \Lambda\frac{\lambda_s}{\rho_s C_s w}\frac{Aw^2\rho_s C_s}{L^2\dot{m}_f c_p} \\ &= \Lambda\frac{\lambda_s}{w\alpha}\frac{\alpha A}{\dot{m}_f c_p}\left(\frac{w}{L}\right)^2 = \frac{1}{\mathrm{Bi}}\Lambda^2\left(\frac{w}{L}\right)^2\end{aligned} \tag{3.21b}$$

indicating again the strong influence of Λ and w/L upon the likely effect of longitudinal conductivity. The differential equations (3.7) and (3.8) are reduced to the simple form

$$\frac{\partial T_f}{\partial \xi} = T_s - T_f \tag{3.22}$$

$$\frac{\partial T_s}{\partial \eta} = T_f - T_s + \gamma\Lambda\frac{\partial^2 T_s}{\partial \xi^2} \tag{3.23a}$$

Alternatively, this equation can be written in the form

$$\frac{\partial T_s}{\partial \eta} = T_f - T_s + \frac{1}{\mathrm{Bi}}\Lambda^2\left(\frac{w}{L}\right)^2\frac{\partial^2 T_s}{\partial \xi^2} \tag{3.23b}$$

3.6 Analysis of the Effect of Longitudinal Conduction

Handley and Heggs examined this problem but considered that the dimensionless group $\gamma\Lambda$ did not involve bed length, L. This arose as a consequence of their choice of a distance dimensionless parameter. Their results are most interesting and cover the range $1 \leq \Lambda \leq 20$. For small values of reduced length, $\Lambda \approx 1$, there is little difference between the exit fluid temperatures from the bed computed with the axial conduction model and the simple Schumann model, even for γ as large as 10. As Λ increases, smaller values of the Tipler factor, γ, are required for such a good correlation. Thus for $\Lambda = 16$ and $\Lambda = 20$, a good correlation between models is realized with $\gamma = 10^{-2}$. This would be anticipated from consideration of equation (3.23). Handley and Heggs conclude that for $\Lambda > 4$, longitudinal (axial) conductivity can be safely ignored provided $\Lambda\gamma < 0.1$. Typical values of γ lie in the range $10^{-5} < \gamma < 10^{-3}$ so that, as a rule of thumb, caution about the effect of axial thermal conductivity should be exercised should $\Lambda > 10^2$.

In Figure 3.1 are displayed the so-called *breakthrough curves*, the variations of dimensionless exit fluid temperature, $T_{f,x}$, as a function of dimensionless time, η, as offered by Handley and Heggs in their paper. In this figure, it is assumed that $T_{s,0} = 0$ and $T_{f,in} = 1$. These curves appear to indicate that initially, longitudinal conduction accelerates the transmission

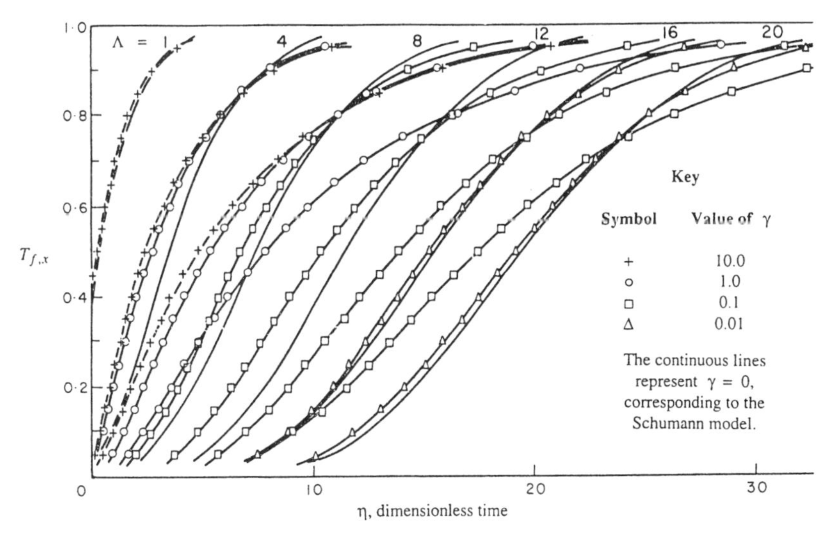

Figure 3.1: Variation of fluid exit temperature, $T_{f,x}$ with dimensionless time, η, for the axial conduction model with adiabatic boundary conditions. (Handley and Heggs [5].)

of heat down the bed, that is, it reduces the time taken for the fluid exit temperature to attain a particular value. However, towards the end of the single blow, the reverse appears to happen and the dimensionless fluid exit temperature moves more slowly toward thc final value of 1 the more significant axial conduction becomes.

Presumably, due to axial conduction, the solid temperature has risen more quickly initially and, as a consequence, toward the end of the single blow, the temperature difference between gas and solid is smaller than it might otherwise have been in the absence of longitudinal conduction. This temperature difference is the driving force for heat transfer between the gas and the solid and its reduction makes smaller the *rate* of heat transfer. In addition, as $T_{f,x} \to 1$ so the dimensionless temperature of the solid in the storage unit as a whole also approaches its final value of 1, thereby making the temperature gradient down the bed smaller and hence reducing any possible effect of longitudinal conduction.

3.7 Axial Conductivity in Packed Beds

Schotte [7] points out that it is generally recognized that packed beds, for example, beds of spheres, have much lower conductivities than might be predicted by the thermal conductivity alone of the material of which the particles in the bed are composed. He indicates that the conduction process is extremely complex, involving the thermal conductivity of both gas and solid, being a function of these two quantities and the void fraction of the bed. Solid conduction alone is restricted simply because heat is transferred from one sphere to other spheres, for example, only at the points of contact. At higher temperatures, radiation between particles enhances any axial conductivity effect, although convective heat transfer still dominates except at very low gas velocities. The overall conclusion must be that likely values of the Tipler factor γ will be even lower than might be predicted for solid slabs, hollow cylinders, or similar heat storage arrangements and that longitudinal conductivity in packed beds is most likely to be unimportant.

3.8 A Comparison Between the Models Representing Latitudinal Conduction

It is now considered that axial conduction in the packing is negligible and we focus upon the effect of *latitudinal* or *transverse* thermal conduction within the solid material, in a direction perpendicular to gas flow. We regard the transverse finite-conductivity model as ideal because it makes fewest assumptions about the effect of this conductivity. This model is compared with two approximating models that were introduced in Chapter 2, namely the Schumann model, in which transverse conduction is regarded as infinitely large and therefore ignored, and the modified heat transfer coefficient model, where an attempt is made to retain the simplicity of the Schumann equations (2.7) and (2.8) while attempting to represent a finite latitudinal conductivity in the packing of the storage unit.

However, to compare the transverse finite-conductivity model with the Schumann model as embodied in equations (2.7) and (2.8) and with the modified heat transfer coefficient model (see equations (2.60)), it is necessary that the three models share the same dimensionless time scale, η. The system of equations (3.1) and (3.6) is simplified using

$$\xi = \frac{\alpha A}{\dot{m}_f c_p L} y \qquad \eta = \frac{\alpha A}{M C_s}\left(\tau - \frac{m_f y}{\dot{m}_f L}\right) \qquad X = \frac{x}{w} \tag{3.24}$$

It can be shown that this dimensionless time variable, η, is related to the Fourier number:

$$\eta = \mathrm{Bi} \times \mathrm{Fo}$$

Equations (3.8), (3.11), and (3.12) remain unchanged. However, equation (3.6) is developed in the following manner. As previously,

$$\frac{\partial^2 T}{\partial x^2} = \frac{1}{w^2} \frac{\partial^2 T}{\partial X^2}$$

The left-hand side of the equation takes the form

$$\frac{\partial T}{\partial \tau} = \frac{\alpha A}{M C_s} \frac{\partial T}{\partial \eta}$$

It follows that substituting in equation (3.6) yields

$$\frac{\partial T}{\partial \eta} = \frac{\kappa_s}{w^2} \frac{M C_s}{\alpha A} \frac{\partial^2 T}{\partial X^2}$$

But $M = A w \rho_s$ for slabs, where ρ_s = density of the packing material. It is recalled that $\kappa_s = \lambda_s/(\rho_s C_s)$ from which it will be seen that

$$\frac{\kappa_s}{w^2} \frac{M C_s}{\alpha A} = \frac{\lambda_s}{\rho_s C_s w^2} \frac{A w \rho_s C_s}{\alpha A} = \frac{\lambda_s}{\alpha w} = \frac{1}{\mathrm{Bi}}$$

where Bi is again the Biot modulus. Equation (3.6) now takes the form

$$\frac{\partial T}{\partial \eta} = \frac{1}{\mathrm{Bi}} \frac{\partial^2 T}{\partial X^2} \tag{3.25}$$

In effecting this comparison between models, be it sufficient here to mention that the finite-difference method used here for the solution of equations (3.8) and (3.25) with the linking boundary conditions is a development by Evans [8] of a method proposed by Willmott [9] in 1969. This model has been used to compute the breakthrough curves for different values of reduced length, Λ and Biot modulus, Bi. These results are presented in tabular form in Tables 3.1 to 3.4. Sadly, we have not been able to reproduce the tables published by Schmidt and Szego [4] where their parameter G^+/V^+ can be shown to be the same as Λ/Bi. The present author is satisfied with the correctness of the methodology and software used [8, 9]. It must be assumed that problems arose in the generation of the Schmidt–Szego results as a consequence of their inclusion of a longitudinal conduction term, however small, in their model and/or the method of computation they chose to employ. The results offered here only agree with those of Schmidt and Szego [4] to one or two significant figures.

In the Schumann model, it is assumed that $\lambda_s = \infty$ and therefore that $\mathrm{Bi} = 0$. It is expected therefore that this model will satisfactorily approximate the finite-conductivity model only for small values of the Biot modulus, Bi. That our methodology and software are correct is illustrated by the fact that, for example, Table 3.1 (finite-conductivity model with $\mathrm{Bi} = 10^{-2}$) and Table 2.2 (Schumann model) show that the Schumann model does indeed satisfactorily approximate the finite-conductivity model for a small value of Biot modulus. On the other hand, this does confirm the assertion of Schmidt and Szego [4] that the Schumann model can be used when $\mathrm{Bi} < 10^{-1}$.

In the so-called *Hausen model*, in which a modified heat transfer coefficient is used within the simplified model, as described at the end of Chapter 2, an attempt is made to represent the case where $\lambda \neq \infty$, that is, where a finite value of the solid thermal conductivity is included. Here $\mathrm{Bi} > 0$. The calculated values of the exit fluid temperature calculated using this model are offered in Table 3.5 (for $\mathrm{Bi} = 10^{-1}$), Table 3.6 (for $\mathrm{Bi} = 1.0$), and Table 3.7 (for $\mathrm{Bi} = 10.0$). These can be compared with the corresponding tables, namely Table 3.2 (for $\mathrm{Bi} = 10^{-1}$), Table 3.3 (for $\mathrm{Bi} = 1.0$), and Table 3.4 (for $\mathrm{Bi} = 10.0$), yielded using the full 3-D finite-conductivity model. It is more convenient, however, to make the comparison graphically, in the manner described below.

3.9 Comparison of the Three Models

Schmidt and Szego [4] claim that the largest differences between models appear in the calculation of the breakthrough curves, that is the exit fluid temperature, $T_f(\Lambda, \eta)$ from the storage unit, certainly in the ranges $0 < \Lambda < 20$ and $0 < \eta < 20$. Note that $T_f(\Lambda, \eta)$ is the fluid temperature computed using the 3-D model, that is, the finite-conductivity model where the three dimensions are ξ, η, and X.

We define first the differences

$$\delta_S(\Lambda, \eta) = T_f(\Lambda, \eta) - T_{f,S}(\Lambda, \eta)$$
$$\delta_H(\Lambda, \eta) = T_f(\Lambda, \eta) - T_{f,H}(\Lambda, \eta)$$

for a specified Biot modulus, Bi. Here $T_{f,S}(\Lambda, \eta)$ is the exit fluid temperature from the storage unit, calculated using the Schumann model, and $T_{f,H}(\Lambda, \eta)$ is the corresponding temperature computed using the Hausen, modified heat transfer coefficient model.

We display the regions in the Λ–η plane for which, in Figure 3.2, $\delta_S(\Lambda, \eta) < \varepsilon$ with $\varepsilon = 10^{-2}$, 10^{-3}, and 10^{-4} for $\text{Bi} = 0.1$. We know that

$$\lim_{\eta \to \infty} T_f(\Lambda, \eta) = \lim_{\eta \to \infty} T_{f,S}(\Lambda, \eta) = 1.0$$

so that for a large enough time, η, both models should yield the same exit temperature, 1.0, from the bed. Similarly, for a large enough length, Λ, at the start of the single blow, there will be little or no penetration of the heat front down the length of the bed. Both models should, therefore, yield the same exit temperature of the fluid, 0.0. From Figure 3.2, it will be seen that this is confirmed by the regions at the top left- and the bottom right-hand corners of the region where $\varepsilon = 10^{-4}$.

Down the center of the Λ–η region, running from SW to NE is a narrow region where $\varepsilon = 10^{-3}$: this will correspond to the region around which the two breakthrough curves intersect. In between this central region and the outer NW and SE corners lie two broad bands where $\varepsilon = 10^{-2}$, where the Schumann model achieves less accuracy.

A similar set of regions in the Λ–η plane arises but with poorer values of ε in the case where $\text{Bi} = 1.0$. These are given in Figure 3.3 and indicate that the Schumann model is hardly adequate to represent the finite-conductivity model in this case, over the greater part of the Λ–η plane shown.

Where $\text{Bi} = 10$, there is a poor correspondence between the finite conductivity and the Schumann models and this will be observed simply by comparing Tables 2.2 and 3.4.

The Hausen model yields the predicted improvement over the Schumann model. Indeed, $\varepsilon = 10^{-4}$ over the Λ–η plane for $\text{Bi} = 0.1$. On the other hand, the Hausen model is quite inadequate to represent the finite-conductivity model where $\text{Bi} = 100$ and it is only marginally better when $\text{Bi} = 10$. Here, the regions in the Λ–η plane are those defined by $\delta_H(\Lambda, \eta) < \varepsilon$.

Table 3.1: Dimensionless fluid temperatures, $T_f(\xi, \eta)$,[a] calculated using the 3-D single-blow model with $\mathrm{Bi} = 10^{-2}$

	Dimensionless distance, ξ									
η	1	2	3	4	5	6	7	8	9	10
1	0.6546	0.3949	0.2256	0.1239	0.0660	0.0343	0.0175	0.0088	0.0043	0.0021
2	0.8173	0.6037	0.4152	0.2706	0.1691	0.1022	0.0600	0.0344	0.0193	0.0106
3	0.9059	0.7529	0.5835	0.4273	0.2987	0.2009	0.1307	0.0827	0.0511	0.0309
4	0.9526	0.8517	0.7169	0.5719	0.4354	0.3184	0.2249	0.1540	0.1027	0.0668
5	0.9765	0.9137	0.8147	0.6929	0.5640	0.4413	0.3333	0.2438	0.1734	0.1203
6	0.9885	0.9510	0.8825	0.7874	0.6755	0.5583	0.4459	0.3450	0.2593	0.1899
7	0.9945	0.9728	0.9275	0.8572	0.7661	0.6621	0.5539	0.4495	0.3545	0.2724
8	0.9974	0.9852	0.9563	0.9066	0.8362	0.7490	0.6513	0.5504	0.4525	0.3625
9	0.9988	0.9920	0.9742	0.9403	0.8881	0.8184	0.7348	0.6425	0.5474	0.4550
10	0.9994	0.9958	0.9850	0.9626	0.9253	0.8718	0.8031	0.7227	0.6350	0.5450
11	0.9997	0.9978	0.9914	0.9771	0.9512	0.9114	0.8572	0.7899	0.7124	0.6286
12	0.9999	0.9989	0.9952	0.9861	0.9687	0.9400	0.8985	0.8441	0.7782	0.7033
13	1.0000	0.9994	0.9973	0.9918	0.9802	0.9601	0.9293	0.8866	0.8323	0.7678
14	1.0000	0.9997	0.9985	0.9951	0.9877	0.9739	0.9515	0.9190	0.8757	0.8217
15	1.0000	0.9999	0.9992	0.9972	0.9925	0.9832	0.9674	0.9432	0.9094	0.8655
16	1.0000	0.9999	0.9996	0.9984	0.9954	0.9893	0.9783	0.9608	0.9350	0.9002
17	1.0000	1.0000	0.9998	0.9991	0.9973	0.9933	0.9858	0.9733	0.9542	0.9271
18	1.0000	1.0000	0.9999	0.9995	0.9984	0.9959	0.9908	0.9821	0.9681	0.9476
19	1.0000	1.0000	0.9999	0.9997	0.9990	0.9975	0.9942	0.9881	0.9781	0.9629
20	1.0000	1.0000	1.0000	0.9998	0.9994	0.9984	0.9963	0.9922	0.9852	0.9740

Table 3.1 *(contd.)*

η	Dimensionless distance, ξ									
	11	12	13	14	15	16	17	18	19	20
1	0.0010	0.0005	0.0002	0.0001	0.0000	0.0000	0.0000	0.0000	0.0000	0.0000
2	0.0058	0.0031	0.0016	0.0008	0.0004	0.0002	0.0001	0.0000	0.0000	0.0000
3	0.0183	0.0106	0.0061	0.0034	0.0019	0.0011	0.0006	0.0003	0.0001	0.0001
4	0.0426	0.0266	0.0164	0.0099	0.0059	0.0034	0.0020	0.0011	0.0006	0.0003
5	0.0815	0.0541	0.0352	0.0226	0.0142	0.0088	0.0054	0.0032	0.0019	0.0011
6	0.1358	0.0950	0.0652	0.0439	0.0290	0.0189	0.0122	0.0077	0.0048	0.0030
7	0.2042	0.1497	0.1075	0.0757	0.0524	0.0357	0.0239	0.0158	0.0103	0.0066
8	0.2834	0.2166	0.1621	0.1190	0.0857	0.0607	0.0423	0.0291	0.0197	0.0131
9	0.3693	0.2930	0.2276	0.1734	0.1296	0.0952	0.0688	0.0489	0.0343	0.0237
10	0.4571	0.3751	0.3015	0.2375	0.1836	0.1394	0.1041	0.0766	0.0554	0.0396
11	0.5429	0.4590	0.3803	0.3089	0.2463	0.1929	0.1486	0.1126	0.0840	0.0618
12	0.6231	0.5410	0.4607	0.3848	0.3156	0.2543	0.2015	0.1571	0.1206	0.0912
13	0.6954	0.6182	0.5394	0.4621	0.3889	0.3216	0.2616	0.2094	0.1650	0.1282
14	0.7584	0.6883	0.6138	0.5380	0.4634	0.3926	0.3271	0.2683	0.2167	0.1725
15	0.8119	0.7500	0.6819	0.6099	0.5367	0.4646	0.3959	0.3321	0.2744	0.2235
16	0.8560	0.8030	0.7424	0.6761	0.6064	0.5355	0.4657	0.3989	0.3367	0.2801
17	0.8916	0.8472	0.7948	0.7354	0.6709	0.6032	0.5344	0.4667	0.4017	0.3409
18	0.9195	0.8834	0.8390	0.7872	0.7290	0.6661	0.6002	0.5334	0.4676	0.4043
19	0.9412	0.9122	0.8756	0.8314	0.7801	0.7231	0.6616	0.5975	0.5325	0.4684
20	0.9576	0.9349	0.9052	0.8683	0.8242	0.7736	0.7176	0.6575	0.5950	0.5317

[a]Three-figure accuracy is assured.

Table 3.2: Dimensionless fluid temperatures, $T_f(\xi, \eta)$,[a] calculated using the 3-D single-blow model with $\mathrm{Bi} = 10^{-1}$

	Dimensionless distance, ξ									
η	1	2	3	4	5	6	7	8	9	10
1	0.6574	0.4004	0.2316	0.1291	0.0699	0.0369	0.0192	0.0098	0.0049	0.0024
2	0.8166	0.6054	0.4192	0.2758	0.1742	0.1066	0.0634	0.0369	0.0210	0.0118
3	0.9042	0.7521	0.5848	0.4307	0.3033	0.2058	0.1353	0.0866	0.0541	0.0331
4	0.9510	0.8497	0.7161	0.5730	0.4384	0.3227	0.2296	0.1586	0.1068	0.0703
5	0.9753	0.9115	0.8127	0.6921	0.5650	0.4440	0.3372	0.2484	0.1780	0.1245
6	0.9877	0.9491	0.8802	0.7854	0.6748	0.5592	0.4483	0.3486	0.2637	0.1945
7	0.9940	0.9713	0.9253	0.8547	0.7642	0.6614	0.5548	0.4518	0.3580	0.2765
8	0.9971	0.9841	0.9544	0.9042	0.8337	0.7471	0.6507	0.5511	0.4546	0.3658
9	0.9986	0.9913	0.9727	0.9381	0.8855	0.8159	0.7330	0.6419	0.5482	0.4570
10	0.9993	0.9953	0.9839	0.9608	0.9229	0.8691	0.8007	0.7210	0.6344	0.5457
11	0.9997	0.9975	0.9907	0.9757	0.9491	0.9088	0.8545	0.7875	0.7107	0.6281
12	0.9998	0.9987	0.9947	0.9851	0.9670	0.9377	0.8958	0.8414	0.7758	0.7017
13	0.9999	0.9993	0.9970	0.9910	0.9789	0.9582	0.9268	0.8839	0.8296	0.7655
14	1.0000	0.9997	0.9983	0.9946	0.9867	0.9723	0.9494	0.9164	0.8728	0.8189
15	1.0000	0.9998	0.9991	0.9969	0.9918	0.9820	0.9656	0.9409	0.9067	0.8626
16	1.0000	0.9999	0.9995	0.9982	0.9950	0.9884	0.9769	0.9588	0.9326	0.8975
17	1.0000	1.0000	0.9997	0.9989	0.9969	0.9927	0.9847	0.9717	0.9520	0.9246
18	1.0000	1.0000	0.9999	0.9994	0.9982	0.9954	0.9900	0.9808	0.9663	0.9454
19	1.0000	1.0000	0.9999	0.9997	0.9989	0.9972	0.9936	0.9872	0.9767	0.9609
20	1.0000	1.0000	1.0000	0.9998	0.9994	0.9982	0.9959	0.9915	0.9841	0.9724

Table 3.2 *(contd.)*

	Dimensionless distance, ξ									
η	11	12	13	14	15	16	17	18	19	20
1	0.0012	0.0006	0.0003	0.0001	0.0000	0.0000	0.0000	0.0000	0.0000	0.0000
2	0.0065	0.0035	0.0019	0.0010	0.0005	0.0003	0.0001	0.0001	0.0000	0.0000
3	0.0199	0.0118	0.0068	0.0039	0.0022	0.0012	0.0007	0.0004	0.0002	0.0001
4	0.0453	0.0287	0.0178	0.0109	0.0066	0.0039	0.0023	0.0013	0.0008	0.0004
5	0.0852	0.0571	0.0376	0.0244	0.0155	0.0098	0.0061	0.0037	0.0022	0.0013
6	0.1402	0.0989	0.0685	0.0466	0.0312	0.0205	0.0133	0.0086	0.0054	0.0034
7	0.2086	0.1541	0.1115	0.0792	0.0553	0.0380	0.0258	0.0172	0.0113	0.0074
8	0.2874	0.2210	0.1665	0.1231	0.0894	0.0639	0.0449	0.0311	0.0213	0.0144
9	0.3724	0.2969	0.2319	0.1777	0.1338	0.0990	0.0721	0.0517	0.0366	0.0256
10	0.4591	0.3781	0.3052	0.2417	0.1879	0.1436	0.1080	0.0800	0.0584	0.0421
11	0.5435	0.4609	0.3831	0.3126	0.2504	0.1972	0.1528	0.1166	0.0876	0.0650
12	0.6225	0.5416	0.4624	0.3876	0.3191	0.2583	0.2057	0.1613	0.1246	0.0949
13	0.6938	0.6176	0.5400	0.4638	0.3916	0.3251	0.2656	0.2136	0.1692	0.1322
14	0.7562	0.6868	0.6133	0.5385	0.4651	0.3951	0.3305	0.2722	0.2208	0.1767
15	0.8092	0.7478	0.6804	0.6094	0.5372	0.4662	0.3984	0.3354	0.2782	0.2276
16	0.8531	0.8003	0.7402	0.6747	0.6059	0.5360	0.4672	0.4014	0.3399	0.2838
17	0.8887	0.8444	0.7921	0.7333	0.6695	0.6027	0.5349	0.4682	0.4041	0.3440
18	0.9169	0.8805	0.8362	0.7846	0.7269	0.6647	0.5998	0.5340	0.4690	0.4066
19	0.9388	0.9095	0.8727	0.8285	0.7776	0.7210	0.6603	0.5971	0.5330	0.4698
20	0.9555	0.9324	0.9024	0.8653	0.8213	0.7710	0.7156	0.6563	0.5946	0.5322

[a]Three-figure accuracy is assured.

Table 3.3: Dimensionless fluid temperatures, $T_f(\xi, \eta)$,[a] calculated using the 3-D single-blow model with Bi = 1.0

	Dimensionless distance, ξ									
η	1	2	3	4	5	6	7	8	9	10
1	0.6867	0.4510	0.2866	0.1775	0.1076	0.0641	0.0376	0.0218	0.0125	0.0071
2	0.8144	0.6233	0.4557	0.3216	0.2205	0.1477	0.0969	0.0625	0.0397	0.0249
3	0.8913	0.7479	0.5985	0.4609	0.3438	0.2497	0.1772	0.1233	0.0843	0.0567
4	0.9370	0.8347	0.7117	0.5845	0.4648	0.3594	0.2711	0.2002	0.1451	0.1033
5	0.9637	0.8934	0.7976	0.6879	0.5751	0.4577	0.3710	0.2878	0.2189	0.1635
6	0.9793	0.9322	0.8606	0.7707	0.6707	0.5684	0.4701	0.3802	0.3013	0.2344
7	0.9883	0.9574	0.9056	0.8348	0.7501	0.6575	0.5632	0.4720	0.3877	0.3125
8	0.9934	0.9736	0.9369	0.8829	0.8138	0.7335	0.6470	0.5590	0.4736	0.3939
9	0.9963	0.9837	0.9584	0.9183	0.8635	0.7963	0.7199	0.6384	0.5555	0.4749
10	0.9979	0.9901	0.9729	0.9437	0.9014	0.8467	0.7814	0.7084	0.6311	0.5526
11	0.9988	0.9940	0.9825	0.9617	0.9298	0.8863	0.8320	0.7685	0.6986	0.6348
12	0.9994	0.9964	0.9888	0.9742	0.9506	0.9168	0.8726	0.8189	0.7573	0.6900
13	0.9996	0.9978	0.9929	0.9828	0.9656	0.9398	0.9047	0.8602	0.8073	0.7474
14	0.9998	0.9987	0.9955	0.9886	0.9763	0.9570	0.9295	0.8935	0.8489	0.7968
15	0.9999	0.9992	0.9972	0.9925	0.9838	0.9695	0.9485	0.9197	0.8830	0.8386
16	0.9999	0.9996	0.9983	0.9951	0.9891	0.9787	0.9627	0.9401	0.9104	0.8733
17	1.0000	0.9997	0.9989	0.9969	0.9926	0.9852	0.9732	0.9558	0.9321	0.9016
18	1.0000	0.9998	0.9993	0.9980	0.9951	0.9898	0.9810	0.9677	0.9490	0.9243
19	1.0000	0.9999	0.9996	0.9987	0.9968	0.9930	0.9866	0.9766	0.9621	0.9424
20	1.0000	0.9999	0.9998	0.9992	0.9979	0.9953	0.9906	0.9832	0.9721	0.9565

Table 3.3 *(contd.)*

η	Dimensionless distance, ξ									
	11	12	13	14	15	16	17	18	19	20
1	0.0040	0.0022	0.0012	0.0007	0.0004	0.0002	0.0001	0.0000	0.0000	0.0000
2	0.0154	0.0094	0.0057	0.0034	0.0020	0.0012	0.0007	0.0004	0.0002	0.0001
3	0.0376	0.0247	0.0159	0.0102	0.0065	0.0041	0.0025	0.0016	0.0010	0.0006
4	0.0725	0.0501	0.0342	0.0231	0.0154	0.0102	0.0066	0.0043	0.0028	0.0018
5	0.1201	0.0869	0.0621	0.0437	0.0305	0.0210	0.0143	0.0097	0.0065	0.0043
6	0.1793	0.1350	0.1002	0.0734	0.0531	0.0380	0.0269	0.0188	0.0131	0.0090
7	0.2476	0.1931	0.1483	0.1124	0.0841	0.0622	0.0454	0.0329	0.0235	0.0167
8	0.3220	0.2590	0.2052	0.1603	0.1237	0.0942	0.0709	0.0528	0.0389	0.0284
9	0.3992	0.3302	0.2690	0.2161	0.1712	0.1340	0.1037	0.0793	0.0600	0.0449
10	0.4761	0.4037	0.3373	0.2778	0.2258	0.1811	0.1436	0.1126	0.0873	0.0670
11	0.5501	0.4771	0.4077	0.3436	0.2857	0.2345	0.1902	0.1525	0.1210	0.0949
12	0.6194	0.5480	0.4780	0.4113	0.3492	0.2928	0.2425	0.1986	0.1608	0.1289
13	0.6824	0.6146	0.5461	0.4788	0.4144	0.3543	0.2993	0.2499	0.2063	0.1686
14	0.7385	0.6757	0.6104	0.5444	0.4795	0.4173	0.3588	0.3051	0.2566	0.2135
15	0.7874	0.7305	0.6697	0.6066	0.5428	0.4801	0.4198	0.3630	0.3105	0.2628
16	0.8291	0.7787	0.7233	0.6642	0.6031	0.5415	0.4807	0.4222	0.3668	0.3154
17	0.8642	0.8203	0.7708	0.7167	0.6593	0.6000	0.5402	0.4813	0.4243	0.3704
18	0.8932	0.8557	0.8122	0.7635	0.7107	0.6548	0.5971	0.5391	0.4818	0.4263
19	0.9168	0.8853	0.8477	0.8047	0.7568	0.7051	0.6506	0.5945	0.5380	0.4823
20	0.9358	0.9096	0.8778	0.8403	0.7977	0.7506	0.7000	0.6468	0.5921	0.5371

[a]Three-figure accuracy is assured.

Table 3.4: Dimensionless fluid temperatures, $T_f(\xi, \eta)$,[a] calculated using the 3-D single-blow model with Bi = 10.0

	Dimensionless distance, ξ									
η	1	2	3	4	5	6	7	8	9	10
1	0.8375	0.6912	0.5628	0.4523	0.3592	0.2819	0.2189	0.1682	0.1280	0.0964
2	0.8807	0.7682	0.6638	0.5686	0.4829	0.4067	0.3399	0.2818	0.2320	0.1896
3	0.9037	0.8101	0.7207	0.6365	0.5581	0.4862	0.4207	0.3618	0.3093	0.2628
4	0.9206	0.8411	0.7630	0.6875	0.6155	0.5478	0.4846	0.4264	0.3731	0.3248
5	0.9342	0.8663	0.7978	0.7300	0.6639	0.6003	0.5399	0.4830	0.4299	0.3808
6	0.9455	0.8874	0.8274	0.7666	0.7061	0.6468	0.5892	0.5341	0.4819	0.4327
7	0.9548	0.9052	0.8527	0.7984	0.7432	0.6881	0.6338	0.5809	0.5300	0.4814
8	0.9625	0.9202	0.8744	0.8259	0.7758	0.7250	0.6740	0.6237	0.5745	0.5269
9	0.9689	0.9329	0.8929	0.8498	0.8045	0.7578	0.7103	0.6627	0.6156	0.5694
10	0.9742	0.9435	0.9087	0.8705	0.8297	0.7870	0.7429	0.6982	0.6534	0.6090
11	0.9786	0.9525	0.9222	0.8885	0.8518	0.8129	0.7722	0.7304	0.6881	0.6456
12	0.9823	0.9600	0.9338	0.9040	0.8712	0.8358	0.7984	0.7596	0.7197	0.6794
13	0.9853	0.9664	0.9437	0.9174	0.8881	0.8561	0.8219	0.7859	0.7486	0.7104
14	0.9878	0.9718	0.9521	0.9290	0.9029	0.8740	0.8428	0.8096	0.7749	0.7390
15	0.9899	0.9763	0.9593	0.9390	0.9158	0.8898	0.8614	0.8309	0.7987	0.7651
16	0.9916	0.9801	0.9654	0.9476	0.9270	0.9037	0.8780	0.8501	0.8202	0.7889
17	0.9931	0.9833	0.9706	0.9551	0.9368	0.9159	0.8927	0.8672	0.8397	0.8106
18	0.9942	0.9859	0.9750	0.9615	0.9453	0.9267	0.9057	0.8825	0.8572	0.8303
19	0.9952	0.9882	0.9788	0.9670	0.9527	0.9361	0.9172	0.8961	0.8730	0.8481
20	0.9960	0.9901	0.9820	0.9717	0.9592	0.9444	0.9274	0.9083	0.8872	0.8643

Table 3.4 *(contd.)*

η	Dimensionless distance, ξ									
	11	12	13	14	15	16	17	18	19	20
1	0.0720	0.0533	0.0391	0.0285	0.0206	0.0148	0.0105	0.0074	0.0052	0.0036
2	0.1539	0.1241	0.0994	0.0791	0.0626	0.0492	0.0385	0.0299	0.0231	0.0178
3	0.2221	0.1866	0.1560	0.1297	0.1072	0.0882	0.0723	0.0589	0.0478	0.0386
4	0.2814	0.2425	0.2081	0.1777	0.1511	0.1279	0.1078	0.0904	0.0756	0.0629
5	0.3357	0.2947	0.2575	0.2241	0.1942	0.1677	0.1442	0.1235	0.1054	0.0896
6	0.3869	0.3444	0.3053	0.2696	0.2371	0.2078	0.1814	0.1578	0.1368	0.1182
7	0.4353	0.3921	0.3517	0.3143	0.2798	0.2482	0.2194	0.1933	0.1697	0.1486
8	0.4812	0.4378	0.3967	0.3581	0.3221	0.2887	0.2579	0.2296	0.2038	0.1803
9	0.5246	0.4813	0.4400	0.4008	0.3637	0.3290	0.2966	0.2665	0.2388	0.2133
10	0.5653	0.5227	0.4816	0.4421	0.4045	0.3688	0.3352	0.3037	0.2743	0.2470
11	0.6033	0.5618	0.5212	0.4819	0.4440	0.4078	0.3733	0.3407	0.3100	0.2813
12	0.6388	0.5985	0.5588	0.5200	0.4822	0.4458	0.4108	0.3774	0.3457	0.3158
13	0.6718	0.6330	0.5944	0.5562	0.5189	0.4825	0.4474	0.4135	0.3812	0.3503
14	0.7023	0.6651	0.6278	0.5906	0.5540	0.5180	0.4829	0.4488	0.4160	0.3846
15	0.7304	0.6950	0.6591	0.6232	0.5874	0.5520	0.5172	0.4832	0.4502	0.4183
16	0.7563	0.7227	0.6884	0.6538	0.6190	0.5844	0.5502	0.5165	0.4836	0.4515
17	0.7800	0.7483	0.7157	0.6825	0.6489	0.6153	0.5818	0.5486	0.5159	0.4839
18	0.8017	0.7719	0.7410	0.7093	0.6771	0.6445	0.6119	0.5793	0.5471	0.5153
19	0.8216	0.7936	0.7645	0.7344	0.7035	0.6722	0.6405	0.6088	0.5771	0.5458
20	0.8396	0.8135	0.7861	0.7576	0.7282	0.6982	0.6676	0.6368	0.6059	0.5751

[a]Three-figure accuracy is assured.

Table 3.5: Dimensionless fluid temperatures, $T_f(\xi, \eta)$,[a] calculated using the Hausen $\bar{\alpha}$ 2-D single-blow model with $\mathrm{Bi} = 10^{-1}$

	Dimensionless distance, ξ									
η	1	2	3	4	5	6	7	8	9	10
1	0.6573	0.4004	0.2317	0.1291	0.0699	0.0370	0.0192	0.0098	0.0049	0.0025
2	0.8166	0.6054	0.4192	0.2758	0.1743	0.1066	0.0635	0.0369	0.0211	0.0118
3	0.9042	0.7520	0.5848	0.4307	0.3033	0.2059	0.1354	0.0866	0.0542	0.0332
4	0.9510	0.8497	0.7160	0.5729	0.4384	0.3227	0.2296	0.1587	0.1069	0.0703
5	0.9753	0.9115	0.8127	0.6921	0.5650	0.4440	0.3372	0.2484	0.1781	0.1246
6	0.9877	0.9491	0.8801	0.7854	0.6747	0.5592	0.4483	0.3486	0.2637	0.1945
7	0.9940	0.9713	0.9253	0.8547	0.7642	0.6614	0.5547	0.4518	0.3580	0.2765
8	0.9971	0.9841	0.9544	0.9041	0.8337	0.7471	0.6507	0.5511	0.4546	0.3657
9	0.9986	0.9914	0.9727	0.9381	0.8855	0.8159	0.7330	0.6418	0.5482	0.4570
10	0.9993	0.9954	0.9840	0.9609	0.9229	0.8691	0.8007	0.7210	0.6344	0.5456
11	0.9997	0.9975	0.9907	0.9757	0.9491	0.9088	0.8544	0.7874	0.7107	0.6280
12	0.9998	0.9987	0.9947	0.9851	0.9670	0.9377	0.8958	0.8414	0.7758	0.7017
13	0.9999	0.9993	0.9970	0.9910	0.9789	0.9582	0.9268	0.8838	0.8296	0.7654
14	1.0000	0.9997	0.9983	0.9947	0.9867	0.9724	0.9494	0.9164	0.8728	0.8189
15	1.0000	0.9998	0.9991	0.9969	0.9918	0.9820	0.9656	0.9409	0.9067	0.8626
16	1.0000	0.9999	0.9995	0.9982	0.9950	0.9885	0.9770	0.9588	0.9326	0.8974
17	1.0000	1.0000	0.9997	0.9990	0.9970	0.9927	0.9848	0.9717	0.9520	0.9246
18	1.0000	1.0000	0.9999	0.9994	0.9982	0.9954	0.9901	0.9808	0.9664	0.9454
19	1.0000	1.0000	0.9999	0.9997	0.9989	0.9972	0.9936	0.9872	0.9767	0.9609
20	1.0000	1.0000	1.0000	0.9998	0.9994	0.9983	0.9959	0.9915	0.9841	0.9724

Table 3.5 *(contd.)*

	Dimensionless distance, ξ									
η	11	12	13	14	15	16	17	18	19	20
1	0.0012	0.0006	0.0003	0.0001	0.0001	0.0000	0.0000	0.0000	0.0000	0.0000
2	0.0065	0.0035	0.0019	0.0010	0.0005	0.0003	0.0001	0.0001	0.0000	0.0000
3	0.0199	0.0118	0.0069	0.0039	0.0022	0.0013	0.0007	0.0004	0.0002	0.0001
4	0.0453	0.0287	0.0179	0.0109	0.0066	0.0039	0.0023	0.0014	0.0008	0.0004
5	0.0852	0.0572	0.0377	0.0244	0.0156	0.0098	0.0061	0.0037	0.0023	0.0014
6	0.1402	0.0989	0.0685	0.0466	0.0312	0.0206	0.0134	0.0086	0.0054	0.0034
7	0.2087	0.1541	0.1115	0.0793	0.0554	0.0381	0.0258	0.0172	0.0114	0.0074
8	0.2874	0.2210	0.1665	0.1231	0.0894	0.0639	0.0450	0.0312	0.0213	0.0144
9	0.3724	0.2969	0.2319	0.1778	0.1338	0.0990	0.0721	0.0518	0.0366	0.0256
10	0.4591	0.3781	0.3052	0.2417	0.1879	0.1437	0.1081	0.0801	0.0585	0.0421
11	0.5435	0.4608	0.3831	0.3126	0.2504	0.1972	0.1528	0.1166	0.0877	0.0650
12	0.6225	0.5416	0.4624	0.3876	0.3191	0.2583	0.2057	0.1613	0.1246	0.0950
13	0.6938	0.6176	0.5400	0.4638	0.3915	0.3251	0.2655	0.2136	0.1693	0.1323
14	0.7562	0.6867	0.6133	0.5385	0.4650	0.3951	0.3304	0.2721	0.2208	0.1767
15	0.8092	0.7478	0.6804	0.6094	0.5372	0.4662	0.3984	0.3354	0.2782	0.2276
16	0.8531	0.8003	0.7402	0.6747	0.6059	0.5360	0.4672	0.4013	0.3399	0.2838
17	0.8887	0.8443	0.7921	0.7333	0.6695	0.6027	0.5349	0.4681	0.4041	0.3440
18	0.9169	0.8805	0.8361	0.7845	0.7269	0.6647	0.5997	0.5339	0.4690	0.4066
19	0.9388	0.9095	0.8727	0.8285	0.7775	0.7210	0.6603	0.5970	0.5330	0.4698
20	0.9555	0.9324	0.9024	0.8653	0.8213	0.7710	0.7155	0.6562	0.5946	0.5322

[a] Four-figure accuracy is assured.

Table 3.6: Dimensionless fluid temperatures, $T_f(\xi, \eta)$,[a] calculated using the Hausen $\bar{\alpha}$ 2-D single-blow model with Bi = 1.0

	Dimensionless distance, ξ									
η	1	2	3	4	5	6	7	8	9	10
1	0.6837	0.4487	0.2858	0.1778	0.1087	0.0654	0.0389	0.0228	0.0133	0.0077
2	0.8131	0.6215	0.4543	0.3209	0.2205	0.1482	0.0978	0.0635	0.0406	0.0257
3	0.8909	0.7468	0.5971	0.4597	0.3431	0.2495	0.1775	0.1238	0.0850	0.0575
4	0.9369	0.8341	0.7107	0.5833	0.4637	0.3587	0.2709	0.2003	0.1455	0.1039
5	0.9639	0.8932	0.7970	0.6870	0.5742	0.4668	0.3704	0.2875	0.2189	0.1637
6	0.9795	0.9323	0.8603	0.7701	0.6699	0.5675	0.4693	0.3796	0.3010	0.2344
7	0.9884	0.9576	0.9055	0.8345	0.7495	0.6568	0.5623	0.4712	0.3871	0.3122
8	0.9935	0.9737	0.9370	0.8828	0.8134	0.7329	0.6463	0.5582	0.4729	0.3933
9	0.9964	0.9839	0.9585	0.9183	0.8634	0.7959	0.7193	0.6377	0.5548	0.4742
10	0.9980	0.9902	0.9730	0.9438	0.9014	0.8465	0.7810	0.7078	0.6304	0.5520
11	0.9989	0.9941	0.9826	0.9618	0.9299	0.8863	0.8317	0.7681	0.6980	0.6242
12	0.9994	0.9965	0.9889	0.9744	0.9507	0.9168	0.8725	0.8187	0.7569	0.6894
13	0.9997	0.9979	0.9930	0.9829	0.9658	0.9399	0.9047	0.8601	0.8070	0.7470
14	0.9998	0.9988	0.9956	0.9888	0.9765	0.9571	0.9296	0.8934	0.8488	0.7965
15	0.9999	0.9993	0.9973	0.9927	0.9840	0.9697	0.9486	0.9198	0.8829	0.8384
16	0.9999	0.9996	0.9983	0.9952	0.9892	0.9788	0.9628	0.9403	0.9104	0.8732
17	1.0000	0.9998	0.9990	0.9969	0.9928	0.9853	0.9734	0.9560	0.9322	0.9016
18	1.0000	0.9999	0.9994	0.9980	0.9952	0.9899	0.9811	0.9679	0.9492	0.9244
19	1.0000	0.9999	0.9996	0.9988	0.9968	0.9931	0.9867	0.9768	0.9623	0.9425
20	1.0000	1.0000	0.9998	0.9992	0.9979	0.9953	0.9907	0.9833	0.9722	0.9567

Table 3.6 (*contd.*)

η	Dimensionless distance, ξ									
	11	12	13	14	15	16	17	18	19	20
1	0.0044	0.0025	0.0014	0.0008	0.0004	0.0002	0.0001	0.0001	0.0000	0.0000
2	0.0161	0.0099	0.0061	0.0037	0.0022	0.0013	0.0008	0.0005	0.0003	0.0002
3	0.0383	0.0253	0.0165	0.0107	0.0068	0.0043	0.0027	0.0017	0.0011	0.0007
4	0.0731	0.0508	0.0348	0.0236	0.0159	0.0105	0.0070	0.0046	0.0030	0.0019
5	0.1205	0.0875	0.0626	0.0443	0.0310	0.0215	0.0147	0.0100	0.0068	0.0045
6	0.1795	0.1354	0.1007	0.0739	0.0536	0.0385	0.0273	0.0192	0.0134	0.0093
7	0.2475	0.1932	0.1486	0.1128	0.0846	0.0627	0.0459	0.0333	0.0240	0.0171
8	0.3216	0.2589	0.2053	0.1606	0.1240	0.0946	0.0714	0.0533	0.0394	0.0288
9	0.3986	0.3298	0.2688	0.2161	0.1714	0.1343	0.1040	0.0797	0.0604	0.0454
10	0.4754	0.4032	0.3369	0.2776	0.2257	0.1813	0.1439	0.1129	0.0877	0.0674
11	0.5495	0.4765	0.4072	0.3432	0.2855	0.2345	0.1903	0.1527	0.1213	0.0953
12	0.6188	0.5474	0.4774	0.4108	0.3488	0.2926	0.2425	0.1987	0.1610	0.1292
13	0.6819	0.6140	0.5455	0.4782	0.4139	0.3539	0.2990	0.2498	0.2064	0.1687
14	0.7381	0.6752	0.6098	0.5438	0.4790	0.4168	0.3585	0.3049	0.2565	0.2135
15	0.7871	0.7301	0.6692	0.6060	0.5423	0.4796	0.4194	0.3627	0.3102	0.2627
16	0.8289	0.7784	0.7229	0.6638	0.6026	0.5409	0.4802	0.4217	0.3665	0.3152
17	0.8641	0.8201	0.7705	0.7163	0.6588	0.5995	0.5397	0.4808	0.4239	0.3700
18	0.8932	0.8556	0.8120	0.7632	0.7103	0.6543	0.5966	0.5386	0.4813	0.4259
19	0.9169	0.8852	0.8476	0.8045	0.7565	0.7047	0.6501	0.5940	0.5375	0.4818
20	0.9360	0.9097	0.8777	0.8401	0.7974	0.7503	0.6996	0.6463	0.5916	0.5366

[a]Four-figure accuracy is assured.

Table 3.7: Dimensionless fluid temperatures, $T_f(\xi, \eta)$,[a] calculated using the Hausen $\bar{\alpha}$ 2-D single-blow model with Bi = 10.0

	Dimensionless distance, ξ									
η	1	2	3	4	5	6	7	8	9	10
1	0.8322	0.6918	0.5746	0.4768	0.3954	0.3276	0.2712	0.2243	0.1855	0.1532
2	0.8633	0.7433	0.6383	0.5468	0.4675	0.3989	0.3397	0.2888	0.2451	0.2078
3	0.8887	0.7863	0.6928	0.6082	0.5322	0.4643	0.4039	0.3506	0.3035	0.2622
4	0.9094	0.8222	0.7394	0.6619	0.5901	0.5240	0.4638	0.4092	0.3600	0.3158
5	0.9263	0.8522	0.7793	0.7088	0.6416	0.5782	0.5191	0.4644	0.4140	0.3681
6	0.9400	0.8772	0.8132	0.7495	0.6873	0.6272	0.5700	0.5160	0.4654	0.4184
7	0.9512	0.8980	0.8421	0.7849	0.7276	0.6713	0.6165	0.5639	0.5138	0.4666
8	0.9603	0.9153	0.8666	0.8155	0.7633	0.7108	0.6588	0.6082	0.5592	0.5123
9	0.9677	0.9298	0.8875	0.8420	0.7946	0.7460	0.6972	0.6488	0.6014	0.5554
10	0.9738	0.9418	0.9051	0.8649	0.8220	0.7774	0.7319	0.6861	0.6405	0.5958
11	0.9787	0.9517	0.9201	0.8846	0.8460	0.8053	0.7631	0.7200	0.6766	0.6335
12	0.9827	0.9600	0.9327	0.9015	0.8670	0.8300	0.7910	0.7508	0.7098	0.6685
13	0.9859	0.9669	0.9434	0.9160	0.8853	0.8518	0.8160	0.7786	0.7401	0.7009
14	0.9886	0.9726	0.9525	0.9285	0.9012	0.8710	0.8383	0.8037	0.7677	0.7307
15	0.9907	0.9773	0.9601	0.9392	0.9150	0.8878	0.8581	0.8263	0.7928	0.7580
16	0.9924	0.9813	0.9665	0.9483	0.9269	0.9026	0.8757	0.8466	0.8155	0.7830
17	0.9939	0.9845	0.9719	0.9561	0.9373	0.9156	0.8913	0.8647	0.8361	0.8058
18	0.9950	0.9872	0.9765	0.9628	0.9462	0.9269	0.9051	0.8808	0.8546	0.8265
19	0.9960	0.9894	0.9803	0.9685	0.9539	0.9368	0.9172	0.8952	0.8712	0.8452
20	0.9967	0.9913	0.9835	0.9733	0.9606	0.9454	0.9279	0.9080	0.8861	0.8622

Table 3.7 *(contd.)*

η	Dimensionless distance, ξ									
	11	12	13	14	15	16	17	18	19	20
1	0.1265	0.1044	0.0861	0.0710	0.0585	0.0481	0.0396	0.0326	0.0268	0.0220
2	0.1759	0.1486	0.1255	0.1058	0.0891	0.0750	0.0630	0.0529	0.0444	0.0372
3	0.2261	0.1946	0.1672	0.1434	0.1228	0.1050	0.0897	0.0765	0.0651	0.0554
4	0.2764	0.2414	0.2104	0.1829	0.1588	0.1376	0.1190	0.1028	0.0887	0.0764
5	0.3263	0.2885	0.2544	0.2239	0.1966	0.1723	0.1507	0.1316	0.1147	0.0998
6	0.3750	0.3352	0.2988	0.2656	0.2356	0.2086	0.1842	0.1624	0.1429	0.1256
7	0.4223	0.3810	0.3429	0.3077	0.2754	0.2460	0.2192	0.1949	0.1730	0.1533
8	0.4677	0.4257	0.3863	0.3496	0.3155	0.2841	0.2552	0.2287	0.2046	0.1826
9	0.5111	0.4688	0.4287	0.3909	0.3555	0.3225	0.2918	0.2634	0.2373	0.2133
10	0.5522	0.5102	0.4698	0.4314	0.3951	0.3608	0.3287	0.2987	0.2709	0.2451
11	0.5910	0.5496	0.5094	0.4708	0.4339	0.3988	0.3656	0.3343	0.3050	0.2777
12	0.6274	0.5869	0.5473	0.5088	0.4717	0.4361	0.4021	0.3699	0.3394	0.3107
13	0.6614	0.6221	0.5833	0.5453	0.5083	0.4725	0.4381	0.4051	0.3738	0.3440
14	0.6930	0.6552	0.6174	0.5802	0.5435	0.5078	0.4732	0.4399	0.4079	0.3773
15	0.7223	0.6860	0.6496	0.6133	0.5773	0.5420	0.5075	0.4739	0.4416	0.4104
16	0.7493	0.7148	0.6798	0.6446	0.6095	0.5748	0.5406	0.5071	0.4746	0.4431
17	0.7741	0.7414	0.7080	0.6741	0.6401	0.6061	0.5724	0.5393	0.5068	0.4752
18	0.7969	0.7660	0.7343	0.7018	0.6689	0.6359	0.6030	0.5703	0.5381	0.5066
19	0.8177	0.7887	0.7586	0.7277	0.6961	0.6642	0.6322	0.6001	0.5684	0.5371
20	0.8366	0.8095	0.7812	0.7518	0.7217	0.6909	0.6599	0.6287	0.5975	0.5666

[a] Four-figure accuracy is assured.

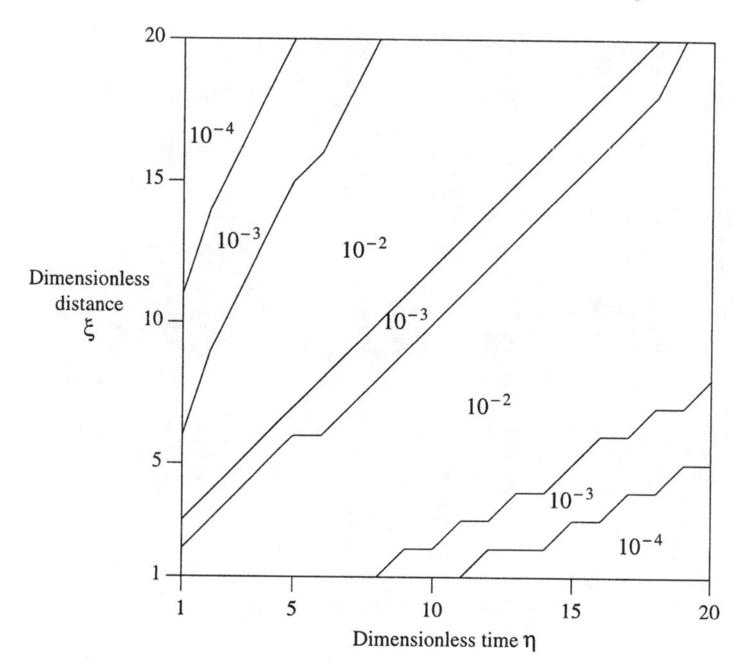

Figure 3.2: 3-D/Schumann model comparison for Bi = 0.1.

The case where Bi = 1.0 is most interesting and is displayed in Figure 3.4. Here the region $\varepsilon \leq 10^{-3}$ extends over the whole Λ–η plane with expected regions where $\varepsilon = 10^{-4}$. This is the case with the exception of a small region where $\varepsilon = 10^{-2}$ for approximately $\Lambda < 5$ and $\eta < 8$ but even this encloses a small $\varepsilon = 10^{-3}$ region. The explanation of the banding remains the same as for the Schumann–finite-conductivity model comparisons. The unusual $\varepsilon = 10^{-2}$ region is probably caused by the inability of the Hausen model to replicate the effect of the development of the transverse solid temperature profile, from an isothermal one, within the packing.

The overall conclusion is that the Hausen lumped heat transfer coefficient model is applicable for Bi ≤ 1.0, whereas the Schumann infinite-transverse - conductivity model should only be used for Bi < 0.1, an improvement by a factor of almost 10^2 of the Hausen over the Schumann model. For Bi > 1.0, it remains important to retain the 3-D transverse-conduction model, although modest approximations can be obtained using the Hausen lumped heat coefficient model, certainly for Bi < 10.

This presentation gives a more detailed overview of the relationship between the finite-conductivity model and the Schumann and Hausen models than is afforded by the broad-sweep conclusions offered by previous authors.

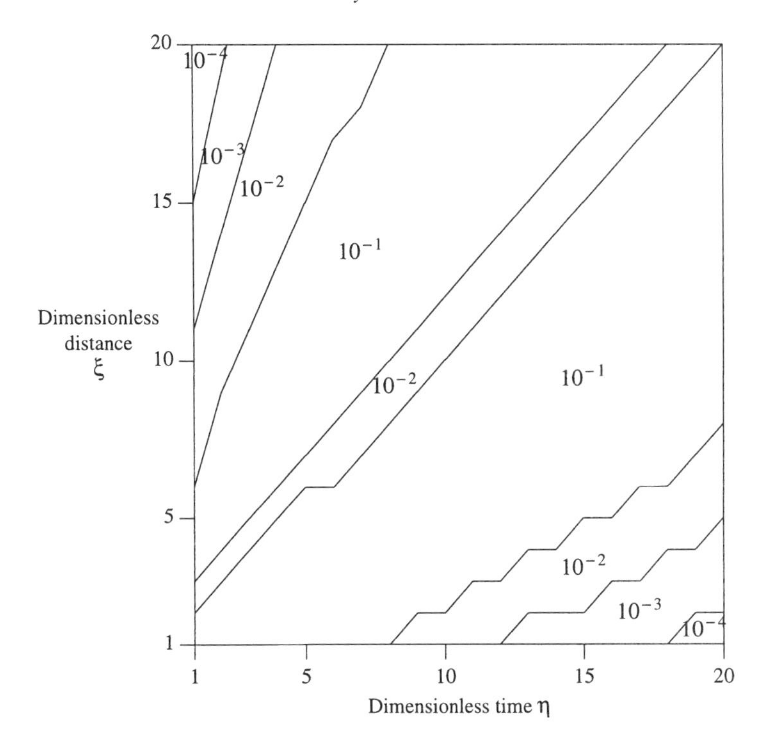

Figure 3.3: 3-D/Schumann model comparison for Bi = 1.0.

3.10 Other Geometries: The Hollow Cylinder

The prediction of the single-blow performance of a heat storage unit whose packing consists of a "bundle" of hollow cylinders is complicated by the presence of two surfaces, one on the inside, the other on the outside of each element of packing of annular cross-section. It is not only necessary to consider the "thickness" of the packing, $r_o - r_i$, but also the ratio r_o/r_i. Here r_i is the inner radius, that is, the radius of the tube down the center of the cylinder, and r_o is the outer radius, that is, the radius of the complete cylinder (Figure 3.5). Schmidt and Szego [10] considered two configurations. In one, the fluid is considered to flow down inside each hollow cylinder and the outside surface is regarded as insulated. In the other, the fluid flows over the outer surface of each of the cylinders and the inner surface is treated as completely insulated. These two geometries are illustrated in Figure 3.5.

Schmidt and Willmott [3] point to the application of the solution to the hollow-cylinder problem to the prediction of several types of heat storage units. In particular, they mention a heat storage exchanger con-

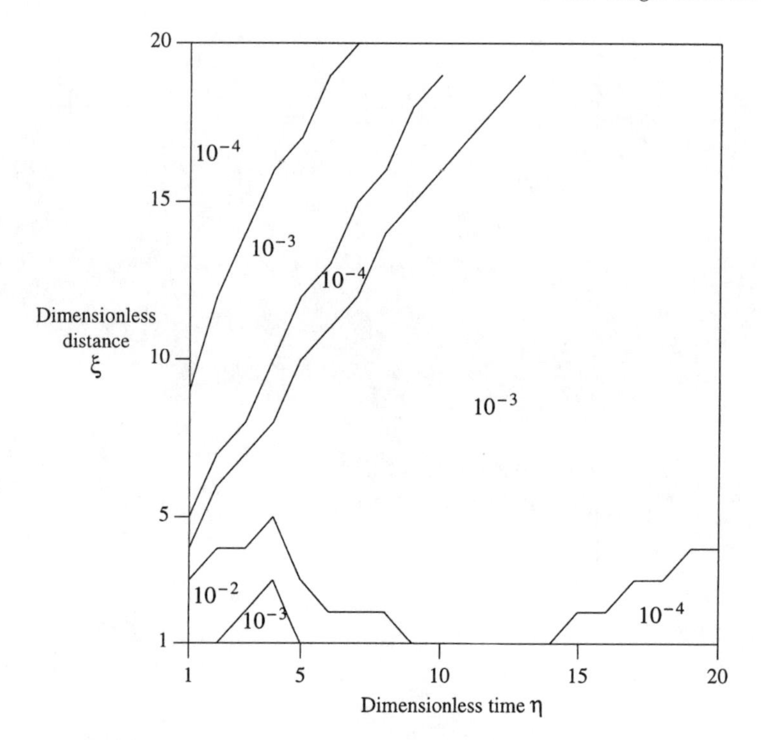

Figure 3.4: 3-D/Hausen model comparison for Bi = 1.0.

structed in a shell-and-tube heat exchanger configuration. More generally, this can be regarded as a "bundle" of hollow cylinders where either the inner or the outer surface of each cylinder is considered to be essentially adiabatic. A third possibility exists where the gas flows over both the inner and outer surfaces. This case will not be considered at this juncture. Hausen [11] considers also the case of the solid cylinder, that is, where $r_i = 0$.

The differential equations that describe the thermal performance of a hollow cylindrical storage unit are similar to those for the plane wall unit. In particular, equation (3.1) again represents the conservation of energy in the moving fluid:

$$\frac{\partial t_f}{\partial y} = \frac{\alpha A}{\dot{m}_f c_p L}(t_{s,o} - t_f) - \frac{m_f}{\dot{m}_f L}\frac{\partial t_f}{\partial \tau} \tag{3.1}$$

The diffusion equation for the solid is slightly different and takes the form shown in equation (3.26):

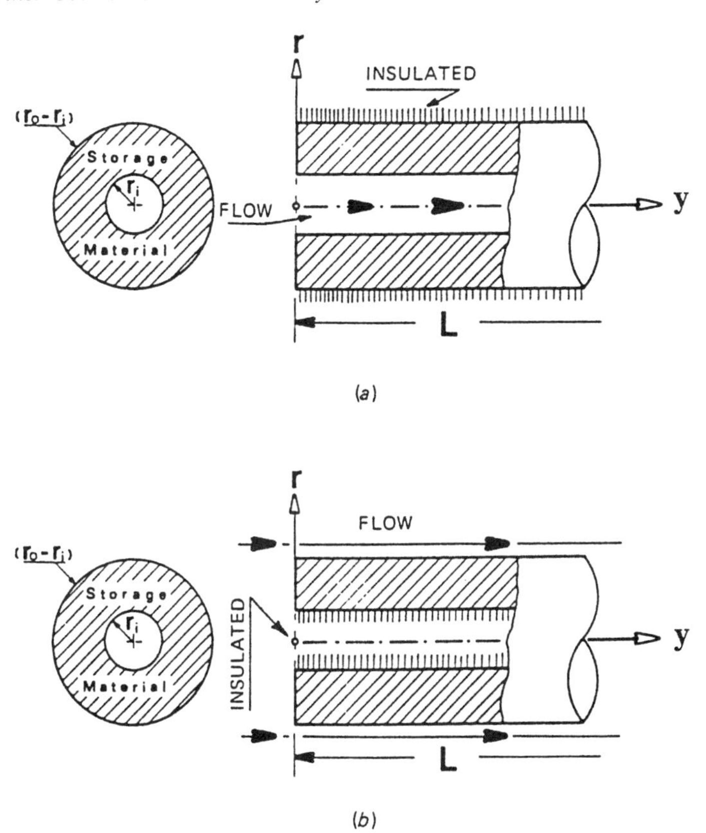

Figure 3.5: Hollow-cylinder configuration with (a) the outside surface insulated and the fluid passing down the center of each cylinder and (b) the inner surface insulated and the fluid passing over the outer surface of each cylinder. (Schmidt and Willmott [3].)

$$\frac{1}{\kappa_s}\frac{\partial t_s}{\partial \tau} = \frac{\partial^2 t_s}{\partial r^2} + \frac{1}{r}\frac{\partial t_s}{\partial r} \tag{3.26}$$

Here the effect of axial conduction is ignored. Also, the linearizing simplifications set out in Chapter 2 regarding the thermophysical properties of the gas and solid (being independent of temperature), the chronological invariance of the entrance gas temperature and the gas flow rate, and the uniformity of the geometrical arrangement of the storage elements (in this case, hollow cylinders) are assumed. Under these assumptions, equations (3.1) and (3.26) can be simplified. In particular, the dimensionless scales set out below are used:

$$\xi = \frac{\alpha A}{\dot{m}_f c_p L} y \qquad \eta = \frac{\alpha A}{MC_s}\left(\tau - \frac{m_f y}{\dot{m}_f L}\right) \qquad z = \frac{r}{R} \tag{3.27}$$

As for the plane wall geometry, equation (3.1) takes the form of equation (3.8):

$$\frac{\partial T_f}{\partial \xi} = T_{s,o} - T_f \tag{3.8}$$

Here $T_{s,o}$ is the dimensionless surface solid temperature. The right-hand side of equation (3.26) takes the form, using the dimensionless temperature, T,

$$\frac{\partial^2 T_s}{\partial r^2} + \frac{1}{r}\frac{\partial T_s}{\partial r} = \frac{1}{R^2}\left(\frac{\partial^2 T_s}{\partial z^2} + \frac{1}{z}\frac{\partial T_s}{\partial z}\right)$$

Equation (3.26) can now be developed to yield

$$\frac{\partial T_s}{\partial \eta} = \kappa_s \frac{Mc_s}{\alpha A}\frac{1}{R^2}\left(\frac{\partial^2 T_s}{\partial z^2} + \frac{1}{z}\frac{\partial T_s}{\partial z}\right) \tag{3.28}$$

We can now set the mass of the packing to be

$$M = \pi(r_o^2 - r_i^2)L\rho_s \tag{3.29}$$

and the heating surface area, A, to be

$$A = 2\pi r_w L \tag{3.30}$$

where r_w is the radius of the "wetted" surface with $r_w = r_i$ if the gas is considered to flow down inside each hollow cylinder, for example. In so doing, we consider the temperature behavior of gas and solid in a single cylinder and then regard the complete storage unit as a bundle of single cylinders that behave identically. The simplifications set down below follow:

$$\frac{\kappa_s}{R^2}\frac{Mc_s}{\alpha A} = \frac{\lambda_s}{\rho_s c_s}\frac{1}{R^2}\frac{\pi(r_o^2 - r_i^2)L\rho_s c_s}{\alpha(2\pi)r_w L} = \frac{\lambda_s}{\alpha R^2}\frac{r_o^2 - r_i^2}{2r_w} \tag{3.31}$$

We next consider that $R = r_o - r_i$ and recall that $r_o^2 - r_i^2 = (r_o + r_i)(r_o - r_i)$. It thus follows that

$$\frac{\lambda_s}{\alpha R^2}\frac{r_o^2 - r_i^2}{2r_w} = \frac{1}{\text{Bi}}\frac{r_o + r_i}{2r_w} \tag{3.32}$$

where we set the Biot modulus

$$\text{Bi} = \frac{\alpha(r_o - r_i)}{\lambda_s}$$

for the hollow-cylinder configuration. Where r_w is equal to r_i, we introduce the "curvature" of the hollow cylinder through the parameter denoted by Schmidt [3] as U^+, where $U^+ = r_i/r_o$. This can be generalized to deal with the case where the inner surface of the hollow cylinder is insulated and the gas flows over the outer surface of each cylinder. We denote the radius of the insulated (adiabatic) surface by r_a and denote

$$U^+ = \frac{r_w}{r_a}$$

From this follows that equation (3.28) can be reduced to the form

$$\frac{\partial T_s}{\partial \eta} = \frac{1}{\mathrm{Bi}} \frac{1 + U^+}{2U^+} \left(\frac{\partial^2 T_s}{\partial z^2} + \frac{1}{z} \frac{\partial T_s}{\partial z} \right) \tag{3.33}$$

which deals with both possible arrangements, that is, the gas flowing down the middle of the cylinder and the gas passing over the outside surface of the cylinder, with the "opposite" surface insulated.

Consider first the case where $r_w = r_i$. At the inner surface, the boundary condition below holds:

$$\frac{\partial T_s}{\partial z} = -\mathrm{Bi}(T_f - T_s) \quad \text{where} \quad z = \frac{r_i}{r_o - r_i} = \frac{r_w}{r_a - r_w} = \frac{U^+}{1 - U^+} \tag{3.34}$$

In the second case, $r_w = r_o$ and

$$z = \frac{r_o}{r_i - r_o} = \frac{r_w}{r_w - r_a} = \frac{U^+}{U^+ - 1}$$

where

$$\frac{\partial T_s}{\partial z} = +\mathrm{Bi}(T_f - T_s) \tag{3.35}$$

at the outer surface of the cylinder. This can be generalized by specifying that

$$\frac{\partial T_s}{\partial z} = \pm\mathrm{Bi}(T_f - T_s) \quad \text{at} \quad z = \frac{U^+}{|1 - U^+|} \tag{3.36}$$

Similarly, the adiabatic surface,

$$\frac{\partial T_s}{\partial z} = 0 \quad \text{at} \quad z = \frac{1}{|1 - U^+|} \tag{3.37}$$

We can rewrite equation (3.33) as

$$\frac{\partial T_s}{\partial \eta} = a\left(\frac{\partial^2 T_s}{\partial z^2} + \frac{1}{z}\frac{\partial T_s}{\partial z}\right)$$

Here,

$$a = \frac{1}{\text{Bi}}\frac{1 + U^+}{2U^+}$$

Mitchell and Griffiths [12] point out that numerical solutions on a rectangular grid are less accurate in the vicinity of the axis of the cylinder (here, should r_i be small) than elsewhere. This is due to the problem of representing $1/z(\partial T_s/\partial z)$ adequately by finite differences where z is small. They go on to suggest that if a new dimensionless variable, w, is introduced by the relation

$$z = 2w^{1/2}$$

then equation (3.33) can be rewritten as

$$\frac{\partial T_s}{\partial \eta} = a\left(w\frac{\partial^2 T_s}{\partial w^2} + \frac{\partial T_s}{\partial w}\right)$$

By an appropriate construction of the space–time (w–η) grid, the solid temperatures can be computed at equally spaced positions in the **z** direction as well as at equally spaced positions in time, η.

Schmidt and Szego [10] report solutions to these equations by a method described previously by them [4]. They restricted themselves to the case where

$$\frac{r_o - r_i}{L} = 10^{-2}$$

thereby minimizing any possible effect of axial conduction. These solutions are presented by Schmidt and Szego [10] and are reproduced here in graphical form as offered by Schmidt and Willmott [3]. Figures 3.6 and 3.8 show the dimensionless fluid exit temperatures displayed as varying with dimensionless time, η, as a function of dimensionless length, Λ, and curvature U^+. Similarly displayed is the dimensionless heat storage, $Q^+(\eta)$, as given in equation (2.44), in Figures 3.9 to 3.11.

The obvious conclusion to be drawn is that the effect of the curvature U^+ is magnified as the Biot modulus increases. The reader is invited to undertake further work in this area and, in particular, to examine the effect of U^+ for small values of Bi and for other values in the range $2 < \text{Bi} < 10$ to provide a more complete picture. Values other than 10^{-2} should also be considered for $(r_o - r_i)/L$.

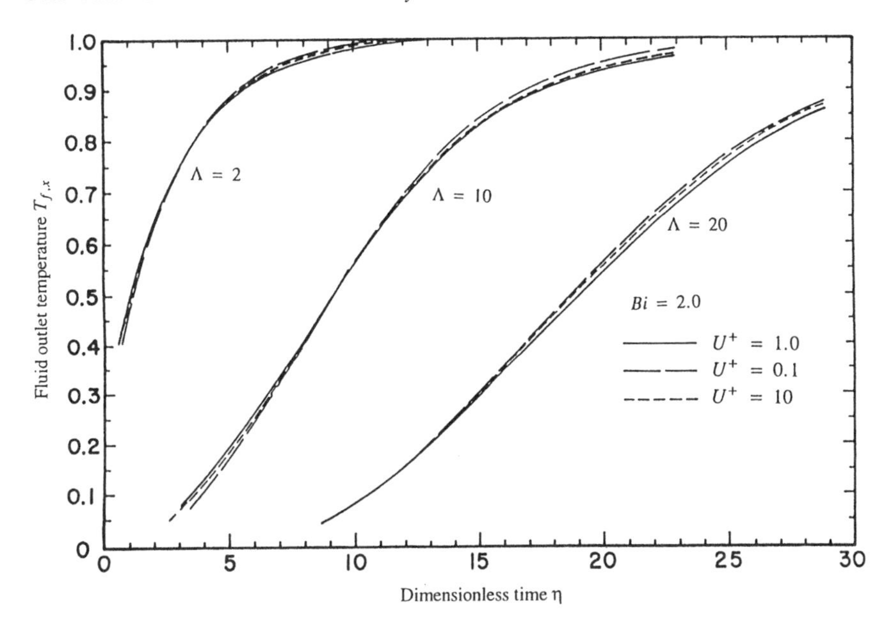

Figure 3.6: Hollow-cylinder configuration. Dimensionless fluid exit temperature. (Schmidt and Szego [10].)

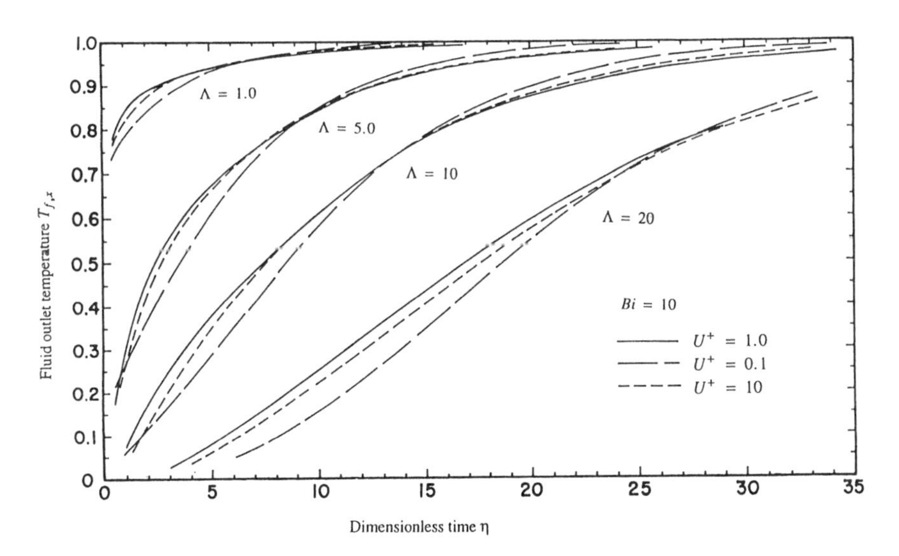

Figure 3.7: Hollow-cylinder configuration. Dimensionless fluid exit temperature. (Schmidt and Szego [10].)

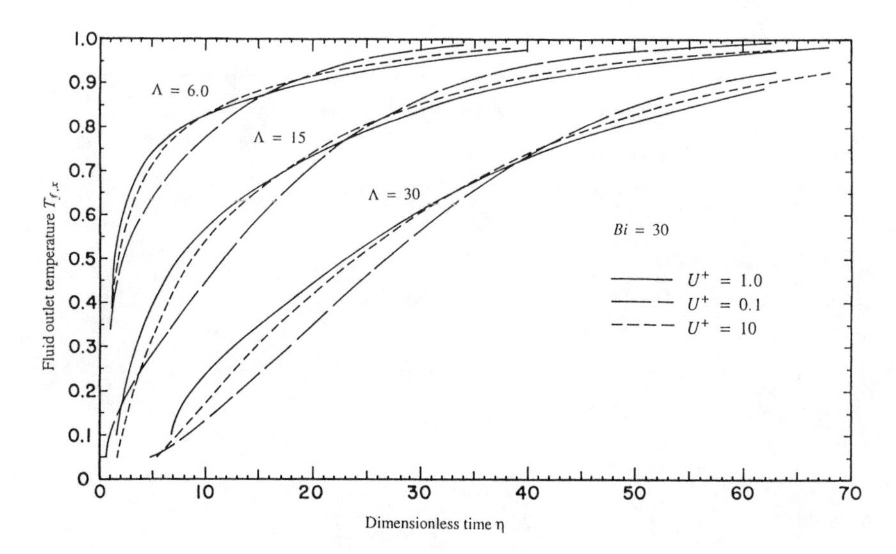

Figure 3.8: Hollow-cylinder configuration. Dimensionless fluid exit temperature. (Schmidt and Szego [10].)

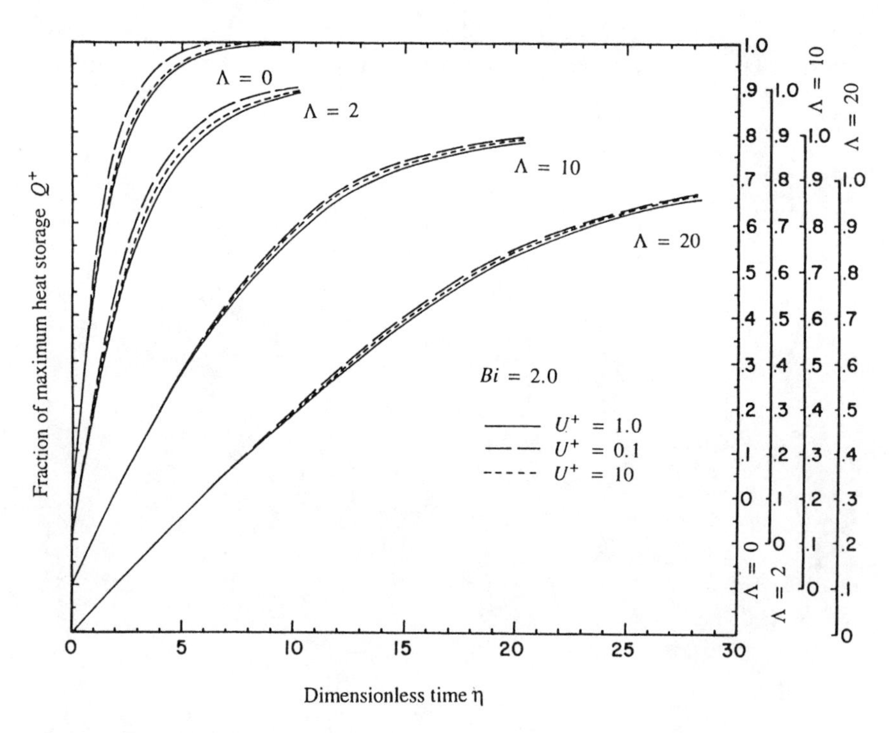

Figure 3.9: Hollow-cylinder configuration. Fraction, Q^+, of maximum heat storage, Q_{max}. (Schmidt and Szego [10].)

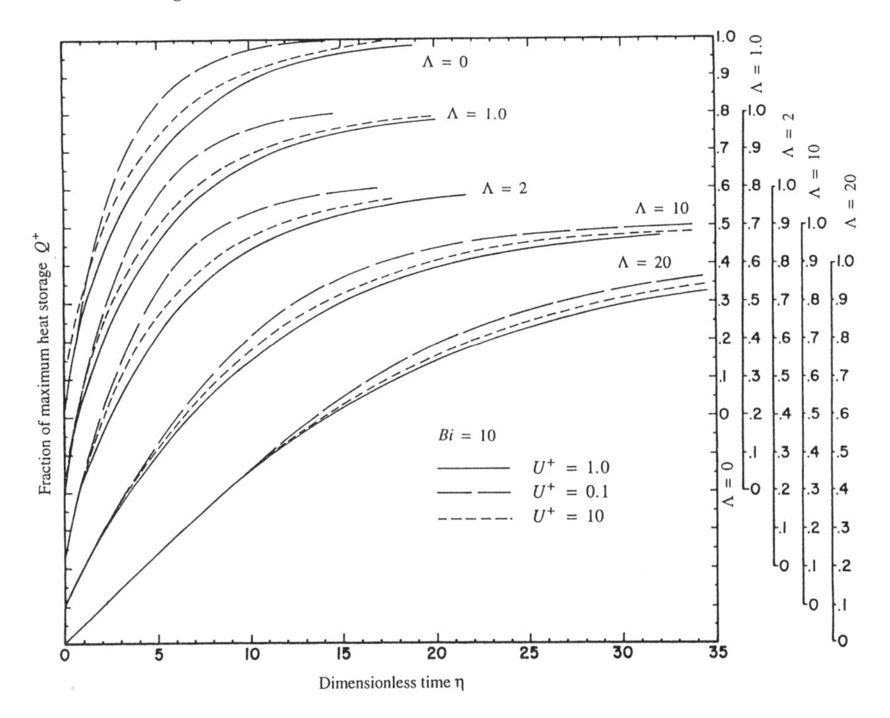

Figure 3.10: Hollow-cylinder configuration. Fraction, Q^+, of maximum heat storage, Q_{max}. (Schmidt and Szego [10].)

3.11 Concluding Remarks

Schmidt and Willmott [3] suggest that the transient response of a heat storage unit, measured in terms of the dimensionless exit fluid temperature, $T_{f,x}$, appears to be nearly independent of the geometry of the packing elements. This appears also to be the case for the dimensionless heat storage, $Q^+(\eta)$. However, as the dimensionless length, Λ decreases, that is, the smaller the unit or the greater the heat capacity flow rate of the fluid, the differences between the $T_{f,x}$ and $Q^+(\eta)$ as computed by the plain-wall model, compared with those calculated using the hollow-cylinder model, for example, become greater. Nevertheless, Schmidt and Willmott conclude that the differences remain small for most practical units.

Schmidt and Szego [10] suggest that the effects of packing geometry be ignored for $0.1 < \mathrm{Bi} < 2.0$ and that the flat-slab model be used to represent hollow-cylinder-type packing for this case. Otherwise, when $\mathrm{Bi} > 2.0$, when,

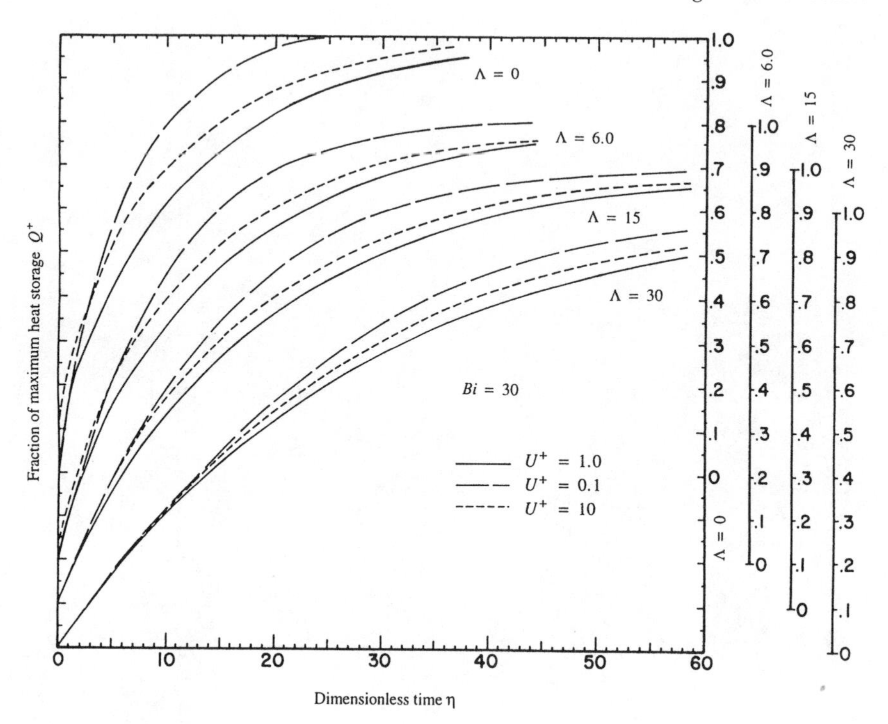

Figure 3.11: Hollow-cylinder configuration. Fraction, Q^+, of maximum heat storage, Q_{max}. (Schmidt and Szego [10].)

for example, the fluid is a liquid, the effects of the curvature should be taken into consideration. The present author suggests that if software is available for the hollow-cylinder geometry of the packing of a unit, it should be used for all cases. Use of a flat-wall model, if only such software is available, for other geometries should be undertaken with caution.

References

1. A. Anzelius, "Heating by Means of a Flowing Medium," *Z.A.M.M.* **6**(4), 291–294 (1926).
2. T. E. W. Schumann, "Heat Transfer: A Liquid Flowing Through a Porous Prism," *J. Franklin Inst.* **208**, 405–416 (1929).
3. F. W. Schmidt, A. J. Willmott, *Thermal Energy Storage and Regeneration*, McGraw-Hill, New York (1981).

4. F. W. Schmidt, J. Szego, "Transient Response of Solid Sensible Heat Thermal Storage Units – Single Fluid," *J. Heat Transfer, Trans. ASME* **98**, 471–477 (1976).
5. D. Handley, P. J. Heggs, "The Effect of Thermal Conductivity of the Packing Material on Transient Heat Transfer in a Fixed Bed," *Int. J. Heat Mass Transfer* **12**, 549–570 (1969).
6. W. Tipler, *A Simple Theory of Heat Exchanger*, Shell Technical Report ICT/14, London (1947).
7. W. Schotte, "Thermal Conductivity of Packed Beds," *A.I.Ch.E. J.* **6**, 63–67 (Mar. 1960).
8. D. J. Evans, "Non-Linear Modelling of Regenerative Heat Exchangers." D. Phil. thesis, University of York (Mar. 1997).
9. A. J. Willmott, "The Regenerative Heat Exchanger Computer Representation," *Int. J. Heat Mass Transfer* **12**, 997–1014 (1969).
10. F. W. Schmidt, J. Szego, "Transient Response of a Hollow Cylindrical-Cross-Section Solid Sensible Heat Storage Unit – Single Fluid," *J. Heat Transfer, Trans. ASME* **100**, 737–739 (1978).
11. H. Hausen, *Heat Transfer in Counterflow, Parallel Flow and Crossflow*, English translation (1983) edited by A. J. Willmott, McGraw-Hill, New York (originally published 1976).
12. A. R. Mitchell, D. F. Griffiths, *The Finite Difference Method in Partial Differential Equations*, Wiley, Chichester, UK (1980).

Chapter 4

BASIC CONCEPTS IN COUNTERFLOW THERMAL REGENERATORS

4.1 Introduction

Chapters 2 and 3 focused upon the single-blow problem in terms of the spatial and time variation of the temperature of the gas passing over the available heating surface area of a solid with heat capacity, as well as the heat stored in that solid. The latter was considered by way of a heat storage factor, Q^+ (see equation (2.44)), which represents the chronological variation of the proportion of the thermal energy held in the solid at any instant, relative to the thermodynamic maximum heat that could be accumulated. The recovery of heat from the storage medium was considered as another single-blow problem and treated in an analogous manner.

A thermal regenerator can be treated as a solid with thermal capacitance, whose available surface area is submitted to a succession of linked single blows, during which heat is alternately accumulated in the solid packing for a length of time known as the *hot period* and then recovered from the solid during the *cold period*. This view might be acceptable in circumstances where it is required to maximize the amount of heat stored and where the cost of losing the thermal energy that is *not* extracted from the hot gas is of no consequence. An example might be a packed bed in which thermal energy is stored from the warm atmospheric air during the day, to be recovered later in the evening or during the night.

It is more common, however, to regard a thermal regenerator as a *regenerative heat exchanger* used to transfer heat between two fluids (usually gases) and to do so as efficiently as possible. That is minimizing the amount of heat which is *not* extracted from the hot gas thereby maximizing the quantity of thermal energy which is transferred from one gas to the other. This transfer of heat is effected by temporally storing it from the hotter gas in a permeable packing and subsequently passing this thermal energy to the cooler gas. All the surface area of the packing is washed by the hot gas for the duration of the hot period during which sensible heat is stored in the packing. At the conclusion of this period, a *reversal* occurs at which the hot gas is switched off and the cooler gas starts to flow through the same channels over the entire surface area. The thermal energy is recovered and passed to the cold gas during the cold period, at the end of which there is another reversal. This complete process is called a *cycle of operation* the duration of which is known as the *cycle time*.

This view of regenerator performance that emerges in this way is rather different from that of a sensible heat store that is submitted to a single blow. The cycle time can be adjusted (ideally, made as small as possible) to maximize the efficiency of operation. On the other hand, it can be adjusted so that the thermal regenerator is able to meet required operating conditions that may be less than optimal. Typically, a regenerator is attached to another process with a view to reducing losses in thermal energy and minimizing the fuel requirements.

As has been indicated, a cycle of operation consists of a hot, followed by a cold period of operation, together with the necessary reversals. After many cycles of identical operation, the temperature performance of the thermal regenerator in one cycle is identical to that in the next. When this condition is realized, the regenerative heat exchanger is said to have reached *cyclic equilibrium* or *periodic steady state*. Should a step change be introduced to one or more of the operating parameters, in particular, the flow rate and entrance temperature of the fluid, for either period of operation, or the duration of the hot and cold periods, the regenerator undergoes a number of *transient* cycles until the new cyclic equilibrium is reached.

In the most common *counterflow* or *contraflow* regenerator operation, the hot gas passes through the regenerator in the *opposite* direction of the cold fluid. In less efficient *parallel flow* or *coflow*, the hot and cold fluids pass through the channels of the packing in the *same* direction. In this chapter, we shall restrict our consideration to contraflow regenerators. In a later chapter, the analysis of the operation of coflow regenerators will be examined.

The periodic operation of regenerators can exploit the periodic operation of the system to which the exchanger is attached. For example, in hot climates, daytime heat can be stored in a packing by passing the warm atmospheric air through it: this heat can then be recovered by blowing cold nighttime air through the same packing during the evening to provide at least some supplementary warming of the living space in a building. Hausen [1] suggests that the throat and nasal passages act as a regenerator packing in cold weather. When an animal breathes in cold air, it is warmed as it passes through the nose and throat before the air reaches the lungs, thereby protecting the lungs from the effects of cold temperatures. As the animal breathes out, the same passages in the nose and throat are warmed by the air leaving the lungs. Clearly, the temperature of the throat and nasal tissue is also regulated by the flow of blood through them.

In general, however, a continuous supply of heated fluid is required so that the discontinuous operation of the regenerator, which is inherent in its design, must be concealed in some way. Indeed, regenerators appear in two distinct forms, the fixed-bed regenerator and the rotary regenerator.

4.2 Fixed-Bed Regenerators

The most obvious technique for realizing *apparent* continuous operation is to use two or more beds of heat-storing material, operating out of phase with respect to one another, so that while one bed is supplying *heated* fluid, heat is stored in the other bed(s) from the *heating* fluid. A seemingly easy way to do this is by enclosing the set of regenerators so formed within a system of ducts or pipes fitted with valves to facilitate the switching of the regenerators at the end of a period of operation. As one set of valves close, at a reversal, so another set open: the flow of hot gas, for example, is diverted from one regenerator to the other by the closing of such a set of valves and the opening of the other. Simultaneously, the flow of cold gas is switched from the other regenerator in a symmetric fashion (see Figure 4.1).

Regenerators of the fixed-bed type are frequently employed under conditions of high temperature and/or significant pressure differences between the hot and cold gas streams. Cowper stoves (see Figure 4.2) are employed to preheat the blast for the iron-making furnace (the so-called *blast furnace*) and are required to deliver preheated air in the temperature range 900–1300°C. The necessary heat is extracted by the Cowper stove from blast furnace gas (a byproduct of the iron-making process) possibly enriched with another fuel, natural gas, for example, which is burnt in a combustion chamber, the hot products of combustion being passed through the channels of the packing, the "checkerwork" of the Cowper stove. The blast to be preheated is passed through the stove, subsequently, in the counterflow direction.

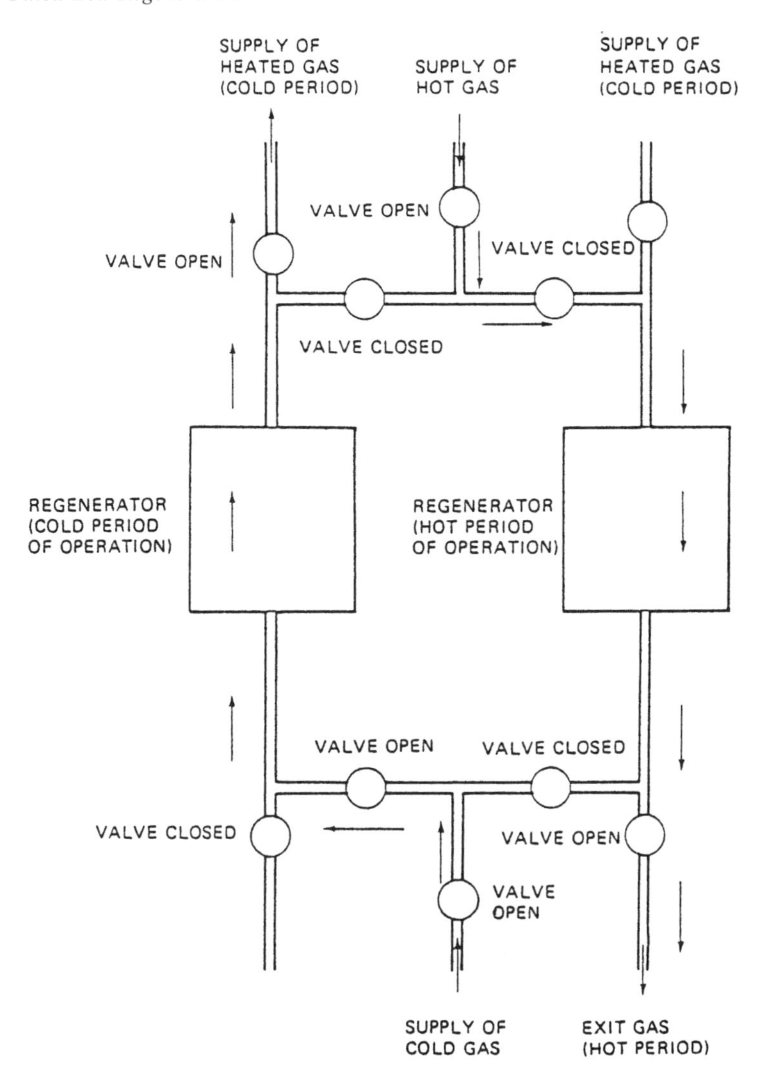

Figure 4.1: Fixed two-bed regenerator system. (Schmidt and Willmott [4].)

In high-temperature regenerators, it is desirable not to have any valves at the hot end of each regenerator. Where this cannot be avoided, as in the case of the Cowper stove, the valves are often expensive, requiring, perhaps, to be water cooled to avoid malfunction at high temperatures. It is often the case, however, that the hot end of the regenerator is attached directly to the

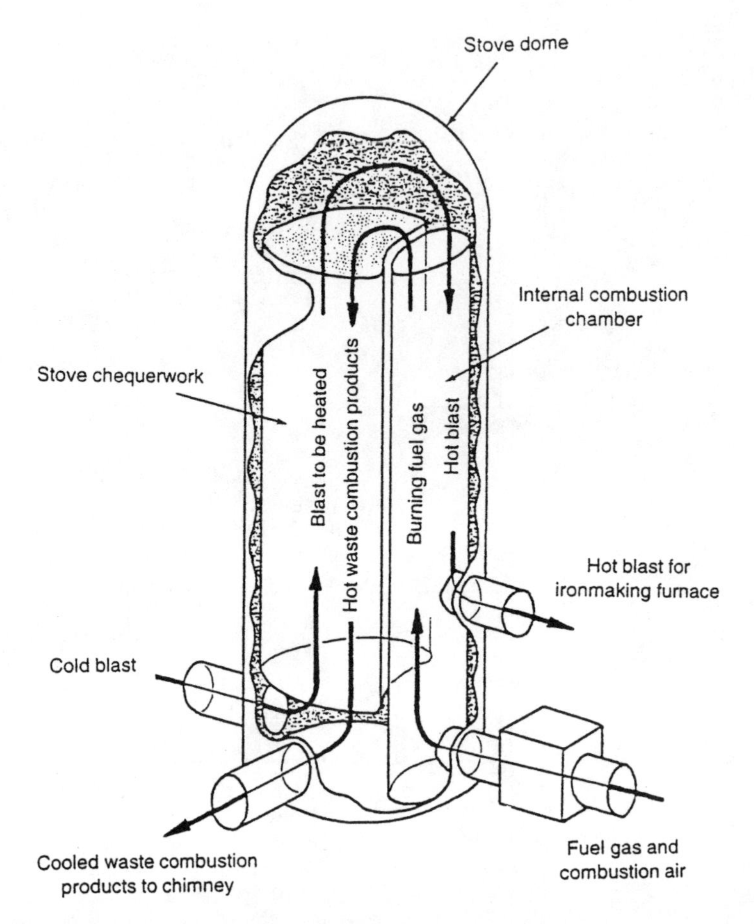

Figure 4.2: Traditional Cowper stove with internal combustion chamber.

furnace or boiler to where preheated air, for the economic combustion of a fuel gas, is supplied directly. From the same position, after a reversal, the hot gas, frequently the waste products of combustion of the fuel, is extracted directly to the regenerator. In this way, valves at the hot end of the regenerator are avoided.

The necessary *suction* of the hot gas through the regenerator is achieved by attaching the exit duct for this gas at the cold end of the regenerator to a chimney, which, if tall enough, will provide the necessary updraft. Valves are employed safely at the cold end of the regenerator to switch the heat exchanger from the chimney exhaust for waste gases to the supply of cold air for the cold period of regenerator operation, or vice versa. A continuous supply of preheated combustion air is achieved by attaching several regenerators to a furnace or boiler, operating as necessary, out of phase with respect to one another.

The *regenerative burner* consists of a compact regenerator, made compact by a packing consisting of ceramic spheres, each of diameter 1–3 cm, coupled directly to a burner lined with refractory material (Figure 4.3). Another arrangement is shown in Figure 4.4. A complete unit consists of two such burners, operating out of phase with respect to one another. Each

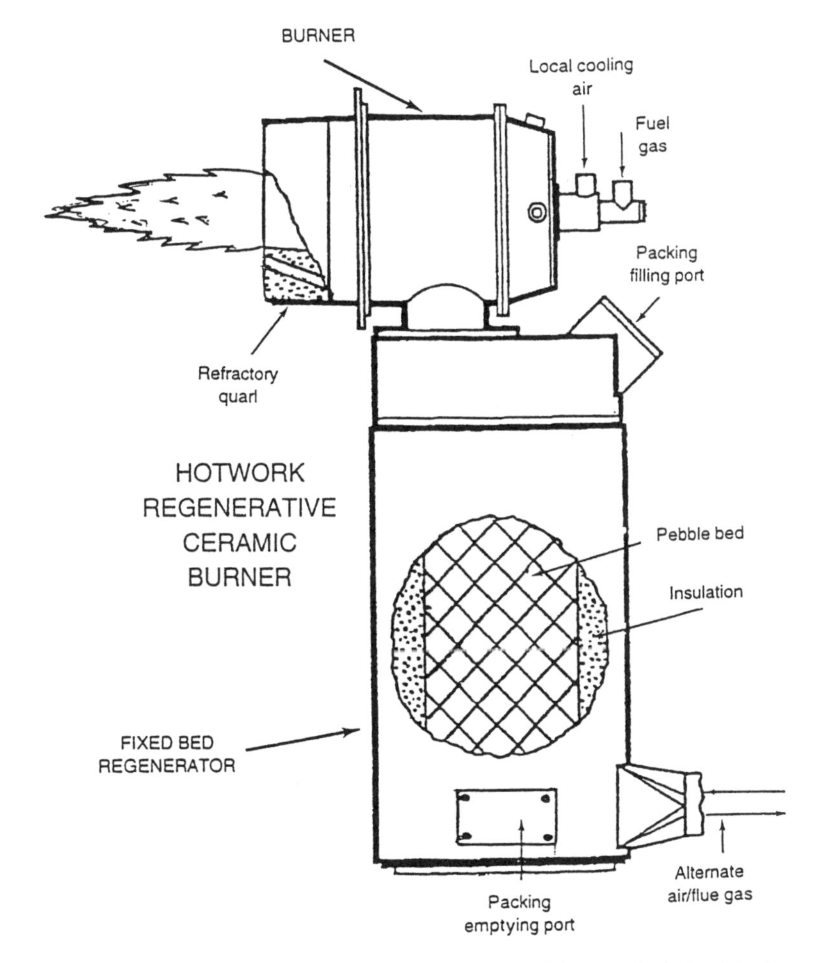

Figure 4.3: Cross-section of regenerative burner with detached fixed-bed regenerator. (Courtesy Hotwork Engineering Ltd., Dewsbury, West Yorkshire, UK.)

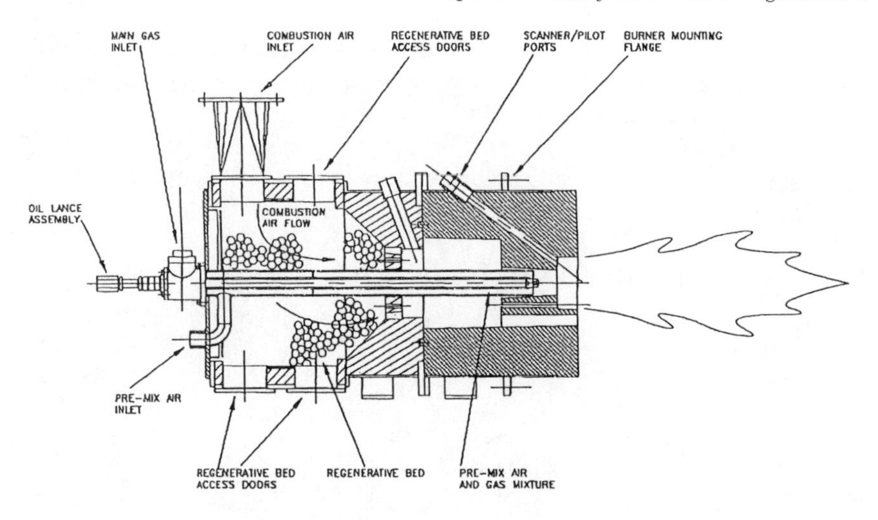

Figure 4.4: Cross-section of dual-fuel, low-NO_x, regenerative integral-bed burner. (Courtesy Stordy Combustion Engineering Ltd., Wolverhampton, UK.)

burner has a regenerator directly linked to it and the two burners are coupled by a reversing valve and the necessary reversal control system, as shown in Figure 4.5. The burners are connected directly to the furnace and, during the hot period of regenerator operation, hot exhaust gases from the furnace are drawn through the burner and then the regenerator to a chimney that affords the necessary suction. In this way, as indicated previously, this eliminates the requirement for valves to operate at the exhaust gas temperature as it leaves the furnace. At the start of the cold period of regenerator operation, the fuel supply is switched to the associated burner and cold air starts to flow through the regenerator to yield preheated combustion air. The cold period concludes by switching off the fuel and combustion air supply to the burner. The *cycle time* is typically 10 min or less. Hot exhaust gas temperatures in the range 1000–1400°C yield combustion air temperatures between 900°C and 1250°C. Around 50% fuel savings are realized compared with firing the fuel gas with ambient combustion air. A large furnace may have several regenerative burner pairs spaced down the length of the unit.

Relatively recently, regenerators have begun to become important as sensible heat storage units. The cycle of operation is exploited as an ideal mechanism to transfer thermal energy in circumstances where the availability of heat or "cold" does not coincide timewise with demand. Here, the thermal storage mechanism within the regenerative cycle is specifically exploited.

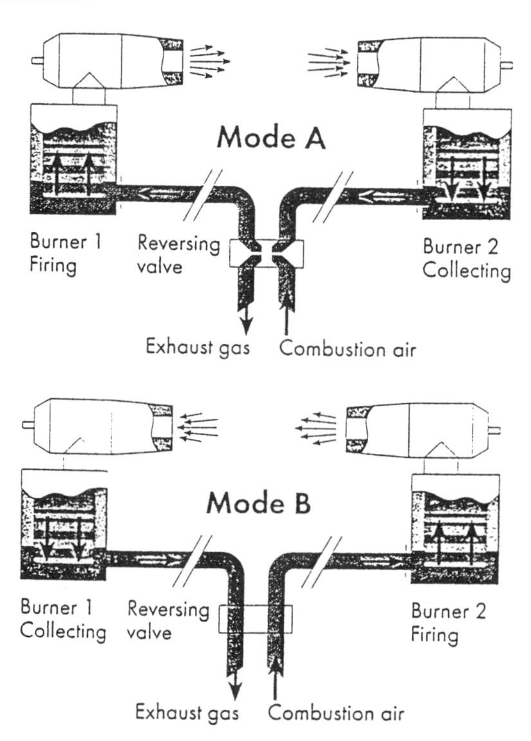

Figure 4.5: Two regenerative burners with necessary valves for continuous firing of a furnace. Mode A = first half of cycle of regenerator operation, Mode B = second half. (Courtesy Hotwork Engineering Ltd., Dewsbury, West Yorkshire, UK.)

4.3 Rotary Regenerators

The Ljungström or rotary regenerator (Figure 4.6) consists of a cylindrical packing through separate sectors of which the hot and cold gases pass at the same time. This cylindrical packing is rotated so that thermal energy, temporally stored in the packing from the hotter gas, is physically moved into the cold gas stream where it is regenerated and the gas heated. It will be seen from Figure 4.6 that this is realized by having the axis of rotation parallel to the direction of flow of both gases, which pass through the heat exchanger in countercurrent mode. The time taken for any section of packing to pass through the hot gas stream corresponds to the *hot period* of fixed-bed operation. Similarly, the *cold period* is the time taken for a section of packing to rotate through the cold gas. A typical industrial rotary regenerator under construction is shown in Figure 4.7.

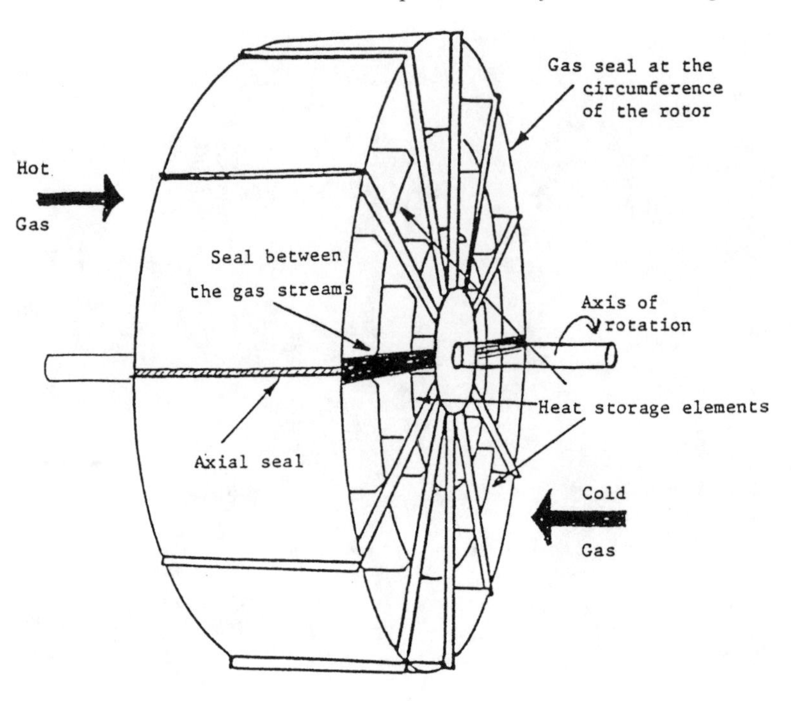

Figure 4.6: Ljungström rotary regenerator.

The same effect is realized in the Rothemühle design of rotary regenerative air preheater. Here the cylindrical packing remains stationary. Alternating hot and cold gas flow is obtained by the rotation of distributing hoods, which are arranged on the top and bottom faces of the cylindrical packing (see Figure 4.8).

In both types of rotary regenerator, considerable attention is paid to providing gas seals that minimize the leakage between hot and cold gas at the gas entrance/exit faces of the heat-storing mass of packing. In the case of the Ljungström form of rotary regenerator, as a section of packing rotates between the gas seals from one gas stream to another, it experiences what is the equivalent of a reversal in the fixed-bed system. Similar reversals are experienced by successive sections of stationary packing as the hoods rotate in the Rothemühle form of rotary regenerator.

These necessary sealing arrangements cannot be expected to endure the harsh temperature conditions experienced at the hot end of fixed-bed regenerators as used in the Siemens type of furnace for glass making, for example. However, under less severe conditions, both forms of rotary regenerator

Figure 4.7: Typical industrial rotary regenerator under construction. (Courtesy Howden-Sirocco Ltd., Glasgow, UK.)

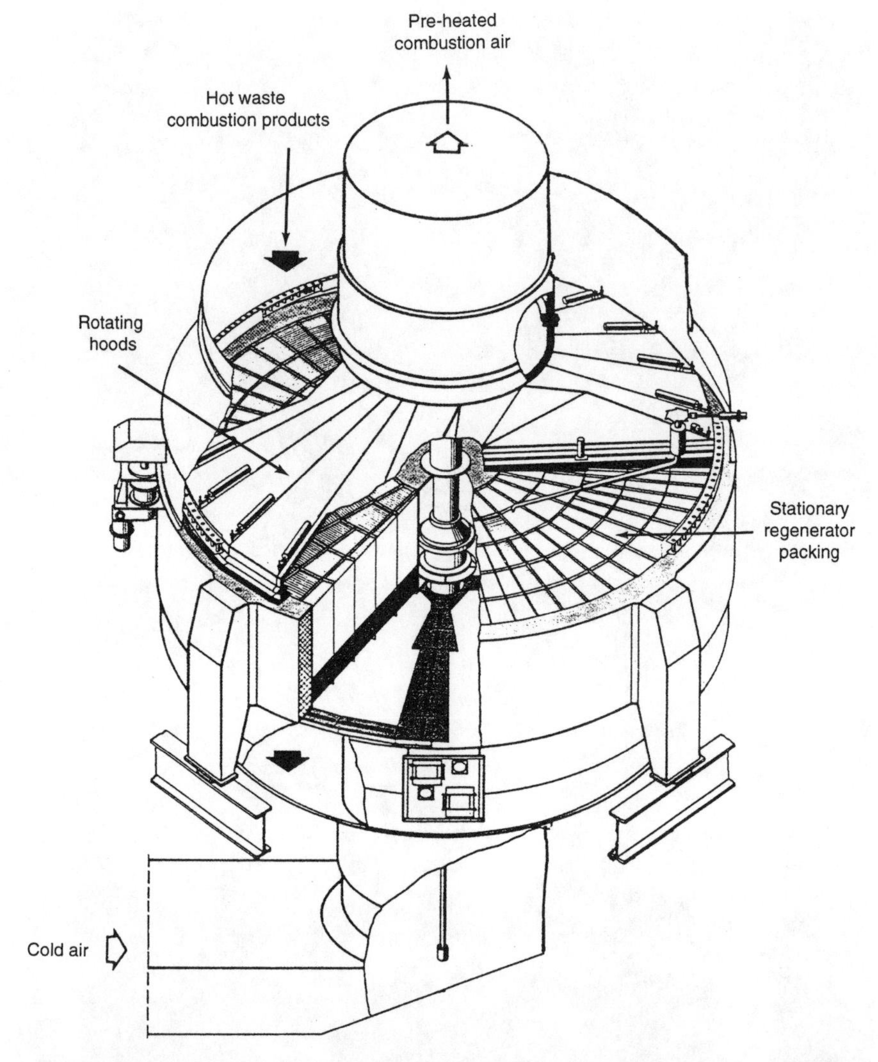

Figure 4.8: Rothemühle rotary regenerative air preheater. (Courtesy Apparatebau Rothemühle, Brandt + Kritzler GmbH, D-57479 Wenden-Rothemühle, Germany.)

prove to be very compact heat exchangers. They are commonly used to recover waste heat from boilers in power stations and aboard ships, from gas turbines to process drying plants. The energy recovered is used to pre-heat the combustion air, effecting significant savings in the fuel gas required for the process under consideration.

The application of rotary regenerators has been extended to recover waste cold (in hot climates) or waste heat (in cold climates) in the air conditioning and ventilating systems of buildings.

4.4 Reversals

By viewing thermal regenerators, in either fixed-bed or rotary form, it might be thought that the necessary switching of the bed, or part of it, from the hot/cold to the cold/hot gas stream (by the opening and closing of valves in the fixed-bed arrangement or by rotating of the faces of the packing beneath the gas seals in either of the rotary arrangements) might be regarded as a nuisance. Indeed Hausen [2] suggests that in ideal circumstances, a regenerator should be used where the plant to which it is attached operates in periodic manner and that the regenerator be allowed to operate in tandem with the plant. This is rarely achieved. Nevertheless, advantage is taken of the alternate passing of the hot and cold gases through the same passages in the packing. Dirt or frozen fluid (in the case of low-temperature applications) deposited by one gas stream can be relied upon to be purged by the other gas stream, thereby avoiding the blocking of the channels that might otherwise occur.

Some applications require that the regenerator(s) be purged before the supply of heated fluid, for example, is switched from one fixed-bed regenerator to the next. In this case, the *cold* period of one regenerator is extended to maintain the supply of heated fluid to the external process to which the set of regenerators is attached. Meanwhile, the *hot* period of the other regenerator is terminated and that regenerator is completely purged before its *cold* period begins. This regenerator then shoulders the burden of supplying of heated fluid from the other regenerator whose "end of cold period reversal" can begin. Such arrangements necessarily complicate the valve and duct facilities associated with a set of fixed-bed regenerators; in addition, a suitable exhaust for fluids purged from the regenerator, not permitted to enter the heated fluid stream, for example, must be provided. Additional sectors (yielding trisector or quadsector arrangements) in rotary regenerators yield the same result to facilitate the purging of gas resident in a sector of packing just prior to a reversal.

Where the fluids are gases, it it not uncommon for the pressure of the cold gas stream, for example, to be significantly higher than that of the hot gas stream. In this case, at the end of a *cold* period, time must be allowed during the reversal for the fixed-bed regenerator to be decompressed before the *hot* period is permitted to begin. Similarly, time must be allowed at the start of a *cold* period for the pressure of the cold gas in the regenerator to build up before the cold period proper can begin. Again, additional valves and pipework must be provided to accommodate these complications. These facilities are required for the operation of Cowper stoves attached to blast furnaces.

4.5 Mathematical Model

Rotary and fixed-bed regenerators used either as heat exchangers or heat storage units can be idealized into a common mathematical model. This can be used to predict the performance of existing systems or of regenerators under design, possibly for new and novel applications. We focus on fixed-bed regenerators and then discuss later the application of the model for rotary regenerators.

The descriptive differential equations in one period of regenerator operations are similar to those developed for the single-blow unit but employing the lumped heat transfer coefficient, $\bar{\alpha}$, to approximate the effect of latitudinal, transverse conduction within the packing. These equations can be developed by consideration of the exchange of heat between the gas flowing through a single channel of the packing and the wall of material through which the channel is driven. The overall behavior of the regenerator is obtained by regarding the packing as a bundle of channels, identical in configuration and operation.

The first differential equation represents the relation between the rate of heat transferred between gas and solid and the rate at which thermal energy is given up/absorbed by the gas flowing through the regenerator:

$$\frac{\partial t_f}{\partial y} = \frac{\bar{\alpha} A}{\dot{m}_f c_p L}(t_s - t_f) - \frac{m_f}{\dot{m}_f L}\frac{\partial t_f}{\partial \tau} \tag{4.1}$$

The other differential equation covers the same rate of heat exchange across the packing surface and the rate at which heat is stored in/recovered from the packing:

$$\frac{\partial t_s}{\partial \tau} = \frac{\bar{\alpha} A}{M C_s}(t_f - t_s) \tag{4.2}$$

Although these equations are identical in form to those used for the single-blow problem, they are applied separately to the hot and cold periods of operation. In regenerators, the initial solid temperature distribution, just after a reversal, $t_s(y, 0)$, is set equal to that at the end of the previous period of operation. (This is in contrast to the single-blow problem where $t_s(y, 0)$ is treated as isothermal.)

Any mathematical representation of regenerator operation requires that two nomenclature problems be resolved:

- It is necessary to distinguish between the hot and cold periods. We use a single prime to denote the cold period, so that $\dot{m}_f'$ is the cold gas flow rate, for example.
- It is important to embody in the model counterflow operation. It is usually specified that the direction y is always measured from the gas entrance to the packing in the direction of gas flow. In this way, equations (4.1) and (4.2) can be applied to both hot and cold periods of operation. Counterflow operation is then embodied in the reversal condition with

$$t_s(y, 0) = t_s'(L - y, P') \tag{4.3}$$

for the start of the hot period and with

$$t_s'(y, 0) = t_s(L - y, P) \tag{4.4}$$

providing the starting conditions at the beginning of the cold period.

The mathematical model embodied in equations (4.1)–(4.4) is based on idealizations of a physical regenerator system. We have already discussed that the packing of the regenerator is regarded as a bundle of channels, identical in configuration and operation. Such an idealization seeks, on the one hand, to provide a generally applicable model. Thus the hot and cold periods, P and P', correspond to the lengths of time the hot and cold gas, respectively, are allowed to pass through the packing of a fixed-bed system. On the other hand, a useful idealization, in order to be mathematically and computationally tractable, must embody assumptions about the physical system. Some of the simplifications add to the generality of the model. Others provide what prove to be acceptable simplifications in the model in the sense that they do not inhibit the ability of the model to represent the physical system to an acceptable degree of accuracy.

In particular, the model assumes that:

- There is a uniform flow of gas across the complete cross-section of the packing for both hot and cold gas streams. It is usually assumed that the gas flow rates do not vary with time.
- There is no axial conduction of heat in the direction of gas flow, either in the solid or the gas.
- The heat transfer coefficients and the thermophysical properties of both gases and solid do not vary spatially or with temperature or time in either the hot or cold period. Nevertheless the parameter values

applicable to the hot period may be different from those in the cold period.

- There are no heat losses from the heat exchanger in either the hot or cold periods of operation.

In many papers, it is assumed also that the gas temperature entering the regenerator, in either period of operation, does not vary with time. That this is the case is not a necessary assumption for the model to retain the linearizations made possible by the third of these assumptions. This linearization takes a similar form to that embodied in the linear single-blow model.

The equations are simplified by the introduction of dimensionless parameters. The dimensionless temperature, T, is defined for regenerators by

$$T = \frac{t - t'_{f,in}}{t_{f,in} - t'_{f,in}} \tag{4.5}$$

and where $t_{f,in}$ and $t'_{f,in}$, are the unchanging temperatures of the gases as they enter the packed bed in the hot and cold periods respectively. Note that T and t in equation (4.5) can refer to either gas or solid temperatures.

From equation (4.5) it follows that

$$T_{f,in} = 1 \quad \text{and} \quad T'_{f,in} = 0$$

on this dimensionless temperature scale.

On the basis of these assumptions given above, it is possible to define the parameters ξ and η in the following way for the hot period:

$$\xi = \frac{\bar{\alpha} A}{\dot{m}_f c_p L} y \tag{4.6}$$

$$\eta = \frac{\bar{\alpha} A}{M C_s} \left(\tau - \frac{m_f y}{\dot{m}_f L} \right) \tag{4.7}$$

As with the single-blow problem, the parameter, ξ, can be viewed as dimensionless distance down the bed from the entrance of the gas and the dimensionless length or *reduced length*, Λ, of the packed bed is given by setting $y = L$ in equation (4.6) in which case, for the hot period:

$$\Lambda = \frac{\bar{\alpha} A}{\dot{m}_f c_p} \tag{4.8}$$

The parameter, η, can be regarded as dimensionless time from the start of the hot period. The dimensionless duration of the hot period or the hot *reduced period*, Π, is developed by setting $y = L$ and $\tau = P$ in equation (4.7). This yields

$$\Pi = \frac{\bar{\alpha} A}{M C_s}\left(P - \frac{m_f}{\dot{m}_f}\right) \tag{4.9}$$

The corresponding parameters Λ' and Π' are defined for the cold period:

$$\Lambda' = \frac{\bar{\alpha}' A}{\dot{m}'_f c'_p} \tag{4.10}$$

$$\Pi' = \frac{\bar{\alpha}' A}{M C'_s}\left(P' - \frac{m'_f}{\dot{m}'_f}\right) \tag{4.11}$$

Equations (4.1) and (4.2) now take the simplified form, as in the case of the single-blow problem,

$$\frac{\partial T_f}{\partial \xi} = T_s - T_f \tag{4.12}$$

$$\frac{\partial T_s}{\partial \eta} = T_f - T_s \tag{4.13}$$

These are applied successively to the hot and then the cold period in a cycle.

The representations of the reversals, which provide the initial conditions for the differential equations (4.12) and (4.13) are given by

$$T_s(\xi, 0) = T'_s(\Lambda'(\Lambda - \xi)/\Lambda, \Pi') \tag{4.14}$$

$$T'_s(\xi', 0) = T_s(\Lambda(\Lambda' - \xi')/\Lambda', \Pi) \tag{4.15}$$

The term

$$\frac{m_f}{\dot{m}_f L}\frac{\partial t_f}{\partial \tau}$$

in equation (4.1) represents the effect of the heat capacity of the gas, resident in the channels of the regenerator, at any instant. The corresponding term $m_f y/(\dot{m}_f L)$ in equation (4.7) represents the delay in the effect of any gas entering the regenerator at a distance that is a proportion y/L of the total length of the channels. At the start of the period under consideration, the gas residing in the channels of the packing from the previous period must be driven out by the incoming new gas. Assuming that no mixing occurs between the old and new gases, $m_f/\dot{m}_f$ is the time taken for the residual gas to be expelled from the regenerator.

In most models, it is assumed that the *reversal effects*, that is, the filling of the channels of the packing with new gas, has no influence upon regenerator performance. Indeed, Hausen [3] goes so far as to include, within an idealized regenerator, pistons that drive out the residual gas at the start of a period and it is assumed that this period only begins when the channels are filled with new gas. This corresponds to solving equations (4.12) and (4.13) from dimensionless time $\eta = 0$, that is, at position y, from when

$$\tau = \frac{m_f y}{\dot{m}_f L}$$

for each period of regenerator operation.

4.6 Modeling Rotary Regenerators

The mathematical representation of rotary regenerators can be realized by modeling a rotary regenerator by an equivalent fixed-bed regenerator. Provided the necessary mapping from rotary to fixed-bed regenerator can be realized, the differential equations (4.1) and (4.2) (or (4.12) and (4.13) in dimensionless form) can be used. It is necessary to make a number of preliminary observations.

- In fixed-bed regenerators, the hot gas flow rate, $\dot{m}_f$, passes through *all* the channels of the packing *at the same time* for the duration of the hot period, P. Similarly, the entire heating surface area, A, of the packing is washed by the cold gas with flow rate, $\dot{m}_f'$, for the duration of the cold period, P'.
- On the other hand, although the entire surface area, A, of the packing of a rotary regenerator is washed first by the hot gas and then by the cold gas in a complete rotation of the packing, the several sections of area enjoy the same cycle of regenerator operation but *not* simultaneously.

The differential equations can be developed by consideration of the process of heat transfer between gas and solid in a single channel. Such a channel experiences a cycle of heating and cooling whether in a fixed-bed regenerator, where the channel remains stationary, or in a rotary regenerator where the channel moves round with the packing.

In a complete cycle of fixed-bed regenerator operation, a mass of hot gas $\dot{m}_f P$ and a mass $\dot{m}_f' P'$ of cold gas pass through the regenerator. In the rotary regenerator, however, the hot and cold gases pass continuously through the heat exchanger so that in a complete cycle, a mass $\dot{m}_f(P + P')$ of hot gas and a mass $\dot{m}_f'(P + P')$ of cold gas (ignoring the times for the reversals) pass through the regenerator. This means that if we represent a rotary regenerator by an equivalent fixed-bed regenerator, we must consider that during the hot period, P, the same mass of gas must flow as passes through the rotary regenerator in a complete cycle, washing the complete surface area, A. The *equivalent* or *apparent* gas flow rates are

$$\dot{m}_f \frac{P+P'}{P} \quad \text{and} \quad \dot{m}_f' \frac{P+P'}{P'}$$

in the hot and cold periods respectively, each of which washes the entire heating surface, A, of the packing. This ensures that the total mass of cold gas, for example, passing through the equivalent fixed-bed regenerator is

$$\dot{m}_f' \frac{P+P'}{P'} P' = \dot{m}_f'(P+P')$$

as required.

In some forms of rotary regenerator, the cross-section of the packing is divided into more than two sectors. Let us consider that there is a third sector (trisector operation) through which a third gas passes with mass flow rate $\dot{m}_f''$. The length of the third period is P''. If we represent this by an equivalent fixed-bed regenerator, then the three apparent gas flow rates are

$$\dot{m}_f \frac{P+P'+P''}{P}, \quad \dot{m}_f' \frac{P+P'+P''}{P'}, \quad \text{and} \quad \dot{m}_f'' \frac{P+P'+P''}{P''}$$

In this way, all the channels of the rotary regenerator are bundled together in such a way as to yield an equivalent fixed-bed regenerator operating with three successive periods of operation, P, P', and P'', respectively.

We define the fractions f and f', where $f+f'=1$, by the relations

$$f = \frac{P}{P+P'} \quad \text{and} \quad f' = \frac{P'}{P+P'}$$

In practice, f is the fraction of the cross-sectional area of the rotary regenerator exposed to the flow of hot gas and f' that exposed to the cold gas provided that the time taken for changeovers can be ignored.

On this basis, we develop the equivalent gas mass flow rates in the form $\dot{m}_f/f$ and $\dot{m}_f'/f'$ for the equivalent fixed-bed regenerator. Where there are two "dead periods" of significant length when the packing passes beneath the gas seals separating the hot gas stream from the cold, when gas neither enters or leaves the packing of the regenerator, then f and f' must be defined in terms of P and P' in the form given above.

The heat transfer coefficients $\bar{\alpha}$ and $\bar{\alpha}'$ are a function of the mass velocities of the gases passing through the individual channels. These velocities are determined from the actual flow rates, $\dot{m}_f$ and $\dot{m}_f'$, and the corresponding free flow areas of cross-section in the rotary regenerator available to the hot and cold gases respectively.

The reduced lengths for the fixed-bed equivalent of a two-sector rotary regenerator are thus

$$\Lambda = \frac{\bar{\alpha} A f}{\dot{m}_f c_p} \quad \text{and} \quad \Lambda' = \frac{\bar{\alpha}' A f'}{\dot{m}_f' c_p'}$$

Confusion sometimes arises from thinking that f is a multiplier of the heating surface area, A. This is not the case because Λ and Λ' would refer to different fixed-bed regenerators, one with a heating surface fA, the other with heating surface $f'A$, which is not intended. The factors $1/f$ and $1/f'$ are multipliers of the mass flow rates, in the manner indicated above.

The term $m_f/\dot{m}_f$ can be rearranged to form L/v where v is the STP velocity of the gas passing through the packing and L/v is the time taken for the gas to pass through the rotating packing. It follows that this is not modified for η (equation (4.7)) or for Π (equation (4.9)). The reduced periods for a trisector regenerator are therefore

$$\Pi = \frac{\bar{\alpha} A}{M C_s}\left(P - \frac{L}{v}\right), \quad \Pi' = \frac{\bar{\alpha}' A}{M C_s'}\left(P' - \frac{L}{v'}\right) \quad \text{and}$$

$$\Pi'' = \frac{\bar{\alpha}'' A}{M C_s''}\left(P'' - \frac{L}{v''}\right)$$

and the equivalent fixed-bed regenerator is then regarded as one undergoing a hot period of dimensionless length Π, followed by two successive cold periods of duration Π' followed by Π''.

One test for the correctness of a computer simulation of a regenerator is one for heat balance. In the rotary regenerator, the gases flow continuously through the regenerator. The rate at which heat is stored in the packing is $\dot{m}_f c_p (t_{f,in} - t_{f,x})$, where $t_{f,x}$ is the chronological average exit gas temperature of the hot gas in the equivalent fixed-bed regenerator. In a trisector regenerator, this is compared with the *sum* of the rates at which heat is extracted by the two cold gas streams, namely $\dot{m}_f' c_p' (t_{f,x}' - t_{f,in}') + \dot{m}_f'' c_p'' (t_{f,x}'' - t_{f,in}')$, where we assume that $t_{f,in}' = t_{f,in}''$.

Modeling a rotary regenerator by way of a fixed-bed regenerator in this way implies a number of assumptions. In particular:

- It is idealized in that although there is a spatial temperature gradient within the gas and solid around the rotor (corresponding to the time-varying temperatures in the equivalent fixed-bed regenerator), there is no heat transfer between channels in a direction perpendicular to gas flow.
- It is further assumed that no mixing of gases between channels occurs as the flow of gas takes place in either the "hot period" or "cold period." In reality, the packing, as it rotates, must disturb the flow of gas through the regenerator.

- Finally, it is assumed that as the gas enters a channel of the packing, it remains there until it leaves the regenerator. That is, the gas is carried round the rotor within the packing. Again, in reality, some of the gas must be "left behind" by the packing as it rotates and there is some migration of gas from one channel to another.

One consequence of these assumptions is that, at cyclic equilibrium, at any level in the rotary regenerator, there is a calculated unchanging spatial variation of temperature, of both gas and solid, around the rotor and this corresponds to the chronological variations at the same level in the equivalent fixed-bed regenerator. In other words, the duration of the hot period, P, is equivalent to the length $2\pi rf$ around the circumference of the cylindrical packing of radius r. Similarly $P' \equiv 2\pi rf'$.

4.7 Discussion of the Design Parameters

The dimensionless parameters ξ and η allow a natural set of design parameters to be evolved, namely the reduced lengths, Λ and Λ', together with the reduced periods, Π and Π' (Table 4.1).

The overall performance is frequently expressed in terms of the thermal ratios η_{REG} and η'_{REG}, where

$$\eta_{REG} = \frac{t_{f,in} - t_{f,x}}{t_{f,in} - t'_{f,in}} \qquad \eta'_{REG} = \frac{t'_{f,x} - t'_{f,in}}{t_{f,in} - t'_{f,in}} \tag{4.16}$$

The problem then arises as to how one should express the relationships

$$\eta_{REG} = \eta_{REG}(\Lambda, \Lambda', \Pi, \Pi') \qquad \eta'_{REG} = \eta'_{REG}(\Lambda, \Lambda', \Pi, \Pi') \tag{4.17}$$

In addition, one seeks, if possible, to give some physical meaning to the dimensionless parameters Λ, Λ', Π, and Π'. We deal first with the latter problem.

Table 4.1: Dimensionless design parameters

$\Lambda = \dfrac{\bar{\alpha} A}{\dot{m}_f c_p}$	$\Lambda' = \dfrac{\bar{\alpha}' A}{\dot{m}'_f c'_p}$
$\Pi = \dfrac{\bar{\alpha} A}{M C_s}\left(P - \dfrac{m_f}{\dot{m}_f}\right)$	$\Pi' = \dfrac{\bar{\alpha}' A}{M C'_s}\left(P' - \dfrac{m'_f}{\dot{m}'_f}\right)$

Consequent upon the evolution of the reduced lengths, Λ and Λ', and the reduced periods, Π and Π', from the dimensionless variables ξ and η, is the convenient separation of the distinct mechanisms of the transfer of heat to/from the gases passing through the packing and the storage of heat within that packing. This proves to be particularly useful in understanding the principles involved in the design and operation of regenerative heat exchangers.

The ability of the regenerator to transfer heat from the gas is proportional to the product $\bar{\alpha}A$. In so discussing the hot period of regenerator operation, we note that the same arguments apply also to the cold period. The corresponding load imposed upon the heat exchanger is proportional to the thermal flow capacity $\dot{m}_f c_p$ of the hot gas. These are combined to yield the reduced length, Λ. The reduced lengths, Λ and Λ', can be regarded as the *thermal sizes* of the regenerator in the hot and cold periods relative to the respective loads, $\dot{m}_f c_p$ and $\dot{m}'_f c'_p$, imposed in those periods.

The thermal ratio, η_{REG}, given by equation (4.16) is defined as the ratio of the actual heat transfer rate in the hot period to the thermodynamically maximum obtainable heat transfer rate in a counterflow regenerator of infinite heat transfer area. Could this maximum rate be achieved, the temperature of the gas leaving the regenerator in the hot period would be equal to the cold period gas temperature at the entrance, $t'_{f,in}$. The cold period thermal ratio, η'_{REG} is defined similarly.

On the dimensionless temperature scale given by equation (4.5), the thermal ratios take the form

$$\eta_{REG} = 1 - T_{f,x} \quad \text{and} \quad \eta'_{REG} = T'_{f,x} \tag{4.18}$$

A regenerator is said to be *symmetric* if $\Lambda = \Lambda'$ and $\Pi = \Pi'$ in which case $\eta_{REG} = \eta'_{REG}$.

One would expect that the larger the thermal size, the greater the efficiency or *effectiveness* of the regenerator. This proves to be the case. Indeed, for a symmetric regenerator, an estimate of the thermal ratio is provided by

$$\eta_{REG} = \frac{\Lambda}{\Lambda + 2} \tag{4.19}$$

This estimate becomes exact for an infinitely small cycle time ($\Pi \to 0$) and is called the *ideal thermal ratio*. Displayed in Figure 4.9 is a graph showing the thermal ratio, η_{REG}, as a function of Λ and Π for the symmetric case. The curve for $\Pi = 0$ is the ideal, computed using equation (4.19). The remaining curves were presented by Schmidt and Willmott [4] having been computed initially by Hausen [2] by solving equations (4.12) and (4.13) subject to the reversal conditions (4.14) and (4.15) and the entrance conditions

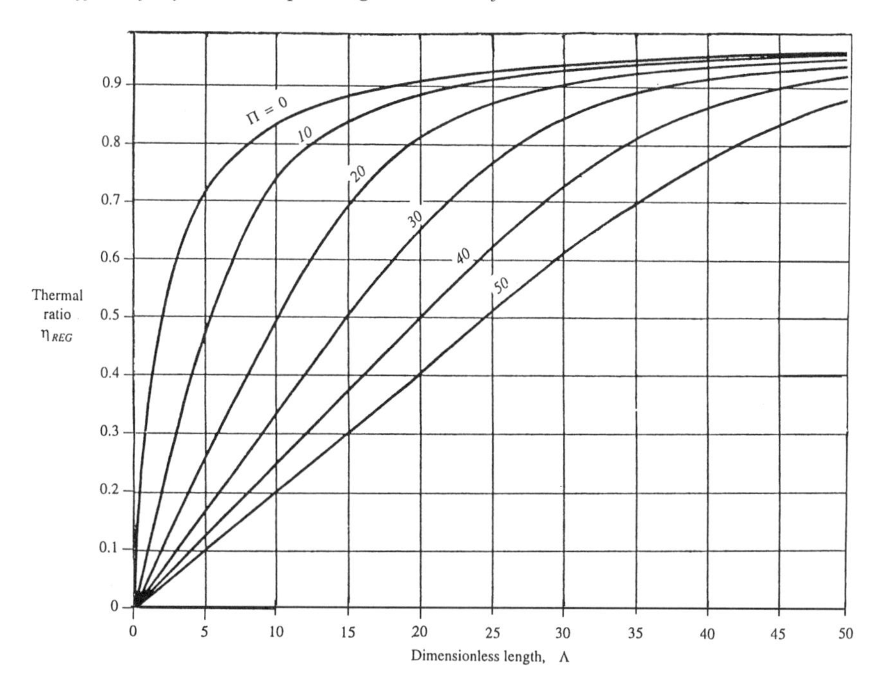

Figure 4.9: Variation of thermal ratio, η_{REG}, as a function of reduced length, Λ and reduced period, Π, for the symmetric case.

$$T_{f,in} = 1 \quad \text{and} \quad T'_{f,in} = 0$$

The method used is similar to that of Iliffe [5], which will be discussed later. Schmidt and Willmott reported that the curves had been recomputed using the method of Nahavandi and Weinstein [6] for some cases and by the method of Willmott [7] for the remainder. Reproduced here is the table of thermal ratios given in the Schmidt and Willmott text (Table 4.2).

4.8 Effect of Cycle Time upon Regenerator Performance

The ideal thermal ratio is a function of reduced length, Λ, alone and does not involve the heat capacity of the regenerator packing. This ideal is calculated for an infinitely short cycle time, in which case $\Pi = \Pi' = 0$.

Again, without loss of generality, we discuss the reduced period for $\Pi > 0$ with respect to the hot period. Any conclusions relate also to the cold period. For finite, nonzero values of reduced period, Π, the storage of heat within the packing becomes important. The heat capacity of the pack-

Table 4.2: Thermal ratio, η_{REG}, as a function of reduced length, Λ, and reduced period, Π for symmetric regenerators

Reduced length, Λ	Reduced period, Π					
	0	10	20	30	40	50
5	0.714	0.469	0.250	0.167	0.125	0.100
10	0.833	0.738	0.494	0.333	0.250	0.200
15	0.882	0.840	0.693	0.498	0.375	0.300
20	0.909	0.886	0.811	0.651	0.505	0.400
25	0.926	0.911	0.871	0.770	0.620	0.500
30	0.936	0.927	0.903	0.845	0.727	0.598
35	0.946	0.939	0.922	0.888	0.810	0.692
40	0.952	0.947	0.935	0.912	0.865	0.773
45	0.957	0.953	0.944	0.928	0.898	0.835
50	0.962	0.958	0.951	0.939	0.909	0.875

ing per period is MC_s/P. The larger the effective interface, $\bar{\alpha}A$, between the gas flowing through the regenerator and the thermal energy storage medium, the greater must be the heat capacity of the packing per period if *overheating* of the packing is not to occur. As period length increases, the thermal ratio becomes smaller than the ideal, as shown in Figure 4.9.

Translated into practical terms, this means that for a given arrangement of the heat storage material, the ratio $\bar{\alpha}A/(MC_s)$ can be computed and this can be matched by a period length, P, to yield as small as feasible a value of reduced period, Π, thereby maximizing the possible value of thermal ratio, η_{REG}. In effect, there is an important relationship between the area-to-mass ratio, A/M, and the cycle time of regenerator operation. This has an important consequence upon the choice of regenerator packings for different applications.

This can be considered in the following way. Economies in regenerator size can be obtained if "thin" packings are used, where the area-to-mass ratio, A/M, is large. Here, small enough values of reduced period, Π, are obtained by operating the regenerator with short cycle times.

On the other hand, the option to use such thin packings might not be available if harsh operating conditions prevail. Here, the regenerator packing might well need to be constructed of suitable materials with a robust geometrical arrangement. As a consequence, the ratio A/M may be relatively small, in which case, small enough values of reduced period, Π, can be realized using long cycle times. This has the advantage that rapid switching of the regenerator is avoided. The significant disadvantage is that in order to generate a large enough heating surface area to realize a big enough value of reduced length, Λ, with a view to achieving a high enough value of thermal ratio, η_{REG}, such regenerators, as will be seen, become massive in construction.

The matter is complicated still further if the process itself of reversing a regenerator is slow. For example, if it is necessary to pressurize the regenerator vessel at the start of the cold period, and then depressurize it at the end of the cold period, as is the case with Cowper stoves used for preheating the blast for ironmaking, then the total cycle time must be long enough for the time necessary for these reversals not to constitute an overlarge proportion of the total cycle time. In this case, the area-to-mass ratio, A/M, must be forced to be small enough to generate sufficiently small values of Π and Π' with the longer hot and cold periods of operation necessary.

4.9 Particular Packings for Different Regenerator Configurations

Very High Temperature Regenerators

Fixed-bed regenerators operating with hot gas inlet temperatures in excess of 1200°C are fitted with packings constructed of fireproof refractories or ceramic materials of special quality, capable of withstanding the effects of any corrosive materials entrained in the hot gas. Thus in glass furnace regenerators, it has been common to use high-alumina packings that are capable of coping with the corrosive effects of lime, potash, silica, sodium sulfate and vanadium, which can be carried over into the regenerator packing from the glass manufacturing process. In Cowper stoves, used to preheat the blast for the iron-making and zinc-smelting processes, the packing is frequently zoned: at the top of the regenerator, materials capable of withstanding the effects of very high temperatures, and further down the regenerator, of high compressive loads, are used, for example, silica. At the bottom of the regenerator, it is imperative to provide materials that possess the mechanical strength and volume stability, able to support the great weight of packing above. Different silica–alumina refractories are often employed in these circumstances (see Figure 4.10).

Not only must the materials of the packings be able to cope with the effects of corrosive materials, so also must the geometry of the packings be arranged that these possibly dirty gases can have free passage through the regenerator. The blocking of the channels must be avoided. In these circumstances, various geometrical arrangements of the refractory materials must be used. "Square chimney" or "closed basket weave" arrangements are frequently used in these circumstances. An example is shown in Figure

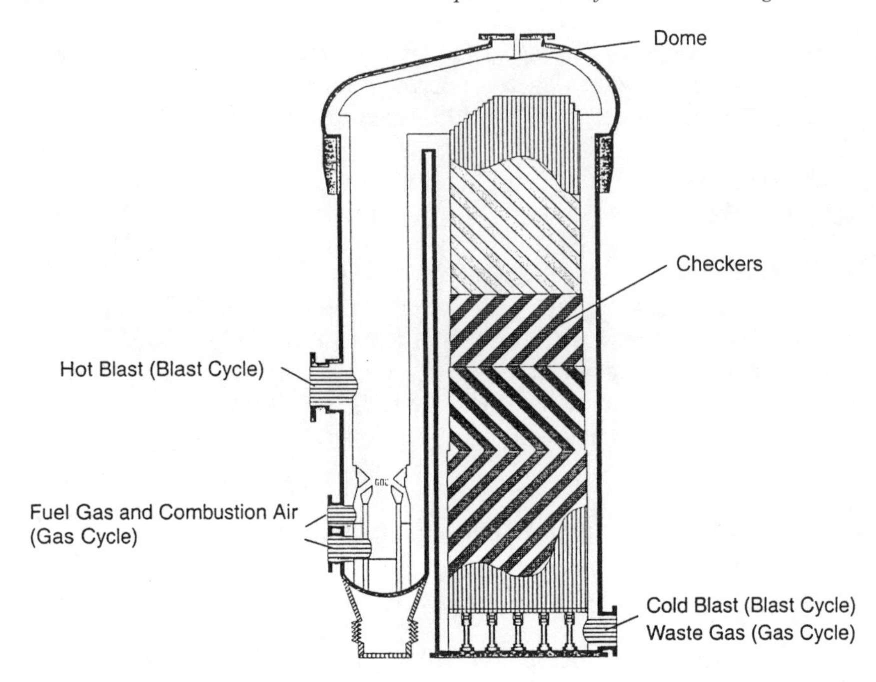

Figure 4.10: Cross-section of Cowper stove with an external combustion chamber. The checkerwork is layered in zones. (Courtesy British Steel, Teesside Technology Centre, Middlesbrough, UK.)

4.11. It can be arranged that the channels are wide enough to provide free passage for dirty gases but for the packing not to be aligned to form chimney passages: the "open basket weave" and "staggered open basket weave" fulfill this role. The channel width can be as large as 200 mm (Figure 4.11).

Where the gases are very hot but relatively clean, as in Cowper stoves, hexagonal bricks are frequently used with passages only 50 mm wide. These passages are formed in the body and on the corners of the bricks. A diagram of the classical Freyn design of checkerwork is shown in Figure 4.12. In this design, these passages are circular, but other shapes are possible. These refractory bricks are arranged in layers in such a way that tubular channels are formed, through which the gases can have a clear passage.

In both kinds of arrangement, the thickness of the packing "behind" the available heating surface area is determined by the mechanical strength required of the packing as well as by the corrosive conditions under which it is required to operate. Under severe conditions, thicknesses as large as 200 mm must be used; in less harsh conditions as can arise in the chemical industry, bricks 50 mm thick may be adequate. This generates a

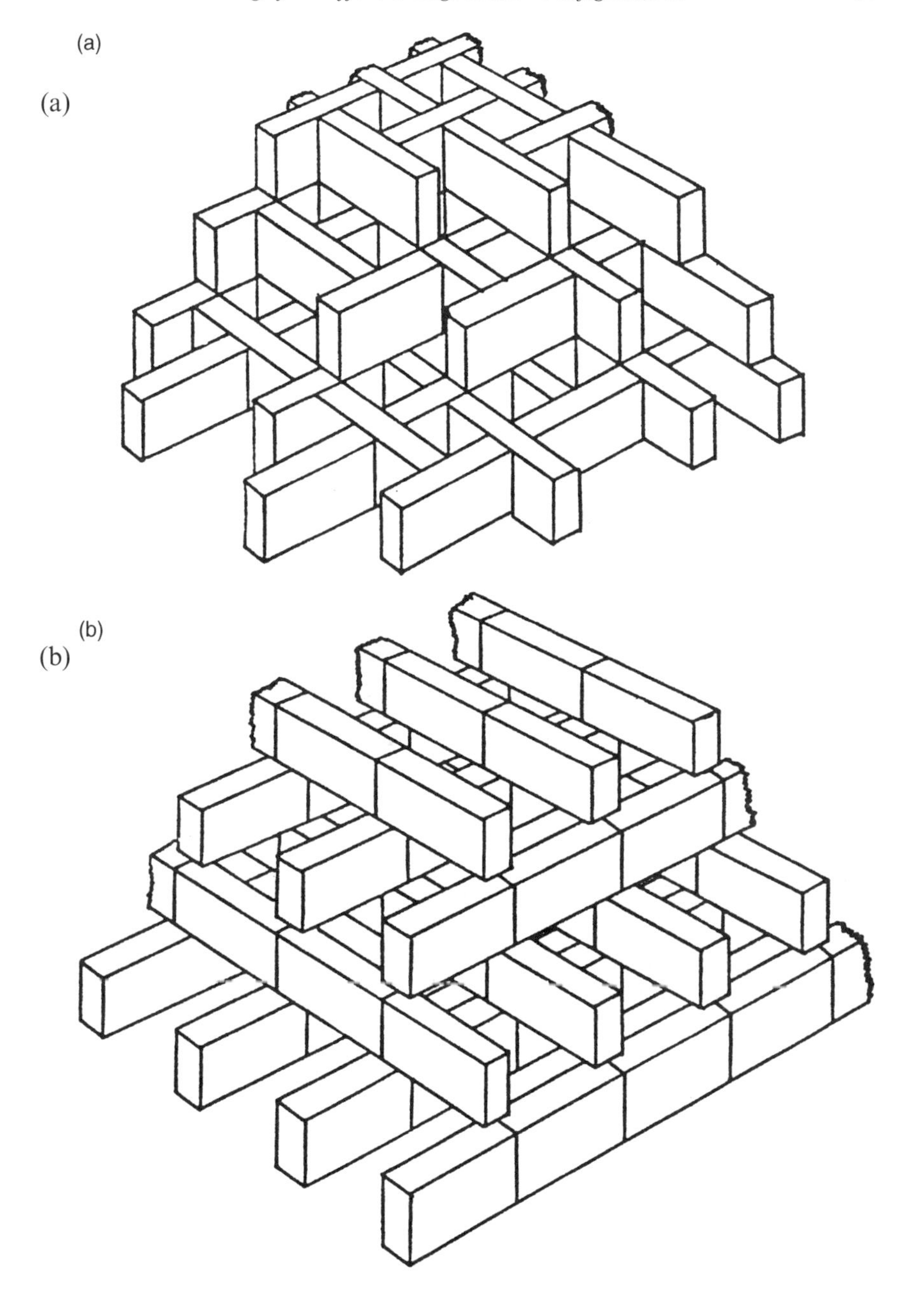

Figure 4.11: Regenerator packings for high temperatures and possibly dirty gases. (a) Square weave arrangement; (b) Staggered basket weave arrangement.

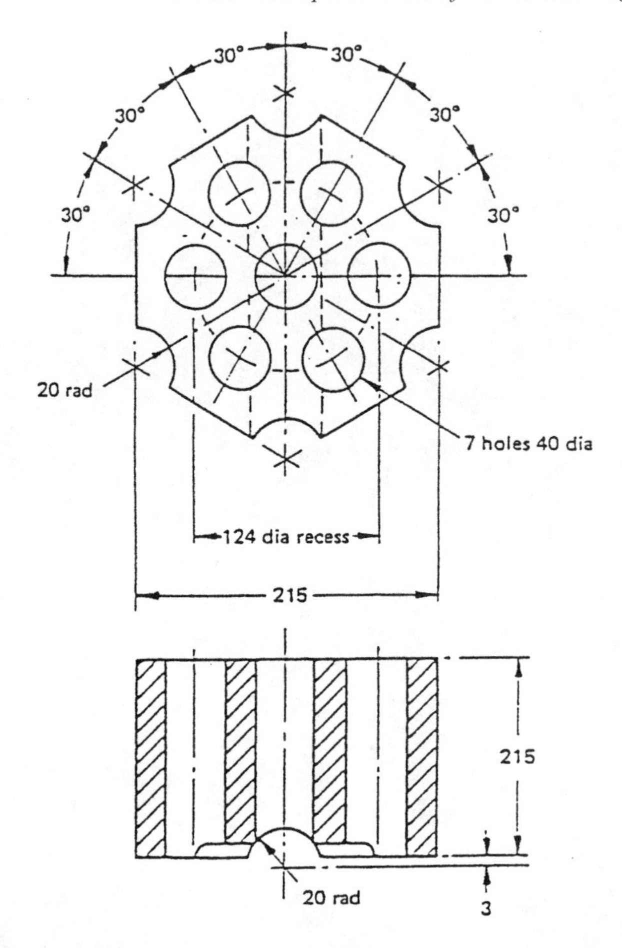

Figure 4.12: Typical design for Freyn-type checkerwork for hot-blast stoves. (Measurements in millimeters.)

packing with a low surface-area-to-volume ratio of the order of 20–30 m^{-1} and hence a low A/M ratio. In order to yield a high enough surface area and thus a correspondingly large enough value of reduced length, Λ, the packing in a Cowper stove is held in a cylindrical vessel, massive in size, of the order of 30 m high and 10 m in diameter, the dimensions of a small church tower! Such regenerators are necessarily expensive.

On the other hand, as has been briefly indicated previously, small enough values of reduced period, Π, can be accommodated in a three-stove arrangement with a hot period, P, in the range 45–110 min and a cold period, P' in the range 25–55 min. Reversals lasting 5–10 min per cycle can be accommodated easily. The Siemens type of regenerator used for glass making operates with a typical cycle time of 40 min.

Reference has been made to the *regenerative burner*, which exploits the fact, among other things, that industrial gases have had to be cleaned in order to meet legal requirements for environment protection. As a consequence, it has been possible to use regenerator packings consisting of ceramic spheres with materials chosen to meet the operating conditions encountered. The spheres are typically 1–3 cm in diameter, yielding area-to-volume ratios of the order of 10 times larger than encountered in the massive regenerators for the glass-making furnace, for example. These ratios, in the range 100–300 m^{-1}, yield small compact regenerators. The particular size is determined by the thermal load, $\dot{m}_f c_p$, that the regenerator is required to support. A bed 0.6 m high and 0.18 m in diameter is not uncommon, although smaller or larger beds can be used for different thermal loads.

Because the ratios $\bar{\alpha}A/(MC_s)$ and $\bar{\alpha}'A/(MC_s')$ will be larger for beds of ceramic spheres with a surface-area-to-volume ratio in the range 100–300 m^{-1}, it is necessary to switch the regenerators far more rapidly than is the case for massive high-temperature regenerators. Indeed, the regenerators and their burners are reversed every 30–180 s.

Moderate Temperature Regenerators

For the Ljungström regenerators, the cylindrical porous packing is rotated around its axis. The packing materials are often fabricated of steel sheets, notched to form a large number of undulating passages. In this way, turbulent flow of the hot and cold gases flowing through the regenerator is promoted, thereby improving the heat transfer characteristics. The metal sheets are arranged radially in removable units holding several such sheets, thereby facilitating rapid and simple maintenance (Figure 4.13).

Such metal sheeting provides a high area-to-volume ratio, in excess of 200 m^{-1}. Nevertheless, they must be constructed in such a way as to be able to withstand the temperatures involved as well as possibly corrosive operating conditions. Where, for example, the hot waste gases have a high SO_2 content, a vitreous enameled heating surface can be employed to protect the steel packing at operating temperatures below the acid dew point of such gases, although this can be prohibitively expensive.

Even higher area-to-volume ratios can be achieved by constructing the regenerator of an assembly of sector-shaped sections of a knitted mesh of wire of another material, depending on the temperature and other operating conditions. For hot-gas entry temperatures of 400°C, stainless-steel mesh can be employed, whereas for temperatures of up to 800°C, ceramic or

Figure 4.13: Modular packing units for rotary regenerators; the units can be withdrawn individually and replaced during servicing. (Typical packing plates shown also.) (Courtesy Howden-Sirocco Ltd., Glasgow, UK.)

alumina fibres have been considered. Other prefabricated heavy-duty ceramic packings can be employed in regenerators required to withstand hot-gas entry temperatures of 800°C or more: here the packing might consist of a honeycomb of ceramic material arranged as alternately flat and wave-shaped layers, as shown in Figure 4.14. Such constructions realize the high surface-area-to-volume ratio necessary to achieve compactness in regenerator construction and, at the same time, allow for a free passage for the flow of the gases through the regenerator. They are also robust enough to survive such tough working temperatures and operating conditions.

Again the ratios $\bar{\alpha}A/(MC_s)$ and $\bar{\alpha}'A/(MC_s')$ will be large in these circumstances. It is not uncommon, therefore, for the packing to be rotated at 2–3 rev/min, yielding hot/cold periods of 30 s or less. The small values

Figure 4.14: Honeycomb of ceramic material for the packing of high-temperature rotary regenerators.

of Π and Π' generated permit regenerator efficiencies of 80% of more to be realized.

Lower Temperature Regenerators

The operation of regenerators at low (ambient or even lower) temperatures permits a good deal of flexibility in the choice of packing materials. Rotary regenerators for air-conditioning applications employ a variety of packings, which include a polyethylene terephthalate film and corrugated, knitted wire mesh. Such packings are wound round the spindle of the rotor yielding "heat wheels" of varying diameters, from 1.25–2.5 m. An example is shown in Figure 4.15. Corrugated aluminum sheets are sometimes used as are various honeycomb arrangements. A variety of packings have been developed to recover both latent heat and specific heat from one of the gases. Included in these are packings that are nonmetallic and fibrous:

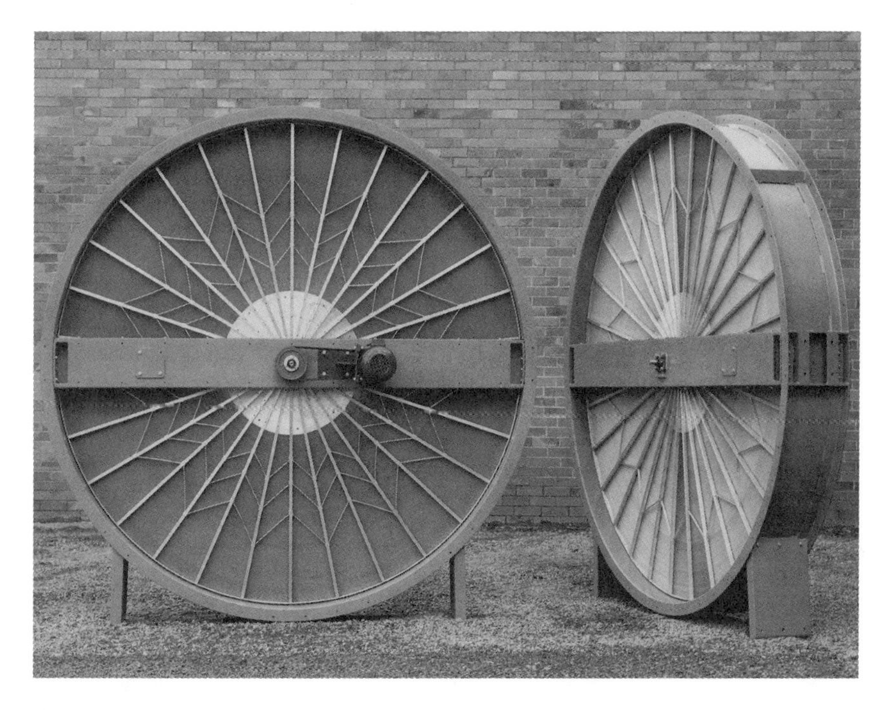

Figure 4.15: Rotors for a low-temperature regenerator. Here the packing is made of polyethylene terephthalate film. (Courtesy Rotary Heat Exchangers Pty, Bayswater, Victoria, Australia.)

they can absorb moisture on the one hand, but are inert to bacterial contamination on the other.

For very low temperature operation, fixed-bed regenerators are frequently used where packings consist of beds of basalt or flint chips, or simply gravel. Corrugated aluminum sheets are sometimes used where corrugations run in alternate directions between the sheets, which are laid on top of one another, generating fine, intersecting channels for the free passage of the gases. Such sheet arrangements can prove to be very expensive.

4.10 Imbalance in Regenerator Performance

A regenerator is said to be *unbalanced* should

$$\frac{\Pi}{A} \neq \frac{\Pi'}{\Lambda'} \tag{4.20}$$

which implies that

$$\dot{m}_f c_p \left(P - \frac{L}{v} \right) \neq \dot{m}_f' c_p' \left(P' - \frac{L}{v'} \right) \tag{4.21a}$$

or, ignoring the L/v terms which are frequently negligible,

$$\dot{m}_f c_p P \neq \dot{m}_f' c_p' P' \tag{4.21b}$$

This is the general and most common case in practice. The regenerator is said to be *balanced* if

$$\frac{\Pi}{\Lambda} = \frac{\Pi'}{\Lambda'} = k \tag{4.22}$$

where k is a constant. It will be seen that the *symmetric* regenerator is the special case where $k = 1$. We note that for the balanced case

$$\dot{m}_f c_p P = \dot{m}_f' c_p' P' \quad \text{and} \quad \eta_{REG} = \eta_{REG}' \tag{4.23}$$

To give a "thumbnail" overall picture of a particular regenerator, it is convenient to approximate an unsymmetric ($k \neq 1$) but balanced regenerator by an equivalent symmetric regenerator. Hausen [2] suggested that the parameters Λ, Λ', Π, and Π' be combined to yield just two parameters, namely Λ_H and Π_H for a symmetric regenerator. These are defined by

$$\frac{2}{\Pi_H} = \frac{1}{\Pi} + \frac{1}{\Pi'} \tag{4.24}$$

$$\frac{2}{\Lambda_H} = \frac{1}{\Pi_H} \left(\frac{\Pi}{\Lambda} + \frac{\Pi'}{\Lambda'} \right) \tag{4.25}$$

where Π_H is a *harmonic mean* and Λ_H is a *weighted harmonic mean*. If the thermal ratio for the equivalent symmetric regenerator is η^H_{REG}, then if it is assumed that the heat exchanged in the balanced regenerator is the same as that exchanged in its symmetric equivalent, then

$$\frac{\Pi}{\Lambda}\eta_{REG} = \frac{\Pi'}{\Lambda'}\eta'_{REG} = \frac{\Pi_H}{\Lambda_H}\eta^H_{REG} \tag{4.26}$$

This proposal of Hausen was verified as acceptable by Iliffe [5] for the cases where $3 \leq \Lambda' \leq 18$ and $3 \leq \Pi' \leq 18$ with $k = 2$ and $k = 3$.

Hausen suggested that this use of harmonic means could be extended to the unbalanced case. The application of equation (4.26) for this in order to compute η_{REG} and η'_{REG} from a calculated value of η^H_{REG} is not very precise for unbalanced regenerators. In any event, the direct calculation of an unbalanced regenerator presents no problems. Nevertheless, the parameters Λ_H and Π_H can be used to give a feel for the size and operation of a particular configuration.

The main problem is that, as they stand, the parameters Λ, Λ', Π, and Π' seem to give little direct information about the operation of a regenerator for *unbalanced* conditions. In addition, when $\eta_{REG} \neq \eta'_{REG}$, there is no single measure available for the thermal effectiveness of the heat exchanger. Baclic [8] approaches this problem in the following manner. He introduces the idea of a *weak* stream of gas, or, more precisely a *weak period*, which is characterized by $(\dot{m}_f c_p P)_{min}$, where

$$(\dot{m}_f c_p P)_{min} = \min(\dot{m}_f c_p P, \dot{m}'_f c'_p P') \tag{4.27}$$

Baclic then defines the hypothetical quantity of heat, Q_{max}, in terms of the maximum possible enthalpy change in the system, namely the enthalpy change of the weak stream undergoing the maximum possible temperature change, $t_{f,in} - t'_{f,in}$.

$$Q_{max} = (\dot{m}_f c_p P)_{min}(t_{f,in} - t'_{f,in}) \tag{4.28}$$

Baclic is then able to provide a single measure of thermal effectiveness, ε, by the equation

$$\varepsilon = \frac{Q_{act}}{Q_{max}} \tag{4.29}$$

where Q_{act} is the actual heat transferred from the hot to the cold gas. This is given by

$$Q_{act} = \dot{m}_f c_p P(t_{f,in} - t_{f,x}) = \dot{m}'_f c'_p P'(t'_{f,x} - t'_{f,in}) \tag{4.30}$$

It follows that

$$\varepsilon = \frac{\dot{m}_f c_p P(t_{f,in} - t_{f,x})}{(\dot{m}_f c_p P)_{min}(t_{f,in} - t'_{f,in})}$$
$$= \frac{\dot{m}'_f c'_p P'(t'_{f,x} - t'_{f,in})}{(\dot{m}_f c_p P)_{min}(t_{f,in} - t'_{f,in})} \tag{4.31}$$

Recalling the definition in equation (4.5) of dimensionless temperature, T, for regenerators, equation (4.31) can be rewritten in the form

$$\varepsilon = \frac{\dot{m}_f c_p P}{(\dot{m}_f c_p P)_{min}}(T_{f,in} - T_{f,x}) = \frac{\dot{m}_f c_p P}{(\dot{m}_f c_p P)_{min}}(1 - T_{f,x})$$
$$= \frac{\dot{m}'_f c'_p P'}{(\dot{m}_f c_p P)_{min}}(T'_{f,x} - T'_{f,in}) = \frac{\dot{m}'_f c'_p P'}{(\dot{m}_f c_p P)_{min}} T'_{f,x} \tag{4.32}$$

Baclic and Dragutinovic [9] extended this analysis by first recalling the concept of the *utilization factor*, $U = \Pi/\Lambda$, introduced by Johnson [10] in 1948. They point out that U does not involve the heat transfer coefficients, α and α'. Neither does it involve the heating surface area. Rather, it represents the ratio of the heat capacity of the fluid that passes through the regenerator in a period to the heat capacity of the packing. We see that

$$U = \frac{\Pi}{\Lambda} = \frac{\dot{m}_f c_p P}{MC_s} \tag{4.33}$$

which implies that U measures the utilization of the heat capacity of the packing. Baclic and Dragutinovic define U_{min} by

$$U_{min} = \min(U, U') \tag{4.34}$$

which is associated with the weaker period. Similarly U_{max}, where

$$U_{max} = \max(U, U') \tag{4.35}$$

is associated with the stronger period. Now if Δt_s is the average temperature swing of the packing, where

$$\Delta t_s = \frac{1}{L}\int_0^L (t_s(y, P) - t_s(y, 0))\, dy$$

then

$$MC_s\, \Delta t_s = \dot{m}_f c_p P(t_{f,in} - t_{f,x}) = \dot{m}'_f c'_p P'(t'_{f,x} - t'_{f,in})$$

in which case

$$U = \frac{\dot{m}_f c_p P}{MC_s} = \frac{\Delta t_s}{t_{f,in} - t_{f,x}} \tag{4.36}$$

with a similar expression for U'. It follows that

$$U_{min} = \frac{\text{mean packing temperature swing}}{\text{time mean gas temperature change}}$$

for the weaker period. The aim is not to make Δt_s as large as possible, rather the reverse: make the regenerator as effective as possible with as small a solid temperature swing as feasible. For favorable operation, one would expect that $U_{min} < 1$. A bridge between U and U' is afforded by the imbalance factor, β, introduced by Schmidt and Willmott [4] in a form which is essentially

$$\beta = \frac{U_{min}}{U_{max}} = \frac{(\dot{m}_f c_p P)_{min}}{(\dot{m}_f c_p P)_{max}} \leq 1 \tag{4.37}$$

if we assume that $C_s = C_s'$. In the case where the weak period is the cold period, that is $(\dot{m}_f c_p P)_{min} = (\dot{m}_f' c_p' P')$, then from equation (4.31) we note that

$$\varepsilon = \eta_{REG}' \quad \text{and} \quad \varepsilon = \frac{1}{\beta}\eta_{REG} \tag{4.38}$$

Similarly, if the hot period is the weaker one, then

$$\varepsilon = \frac{1}{\beta}\eta_{REG}' \quad \text{and} \quad \varepsilon = \eta_{REG} \tag{4.39}$$

Example 4.1 The descriptive data for each channel of a particular regenerator are considered:

Available heating surface area $A = 31.40\,\text{m}^2$
Corresponding heat storing mass $M = 4898\,\text{kg}$
Specific heat of the packing $C = 0.92\,\text{kJ/(kg\,°C)}$
Specific heat of gas in both periods $c_p = 1.011\,\text{kJ/(kg\,°C)}$

The operation is as follows:

Gas mass flow per channel (hot period) $\dot{m}_f = 0.156\,\text{kg/s}$
Gas mass flow per channel (cold period) $\dot{m}_f' = 0.078\,\text{kg/s}$
Heat transfer coefficient (hot period) $\bar{\alpha} = 50.23\,\text{W/(m}^2\,°\text{C)}$
Heat transfer coefficient (cold period) $\bar{\alpha}' = 50.23\,\text{W/(m}^2\,°\text{C)}$
Duration of the hot period $P = 3600\,\text{s}$
Duration of the cold period $P = 10\,800\,\text{s}$

The dimensionless parameters are

$$\text{Reduced length (hot period)} \quad \Lambda = \frac{\bar{\alpha}A}{\dot{m}_f c_p} = \frac{(50.23)(31.4)}{(0.156)(1011.0)} = 10.0$$

$$\text{Reduced length (cold period)} \quad \Lambda' = \frac{\bar{\alpha}'A}{\dot{m}_f' c_p'} = \frac{(25.11)(31.4)}{(0.078)(1011.0)} = 10.0$$

$$\text{Reduced period (hot period)} \quad \Pi = \frac{\bar{\alpha}AP}{MC_s} = \frac{(50.23)(31.4)(3600)}{(4898.4)(920.0)} = 1.26$$

$$\text{Reduced period (cold period)} \quad \Pi' = \frac{\bar{\alpha}'AP'}{MC_s'} = \frac{(25.11)(31.4)(10800)}{(4898.4)(920.0)} = 1.88$$

The harmonic means parameters are

$$\frac{1}{\Pi_H} = \frac{1}{2}\left(\frac{1}{\Pi} + \frac{1}{\Pi'}\right) = \frac{1}{2}\left(\frac{1}{1.26} + \frac{1}{1.88}\right) = 0.6628$$
$$\Pi_H = 1.51$$

$$\frac{1}{\Lambda_H} = \frac{1}{2\Pi_H}\left(\frac{\Pi}{\Lambda} + \frac{\Pi'}{\Lambda'}\right) = \frac{1}{(2)(1.51)}\left(\frac{1.26}{10.0} + \frac{1.88}{10.0}\right) = 0.10397$$
$$\Lambda_H = 9.6178$$

The utilization factors are given as follows:

$$U = \frac{\Pi}{\Lambda} = \frac{1.26}{10.0} = 0.126 = U_{min}$$
$$U' = \frac{\Pi'}{\Lambda'} = \frac{1.88}{10.0} = 0.188 = U_{max}$$

The degree of imbalance, β, is computed:

$$\beta = \frac{U_{min}}{U_{max}} = \frac{0.126}{0.188} = 0.67$$

For $\Lambda = \Lambda' = 10.0$ with $\Pi = 1.26$ and $\Pi' = 1.88$ the computed thermal ratios are

$$\eta_{REG} = 0.947 \quad \text{and} \quad \eta'_{REG} = 0.635$$

The weaker period is the hot period (because $U < U'$). It follows that the effectiveness, ε, is given by

$$\varepsilon = \eta_{REG} = 0.947 = \frac{1}{\beta}\eta'_{REG} = \frac{1}{0.67}(0.635) = 0.947$$

For $\Lambda_H = 9.6178$ with $\Pi_H = 1.51$ (equivalent symmetric regenerator) the computed thermal ratio is

$$\eta^H_{REG} = 0.8251$$

From equation (4.26), we deduce that

$$\eta^*_{REG} = \frac{U_H}{U}\eta^H_{REG} = \frac{0.157}{0.126}0.8251 = 1.028$$
$$\eta'^*_{REG} = \frac{U_H}{U'}\eta^H_{REG} = \frac{0.157}{0.188}0.8251 = 0.689$$

where η^*_{REG} and η'^*_{REG} are the estimated thermal ratios. With $\eta^*_{REG} > 1$, this indicates the inadequacy of the harmonic means model for unbalanced regenerators, certainly for $\beta < 0.7$. We note that $\eta'^*_{REG} - \eta'_{REG} = 0.054$, indicating an error of about 8% in η'^*_{REG}.

The heat balance equation (4.30) can be rewritten in dimensionless temperature form by dividing throughout by $(t_{f,in} - t'_{f,in})$. This yields

$$\dot{m}_f c_p P(T_{f,in} - T_{f,x}) = \dot{m}'_f c'_p P'(T'_{f,x} - T'_{f,in}) \tag{4.40}$$

or

$$\dot{m}_f c_p P(1 - T_{f,x}) = \dot{m}'_f c'_p P'\, T'_{f,in} \tag{4.41}$$

recalling that $T_{f,in} = 1$ and $T'_{f,in} = 0$. Equation (4.41) can be itself rewritten as

$$T'_{f,x} = \frac{\dot{m}_f c_p P}{\dot{m}'_f c'_p P'}(1 - T_{f,x}) = \beta(1 - T_{f,x}) \tag{4.42}$$

where the hot period is the weak period. Similarly

$$T'_{f,x} = \frac{1}{\beta}(1 - T_{f,x}) \tag{4.43}$$

where the stronger period is the hot period. Baclic and Dragutinovic [9] suggest, as a consequence,

$$\beta = \frac{\text{total gas temperature change in the stronger period}}{\text{total gas temperature change in the weaker period}}$$

It is tempting to think that the smaller the imbalance factor, $\beta \ll 1$, the greater will be the effectiveness, ε, and that this is desirable. This stems from the arbitrary decision which is, in effect, to define ε by

$$\varepsilon = \max(\eta_{REG}, \eta'_{REG})$$

It is possible equally to define an alternative effectiveness, ε^* by

$$\varepsilon^* = \min(\eta_{REG}, \eta'_{REG}) = \frac{Q_{act}}{Q^*_{max}} \tag{4.44}$$

where Q^*_{max} is the enthalpy change in the *strong* stream undergoing the maximum temperature change, $t_{f,in} - t'_{f,in}$:

$$Q^*_{max} = (\dot{m}_f c_p P)_{max}(t_{f,in} - t'_{f,in}) \tag{4.45}$$

It is certainly not always the case that the effectiveness, ε, should be maximized by minimizing the value of the imbalance factor, β.

Consider a Cowper stove that is required to deliver a flow rate, $\dot{m}'_f$ of hot blast (air) at a temperature $t'_{f,x}$. This implies a required thermal ratio, η'_{REG}, with $\varepsilon = \eta'_{REG}$ and the cold period being the weaker period. This could be realized provided that the flow rate, $\dot{m}_f$, of the products of combustion of the fuel gas was set sufficiently high and, as a consequence, β set sufficiently low. This would yield a low value of $\eta_{REG} = \beta\varepsilon$. But a low value of η_{REG} implies a high value of $t_{f,x}$, the exit temperature of the flue gases to the chimney stack from the bottom of the stove. In other words, a low proportion, η_{REG}, of the heat available in the fuel burnt at the hot-side entrance of the Cowper stove is transferred to the hot blast, the residue being allowed to escape to the atmosphere, "up the chimney."

This situation is only avoided by making the stove large enough so that β can be set to a value in such a way that the hot blast temperature, $t'_{f,x}$, is achieved with the hot-side flow rate, $\dot{m}_f$ set to such a value so that both the proportion, η_{REG}, of heat transferred to the hot blast from the fuel gas is as high as feasible, subject to the constraint that the exit gas temperature, $t_{f,x}$ in the hot period is high enough to ensure sufficient updraft of waste gases up the chimney.

It would be ideal if there were but one measure of effectiveness, ε. However, the price of realizing that effectiveness, suitably measured as $1 - \varepsilon^*$, must also be considered.

Baclic and Dragutinovic [9] finally introduce another parameter, Λ_2, where Λ_2 is the reduced length associated with the stronger period. This enables them to give two more descriptive parameters for thermal regenerators that are unbalanced. These are Λ_1, the reduced length associated with the weaker period, and σ, where

$$\sigma = \frac{\Lambda_1}{\Lambda_2} \tag{4.46}$$

can be regarded as the ratio of the weaker to the stronger period thermal sizes of the regenerator for the unbalanced regenerator. Baclic and Dragutinovic then replace the relationships

$$\eta_{REG} = \eta_{REG}(\Lambda, \Lambda', \Pi, \Pi') \qquad \eta'_{REG} = \eta'_{REG}(\Lambda, \Lambda', \Pi, \Pi') \tag{4.17}$$

by the relationship describing overall regenerator behavior, given as

$$\varepsilon = \varepsilon(U_{min}, \beta, \Lambda_1, \sigma) \tag{4.47}$$

Using the subscript 1 to denote the weaker period and 2 the stronger period, they point out that these parameters can be related to the original Hausen reduced-period parameters by the relationship

$$\Pi_1 = U_{min}\Lambda_1 \qquad \Pi_2 = \frac{U_{min}\Lambda_1}{\beta\sigma}$$

They claim that significant advantage is realized in using the relationship $\varepsilon = \varepsilon(U_{min}, \beta, \Lambda_1, \sigma)$ through the physical meaning corresponding to the limiting cases of each of the parameters. In particular, the following cases are considered.

1. Where $U_{min} \to 0$, this corresponds to the case where the regenerator is reversed so rapidly that the solid temperature does not change. Provided the effect of the reversals is completely ignored, this becomes the case where the hot- and cold-gas spatial temperature distributions do not change from cycle to cycle, nor do they change *within* the appropriate period. This might be called the *recuperative* operation of a regenerator. In this case, the performance of the heat exchanger is a function of the reduced lengths, Λ and Λ', alone.
2. Where $U_{min} \to \infty$, this corresponds to the single-blow problem, provided it can be assumed that if $T_{f,in} = 1$, the solid is initially isothermal and equal to zero in temperature.
3. $\Lambda_1 \to 0$ and $\Lambda_1 \to \infty$ correspond to "short" and "long" regenerators respectively. In the "short" case, no heat transfer takes place. It is difficult to conceive of the case where the regenerator is infinitely long, because no position can be ascribed to the entrance of one of the gases. On the other hand, one can imagine that the infinitely long regenerator is the case where the $T_{f,x} = T'_{f,in} = 0$ and $T'_{f,x} = \beta$, where the hot period is weaker. In the opposite case, $T'_{f,x} = 1$ and $T_{f,x} = 1 - \beta$.
4. Where $\beta \to 0$, the regenerator is completely unbalanced. Baclic and Dragutinovic [9] suggest that this is where no chronological or spatial change in temperature occurs in the stronger period for any value of σ. However, it corresponds also to the case where $(\dot{m}_f c_p P)_{min} \to 0$,

that is, where no gas flows through the regenerator in the weaker period and is a case, therefore, of little physical meaning.

5. The completely balanced case arises when $\beta = 1$. Each gas stream undergoes the same temperature *change* passing through the exchanger.
6. Where $\sigma \to 0$, a completely asymmetric case is implied, where there is no effective weak period.
7. The opposite, completely asymmetric case arises where $\sigma \to \infty$, in which case, the strong period completely disappears.
8. The combination of $\beta = 1$ and $\sigma = 1$ corresponds to the symmetric case.

The present author considers that the Hausen parameters, Λ and Π, are useful in giving an overall picture of a regenerator and its operation, described previously, and provide insights into the mechanisms of heat transfer taking place in a regenerative heat exchanger. Equally, he acknowledges the need to represent imbalance in regenerator operation and accepts that the parameters U_{min}, β, Λ, and σ go some way to meeting this requirement. The concepts of weak and strong periods are especially useful.

Baclic and Dragutinovic [9] offer graphs of effectiveness, ε as a function of the utilization factor, U_{min}, for different values of the reduced length for the weak period. These are reproduced here as Figures 4.16–4.19. They denote U_{min} by U_1.

4.11 Concluding Remarks

This chapter has been concerned with distinguishing between rotary and fixed-bed regenerators, with the modeling of such regenerators and the representing of the effect of imbalance in regenerator operation. It has been seen that through the models and dimensionless parameters used, some rationale emerges in the way regenerators are designed and then operated. In the next chapter, we concern ourselves with the possible methods of solution of the differential equations (4.1) and (4.2) subject to the reversal conditions given by equations (4.3) and (4.4) of which the models are composed. We focus upon the dimensionless forms of these equations, namely equations (4.12) and (4.13) with reversal conditions (4.14) and (4.15). In a later chapter, we deal with nonlinear models where it is allowed that the thermophysical properties of both gases and the solid, as also the heat transfer coefficients, may vary spatially and with time as functions of temperature. The flow rates of the gases can be considered to vary with time as well, if required.

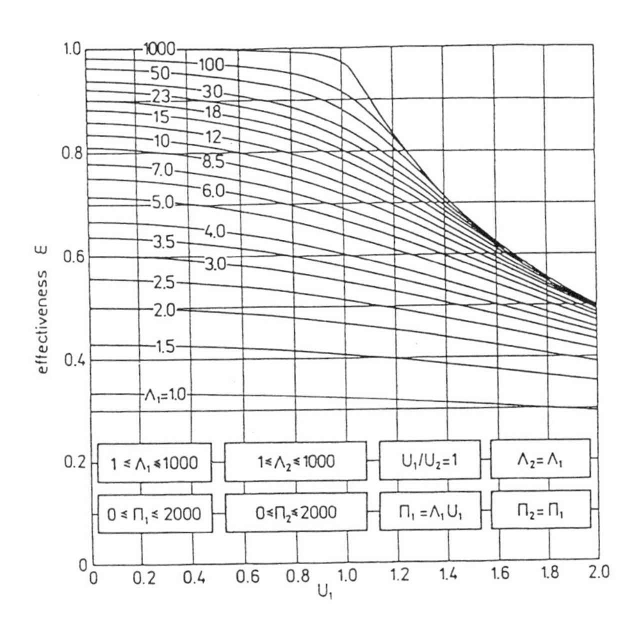

Figure 4.16: Thermal effectiveness, ε, as a function of the utilization factor, U_1, and reduced length, Λ_1, for $\beta = 1$ and $\sigma = 1$. (Baclic and Dragutinovic [9].)

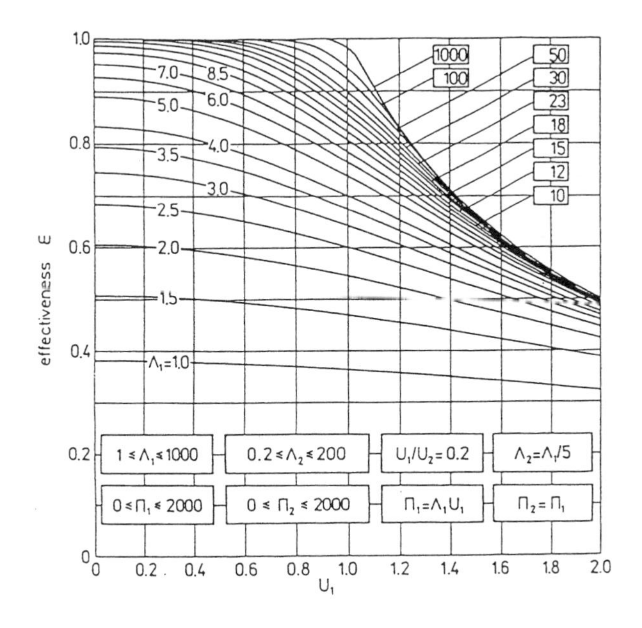

Figure 4.17: Thermal effectiveness, ε, as a function of the utilization factor, U_1, and reduced length, Λ_1, for $\beta = 0.2$ and $\sigma = 5$. (Baclic and Dragutinovic [9].)

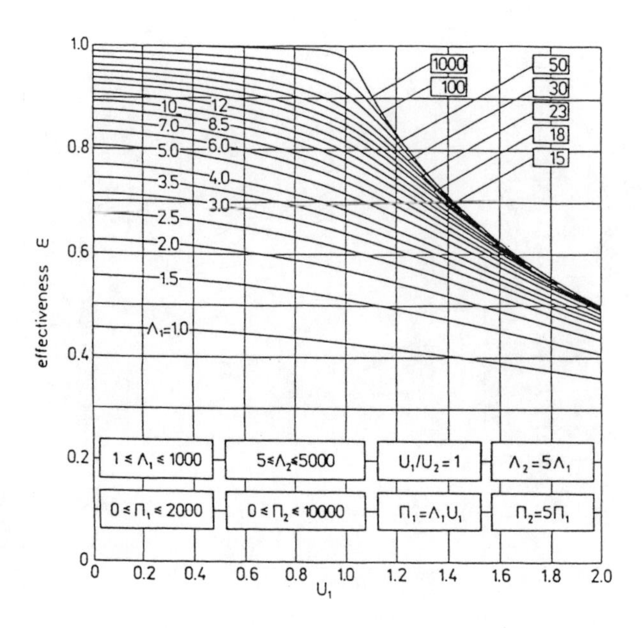

Figure 4.18: Thermal effectiveness, ε, as a function of the utilization factor, U_1, and reduced length, Λ_1, for $\beta = 1$ and $\sigma = 0.2$. (Baclic and Dragutinovic [9].)

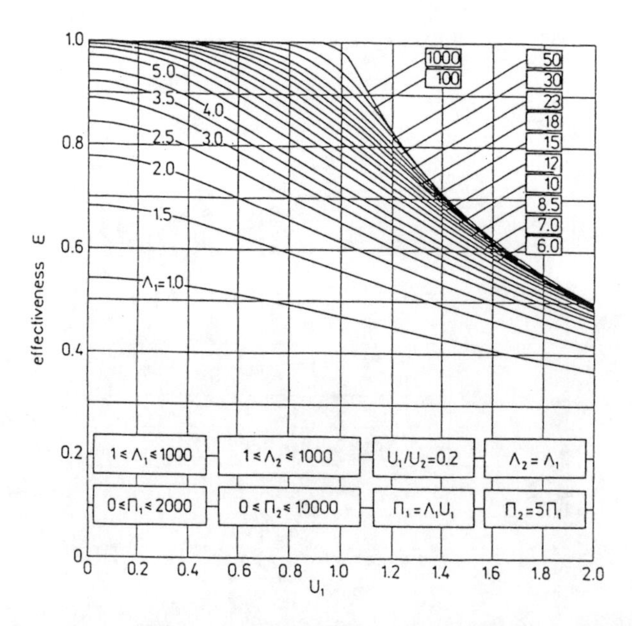

Figure 4.19: Thermal effectiveness, ε, as a function of the utilization factor, U_1, and reduced length, Λ_1, for $\beta = 0.2$ and $\sigma = 1$. (Baclic and Dragutinovic [9].)

References

1. H. Hausen, *Heat Transfer in Counterflow, Parallel Flow and Crossflow*, English translation (1983) edited by A. J. Willmott, McGraw-Hill, New York (originally published 1976).
2. H. Hausen, *Warmeubertragung in Gegenstrom, Gleichstrom und Kreuzstrom*, Springer-Verlag, Berlin (1950).
3. H. Hausen, "Uber die Theorie des Wärmeaustauches in Regeneratoren (The Theory of Heat Exchange in Regenerators)," *Z. angew. Math. Mech.* **9**, 173–200 (June 1929) (RAE Library Translation No. 270, September 1948, W. Shirley).
4. F. W. Schmidt, A. J. Willmott, *Thermal Energy Storage and Regeneration*, McGraw-Hill, New York (1981).
5. C. E. Iliffe, "Thermal Analysis of the Contra-flow Regenerative Heat Exchanger," *Proc. Inst. Mech. Eng.* **159**, 363–372 (1948).
6. A. N. Nahavandi, A. S. Weinstein, "A Solution to the Periodic-Flow Regenerative Heat Exchanger Problem," *Appl. Sci. Res.* **10**, 335–348 (1961).
7. A. J. Willmott, "Digital Computer Simulation of a Thermal Regenerator," *Int. J. Heat Mass Transfer* **7**, 1291–1302 (May 1964).
8. B. S. Baclic, "Misinterpretations of the Diabatic Regenerator Performances," *Int. J. Heat Mass Transfer* **31**, 1605–1611 (1988) (originally published in *Pubs Fac. Tech. Sci. Univ. Novi. Sad.* **16**, 87–100 (1985)).
9. B. S. Baclic, G. D. Dragutinovic, "Asymmetric-Unbalanced Counterflow Thermal Regenerator Problem: Solution by the Galerkin Method and Meaning of Dimensionless Parameters," *Int. J. Heat Mass Transfer* **34**(2), 483–498 (1991).
10. J. E. Johnson, "Regenerator Heat Exchangers for Gas-Turbines," *R.A.E. Report Aero No. 2266 (R&M No. 2630)* **S.D. 27**, 1–72 (May 1948).

Chapter 5

INTRODUCTION TO METHODS FOR SOLVING THE EQUATIONS THAT MODEL COUNTERFLOW REGENERATORS

5.1 Introduction

The cycle of thermal regenerator operation is made up of a hot period followed by a cold period. After a sufficiently large number of identical cycles, the temperature behavior of the regenerator becomes periodic, the period length being the duration of the cycle. The periodic behavior is commonly called the *cyclic equilibrium* or *cyclic* or *periodic steady state*. At cyclic equilibrium, the temperature performance is independent of the initial temperature conditions within the packing. Once, however, this dynamic equilibrium is disturbed by a step change in one or more of the operating parameters, for example, in the hot-/cold-side inlet gas temperature and/or in the hot/cold gas flow rate, then a *transient* behavior will be observed until cyclic steady state is reestablished or until another step change occurs.

The behavior of the regenerative heat exchanger is, therefore, *dynamic*, both in the oscillations in temperature that are inherent in regenerator operation and are, therefore, a feature of cyclic equilibrium, and in the movement from one set of oscillations to another that occurs during transient behavior. This is the basis of the title of this book. This chapter is concerned with an overview of methods that deal with periodic steady-state conditions and, in some cases, with transient conditions also.

The methods of solution of the regenerator problem relate to

$$\frac{\partial T_f}{\partial \xi} = T_s - T_f \tag{5.1}$$

$$\frac{\partial T_s}{\partial \eta} = T_f - T_s \tag{5.2}$$

which are equations (4.12) and (4.13), the simplified form of equations (4.1) and (4.2). These are applied successively to the hot and then the cold period in a cycle. The reversal conditions are given by

$$T_s'(\Lambda'(\Lambda - \xi)/\Lambda, 0) = T_s(\xi, \Pi) \tag{5.3}$$

$$T_s(\Lambda(\Lambda' - \xi')/\Lambda', 0) = T_s'(\xi', \Pi') \tag{5.4}$$

The dimensionless temperature scale, T, is used, which is given by

$$T = \frac{t - t'_{f,in}}{t_{f,in} - t'_{f,in}} \tag{5.5}$$

Here $t_{f,in}$ and $t'_{f,in}$ are the unchanging temperatures of the gases as they enter the regenerator in the hot and cold periods respectively. As pointed out previously, T and t in equation (5.5) can refer to either gas or solid temperatures.

From equation (5.5) it follows that

$$T_{f,in} = 1 \quad \text{and} \quad T'_{f,in} = 0 \tag{5.6}$$

The methods of solution of these equations fall into two distinct classes:

- *Open methods*, in which the gas and solid temperatures, T_f and T_s, are calculated for both periods of operation in successive cycles until the mathematical model attains periodic steady state. The transient behavior is simulated from an arbitrary but, if possible, well-chosen solid temperature distribution, which is provided as an initial condition to the problem.
- *Closed methods*, in which the reversal conditions (5.3) and (5.4) are embodied immediately in the method of solution of the equations of the model. Cyclic steady state is computed directly, without any consideration being taken of any earlier transient cycles.

The closed methods of Hausen [1], Iliffe [2], and Nahavandi and Weinstein [3], as well as later works concerned with closed methods, embody the assumption that $T_{f,in}$ and $T'_{f,in}$ do not vary within each relevant period.

The great advantage of the closed methods lies in the direct calculation of thermal performance at periodic steady state. Some applications of regenerative heat exchangers involve rarely changing operating conditions. Here, modeling of cyclic steady-state performance is sufficient for the design of new units and the prediction of the performance of existing regenerators under possibly new thermal load conditions.

As we shall see in a later chapter, where the surface-to-mass ratio of the packing is large, under typical operating conditions, the duration of any transient conditions from one cyclic equilibrium condition to another will be relatively small. This is the case for most rotary regenerators, for example, when the use of closed methods to predict steady-state behavior is sufficient.

5.2 Open Methods

The most common open methods are the ones in which direct use is made of the differential equations (5.1) and (5.2) by representing them in finite-difference form. In this chapter, we illustrate the principle of an open method by discussing that of Willmott [4] in which it is proposed to integrate these equations using the trapezium rule. The solid and gas temperatures are evaluated at mesh points (i, j) at a distance $i\Delta\xi$ from the gas entrance and at a time $j\Delta\eta$ from the start of the period of operation under consideration. We call $\Delta\xi$ the distance *step length* and $\Delta\eta$ the time step length. This is illustrated in Figure 5.1.

We develop equation (5.1) using the trapezium rule:

$$
\begin{aligned}
T_{f,i,j} &= T_{f,i-1,j} + \frac{\Delta\xi}{2}\left(\left.\frac{\partial T_f}{\partial\xi}\right|_{i,j} + \left.\frac{\partial T_f}{\partial\xi}\right|_{i-1,j}\right) \\
&= T_{f,i-1,j} + \frac{\Delta\xi}{2}\left(T_{s,i,j} - T_{f,i,j} + T_{s,i-1,j} - T_{f,i-1,j}\right) \\
&= A_1 T_{f,i-1,j} + A_2\left(T_{s,i,j} + T_{s,i-1,j}\right) \qquad (5.7)
\end{aligned}
$$

where

$$
A_1 = \frac{2-\Delta\xi}{2+\Delta\xi} \quad \text{and} \quad A_2 = \frac{\Delta\xi}{2+\Delta\xi} \qquad (5.8)
$$

We develop equation (5.2) in a similar fashion:

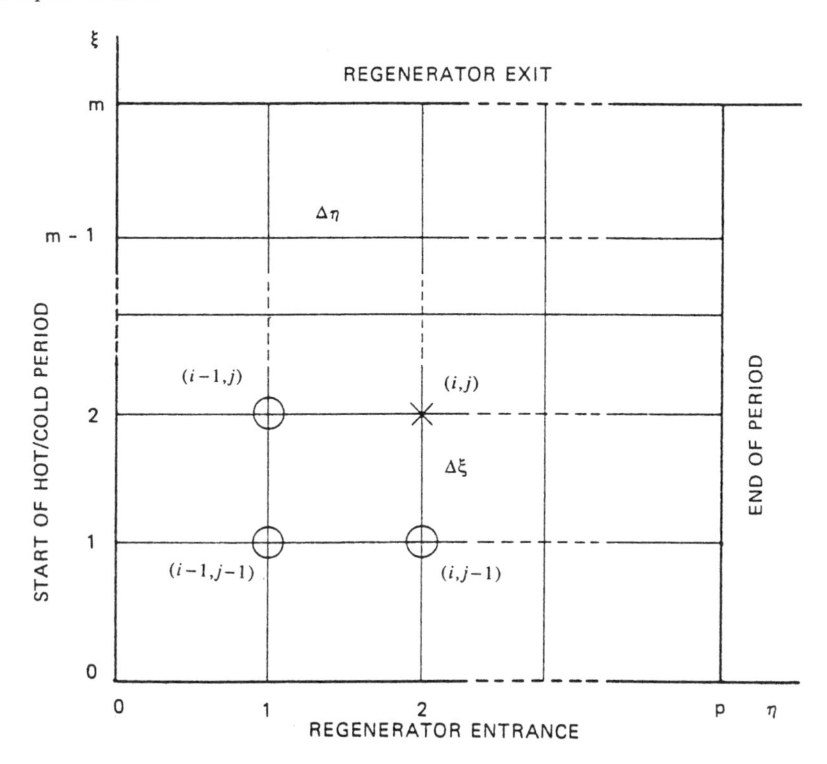

Figure 5.1: Finite-difference mesh for the numerical solution of the regenerator equations. (Schmidt and Willmott [7].)

$$
\begin{aligned}
T_{s,i,j} &= T_{s,i,j-1} + \frac{\Delta\eta}{2}\left(\left.\frac{\partial T_s}{\partial \eta}\right|_{i,j} + \left.\frac{\partial T_s}{\partial \eta}\right|_{i,j-1}\right) \\
&= T_{s,i,j-1} + \frac{\Delta\eta}{2}(T_{f,i,j} - T_{s,i,j} + T_{f,i,j-1} - T_{s,i,j-1}) \\
&= B_1 T_{s,i,j-1} + B_2(T_{f,i,j} + T_{f,i,j-1})
\end{aligned} \tag{5.9}
$$

where

$$
B_1 = \frac{2-\Delta\eta}{2+\Delta\eta} \quad \text{and} \quad B_2 = \frac{\Delta\eta}{2+\Delta\eta} \tag{5.10}
$$

Given the *solid* temperatures at time $j\Delta\eta$, including $j = 0$ at the start of the period, the corresponding gas temperatures for the specified inlet temperature $T_{f,0,j}$ (=1 for the hot period and 0 for the cold period) are calculated for $i = 1, 2, \ldots, m$ using equation (5.7). Note that $m\Delta\xi = \Lambda$, the dimensionless length of the regenerator for the period under consideration. At the start of the period, the solid temperature distribution is given by the distribution at the end of the previous period through the reversal conditions (5.3) and (5.4). At the beginning of the simulation, it is usual to set

$$T_s(\xi, 0) = \tfrac{1}{2}(T_{f,in} + T'_{f,in}) \quad \text{for} \quad 0 \leq \xi \leq \Lambda \tag{5.11}$$

and to use the relations (5.3) and (5.4) at the end of subsequent periods.

At the entrance to the regenerator, the initial solid temperature is given therefore by the reversal conditions, as just described. The solid temperature at this entrance ($i = 0$) is computed using equation (5.9) for $j = 1, 2, \ldots, p$, where $p\Delta\eta = \Pi$, the dimensionless duration of the period under consideration.

For the general position (i, j) with $j > 0$, the solid temperature $T_{s,i,j}$ is computed using a modified form of equation (5.9), by substitution for $T_{f,i,j}$ using equation (5.7). This has the form

$$T_{s,i,j} = K_1 T_{s,i,j-1} + K_2 T_{f,i,j-1} + K_3 T_{s,i-1,j} + K_4 T_{f,i-1,j} \tag{5.12}$$

where

$$K_1 = \frac{B_1}{X}, \quad K_2 = \frac{B_2}{X}, \quad K_3 = \frac{A_2 B_2}{X}, \quad K_4 = \frac{A_1 B_2}{X}, \quad \text{and}$$
$$X = 1 - A_2 B_2$$

Having computed $T_{s,i,j}$ in this manner, the corresponding gas temperature $T_{f,i,j}$ can be calculated immediately using equation (5.7). Thus at time $j\Delta\eta$, the solid and gas temperatures can be calculated down the length of the regenerator for positions $i = 1, 2, \ldots, m$.

Equations (5.1) and (5.2) can thus be integrated, first over the hot period and then the cold period, having applied the reversal conditions (5.3) and (5.4) appropriately. Integration can then be performed over successive cycles in a manner characteristic of an open method. This integration proceeds either from an arbitrary starting position (e.g., equation (5.11)) to cyclic equilibrium, or over a period of transient performance, say from one periodic steady state to another, following one or more step changes in the operating parameters.

5.3 Closed Methods

Many of the closed methods are based on the integral equations developed by Nusselt [5] from the differential equations (5.1) and (5.2). It is useful to begin by introducing the notation

$$F(\xi) = T_s(\xi, 0) \qquad F'(\xi') = T_s'(\xi', 0)$$

with

$$0 \leq \xi \leq \Lambda \quad \text{and} \quad 0 \leq \xi' \leq \Lambda'$$

so that $F(\xi)$ and $F'(\xi')$ are the spatial temperature distributions in the solid packing at the start of the hot and cold periods respectively.

It was left to Iliffe [2] to simplify the original Nusselt equations by the introduction of the dimensionless temperatures defined by equation (5.5). The basic step is to provide an equation that relates $T(\xi, \eta)$, the solid temperature distribution at time η to the initial temperature distribution $F(\xi)$. Nusselt gave this, for the hot period, as

$$T(\xi, \eta) = 1 - e^{-\eta}[1 - F(\xi)] + \int_0^{\xi} \frac{iJ_1\left[2i\sqrt{(\xi - \varepsilon)\eta}\right]}{\sqrt{(\xi - \varepsilon)\eta}} \eta\, e^{-(\xi-\varepsilon)+\eta}[1 - F(\varepsilon)]\, d\varepsilon \tag{5.13}$$

For the cold period, the Nusselt equation is

$$T'(\xi', \eta') = e^{-\eta'} F'(\xi') + \int_0^{\xi'} \frac{iJ_1\left[2i\sqrt{(\xi' - \varepsilon)\eta'}\right]}{\sqrt{(\xi' - \varepsilon)\eta'}} \eta'\, e^{-(\xi'-\varepsilon)+\eta'} F'(\varepsilon)\, d\varepsilon \tag{5.14}$$

Here $iJ_1(iy)$ is a real-valued function with complex argument iy, where $i^2 = -1$ and J_1 is the Bessel function of the first type and of first order.

The notation can be simplified still further for the cases where $\eta = \Pi$ and $\eta' = \Pi'$ using

$$K(\xi - \varepsilon) = \frac{iJ_1\left[2i\sqrt{(\xi - \varepsilon)\Pi}\right]}{\sqrt{(\xi - \varepsilon)\Pi}} \Pi\, e^{-(\xi-\varepsilon)+\Pi} \tag{5.15}$$

and

$$K'(\xi' - \varepsilon) = \frac{iJ_1\left[2i\sqrt{(\xi' - \varepsilon)\Pi'}\right]}{\sqrt{(\xi' - \varepsilon)\Pi'}} \Pi'\, e^{-(\xi'-\varepsilon)+\Pi'} \tag{5.16}$$

This enables equations (5.13) and (5.14) to take simpler forms, namely

$$T_s(\xi, \Pi) = 1 - e^{-\Pi}[1 - F(\xi)] + \int_0^{\xi} K(\xi - \varepsilon)[1 - F(\varepsilon)]\, d\varepsilon \tag{5.17}$$

and

$$T_s'(\xi', \Pi') = e^{-\Pi'} F'(\xi') + \int_0^{\xi'} K'(\xi' - \varepsilon) F'(\varepsilon)\, d\varepsilon \tag{5.18}$$

Starting with an initial temperature distribution, say given by equation (5.11), it is possible to integrate through the hot period using equation (5.17), then to apply the reversal condition (5.3), then to integrate through the cold period using equation (5.18) and finally to apply the reversal condition (5.4), thus yielding the solid temperature distribution, $F(\xi)$, at the start of the next cycle. In this way, another *open method* for solving the equations is available.

However, the reversal conditions can be applied directly within the integral equations to yield

$$F'(\Lambda'(\Lambda - \xi)/\Lambda) = 1 - e^{-\Pi}[1 - F(\xi)] + \int_0^{\xi} K(\xi - \varepsilon)[1 - F(\varepsilon)]\, d\varepsilon \tag{5.19}$$

and

$$F(\Lambda(\Lambda' - \xi')/\Lambda') = e^{-\Pi'} F'(\xi') + \int_0^{\xi'} K'(\xi' - \varepsilon) F'(\varepsilon)\, d\varepsilon \tag{5.20}$$

This pair of equations can then be solved for $F(\xi)$ and $F'(\xi')$ at periodic steady state without computing $F(\xi)$ and $F'(\xi')$ for any prior transient conditions. This is a classical *closed method* for solving the regenerator problem.

In the case where the regenerator and its operation is *symmetric*, when $\Lambda = \Lambda'$ and $\Pi = \Pi'$, the temperature distributions $F(\xi)$ and $F'(\xi')$ are symmetric with respect to one another. This can be written in the form

$$T_{f,in} - T_s(\Lambda - \xi, 0) = T_s'(\xi, \Pi) - T_{f,in}' \quad \text{with} \quad \xi = \xi' \quad \text{when symmetric} \tag{5.21}$$

or, quite simply, recalling equations (5.6),

$$1 - T_s(\Lambda - \xi, 0) = T_s'(\xi, \Pi) \tag{5.22}$$

Substituting equation (5.22) into (5.18), with $\xi = \xi'$, we obtain

$$F'(\Lambda - \xi) + e^{-\Pi} F'(\xi) + \int_0^{\xi} K(\xi - \varepsilon) F'(\varepsilon)\, d\varepsilon = 1 \tag{5.23}$$

At this juncture, it is sufficient to describe the method of Iliffe for the symmetric case, without any loss of generality [2].

The integral in equation (5.23) is approximated by the composite Simpson's rule. Suppose that the temperatures $\{F'_k \mid k = 0, 1, 2, \ldots, m\}$ are equally spaced at positions $\{k\Delta\xi \mid k = 0, 1, 2, \ldots, m\}$ where $m\Delta\xi = \Lambda$. Then, in the simplest case where $k = 2$,

$$\int_0^{2\Delta\xi} K(\xi - \varepsilon)\, F'(\varepsilon)\, d\varepsilon \approx \frac{\Delta\xi}{3}[K(2\Delta\xi - 0)F'(0) + 4K(2\Delta\xi - \Delta\xi)F'(\Delta\xi) + 2K(2\Delta\xi - 2\Delta\xi)F'(2\Delta\xi)] = \frac{\Delta\xi}{3}(K_2F'_0 + 4K_1F'_1 + K_0F'_2) \tag{5.24}$$

where $F'_k = F'(k\Delta\xi)$ and $K_k = K(k\Delta\xi)$.

In the case where $k = 3$, Iliffe used Simpson's *three-eighths* rule:

$$\int_0^{3\Delta\xi} K(\xi - \varepsilon)\, F'(\varepsilon)\, d\varepsilon \approx \frac{3\Delta\xi}{8}\left(K_3F'_0 + 3K_2F'_1 + 3K_1F'_2 + K_0F'_3\right) \tag{5.25}$$

As might be expected, for $k > 2$ and k even, Iliffe used the conventional composite form of Simpson's rule. For $k > 3$ and k odd, Iliffe used

$$\int_0^{k\Delta\xi} K(\xi - \varepsilon)\, F'(\varepsilon)\, d\varepsilon = \int_0^{3\Delta\xi} K(\xi - \varepsilon)\, F'(\varepsilon)\, d\varepsilon + \int_{3\Delta\xi}^{k\Delta\xi} K(\xi - \varepsilon)\, F'(\varepsilon)\, d\varepsilon \tag{5.26}$$

representing the first integral on the right-hand side by the three eighths rule and the second by a composite (if necessary) form of Simpson's rule.

For $k = 1$, Iliffe first evaluated $F_{1/2}$ using a cubic interpolation formula, namely

$$F'_{1/2} = \frac{5F'_0 + 15F'_1 - 5F'_2 + F'_3}{16} \tag{5.27}$$

and then found, using again Simpson's rule,

$$\int_0^{\Delta\xi} K(\xi - \varepsilon)F'(\varepsilon)\, d\varepsilon = \frac{\Delta\xi}{6}\left(K_1F'_0 + 4K_{1/2}F'_{1/2} + K_0F'_1\right) \tag{5.28}$$

If we apply these quadrature formulae to the equation (5.25) for $k = 0, 1, 2, \ldots, m$, the $(0..m) \times (0..m)$ matrix P can be constructed, where

$$P = \Delta\xi \begin{bmatrix} 0 & 0 & 0 & 0 & & & \\ \frac{5K_{1/2} + 4K_1}{24} & \frac{4K_0 + 15K_{1/2}}{24} & \frac{-5K_{1/2}}{24} & \frac{K_{1/2}}{24} & & & \\ \frac{K_2}{3} & \frac{4K_1}{3} & \frac{K_0}{3} & 0 & & & \\ \frac{3K_3}{8} & \frac{9K_2}{8} & \frac{9K_1}{8} & \frac{3K_0}{8} & & & \\ \frac{K_4}{3} & \frac{4K_3}{3} & \frac{2K_2}{3} & \frac{4K_1}{3} & \frac{K_0}{3} & & \\ \cdots & \cdots & \cdots & \cdots & \cdots & \cdots & \cdots \\ \frac{K_m}{3} & \frac{4K_{m-1}}{3} & \frac{2K_{m-2}}{3} & \cdots & \cdots & \frac{4K_1}{3} & \frac{K_0}{3} \end{bmatrix} \tag{5.29}$$

for $m =$ even. With the exception of the first two rows, this matrix P is lower triangular.

The quadrature formulae are then combined in matrix form $\boldsymbol{PF'}$ where $\boldsymbol{F'} = [F'_0, F'_1, F'_2, \ldots, F'_n]^T$. We apply this to equation (5.23) to yield

$$(P + Q + R)\boldsymbol{F'} = \boldsymbol{e} \tag{5.30}$$

with $Q = e^{-\Pi}I$, where $\boldsymbol{e} = [1, 1, \ldots, 1]^T$, where I is the unit (identity) matrix and where $R = [r_{i,j}]$ with $r_{j,m-j} = 1$ for $j = 0, 1, 2, \ldots, m$, and all other elements of R equal to zero. In other words, by the application of Simpson quadrature formulae given above to equation (5.23) at positions $k = 0, 1, 2, \ldots, m$, the matrix equation (5.30) is formed.

Solution of equation (5.30) yields the required vector $\boldsymbol{F'}$. Equation (5.22) can be written in the form

$$F'(\Lambda - \xi) + F(\xi) = 1 \tag{5.31}$$

Using this equation, the vector $\boldsymbol{F}$ can be obtained from $\boldsymbol{F'}$ from $F'_{m-k} + F_k = 1$ for $k = 0, 1, 2, \ldots, m$.

This method is straightforward. It involved, however, Iliffe in a good deal of hand calculation. The work was done in the mid-1940s before computers were generally available. Iliffe mentions the use of "five figure logarithms to evaluate $K(\xi - \varepsilon)$ . . . and a twenty inch slide rule thereafter." He went on to mention that the number of hours required to solve equation (5.30) was equal to the value of m where the vector $\boldsymbol{F'}$ is of size $(0..m)$.

Hausen [6] developed a very similar method to that of Iliffe except that he approached the cases where k is odd and in particular, where $k = 1$ in a different manner. He suggested that where k is odd,

$$\int_0^{k\Delta\xi} K(\xi - \varepsilon)F'(\varepsilon)\,d\varepsilon = \int_0^{(k-1)\Delta\xi} K(\xi - \varepsilon)\,F'(\varepsilon)\,d\varepsilon + \int_{(k-1)\Delta\xi}^{k\Delta\xi} K(\xi - \varepsilon)\,F'(\varepsilon)\,d\varepsilon \tag{5.32}$$

where

$$\int_0^{(k-1)\Delta\xi} K(\xi - \varepsilon)F'(\varepsilon)\,d\varepsilon$$

can be computed using Simpson's rule because $k - 1$ will be even. Hausen makes no use of Simpson's three-eighths rule here. Instead, Hausen proposed to develop a quadratic interpolation formula passing through $K_2F'_{k-2}$, $K_1F'_{k-1}$, and $K_0F'_k$. By integrating this interpolation formula from $(k - 1)\Delta\xi$ to $k\Delta\xi$, Hausen developed the approximation

$$\int_{(k-1)\Delta\xi}^{k\Delta\xi} K(\xi - \varepsilon)F'(\varepsilon)\,d\varepsilon \approx \frac{\Delta\xi}{12}\left(-F'_{k-2}K_2 + 8F'_{k-1}K_1 + 5F'_kK_0\right)$$

Hausen explained that this approximation involved F'_{-1} in the case where $k = 1$. He decided to estimate F'_{-1} by using the quadratic interpolation curve passing through F'_0, F'_1, and F'_2. This yields

$$\int_0^{\Delta\xi} K(\xi - \varepsilon)F'(\varepsilon)\,d\varepsilon \approx \frac{\Delta\xi}{12}\left[F'_0(8K_1 - 3K_2) + F'_1(5K_0 + 3K_2) - F'_2K_2\right]$$

Hausen admits that Iliffe's use of the cubic interpolation to find $F'_{1/2}$ and the use of the accurate Simpson's three-eighths rule is superior to his own approach but suggests that in many cases, the temperature distribution in the solid is all but linear and that, therefore, quadratic interpolation is sufficient. The present author considers that any advantage offered by the method of Iliffe [2] should be exploited in any software developed because cases do exist where as good an accuracy as is feasible should be realized.

5.4 Concluding Remarks

The basic principles embodied in the *open* and *closed methods* for solving the equations that model counterflow regenerators have been set out, using the methods of Willmott [4] and Iliffe [2] by way of illustration. At first sight, it might appear that the closed methods should be superior to the closed ones, if only because the simulation of any transient cycles is avoided.

In later chapters, we shall see that Willmott's open method can be readily adapted to deal with nonlinear problems (where, for example, temperature-dependent thermophysical properties of gas/solid are represented within the model), whereas this is far from easy in the case of closed methods. We shall see also that Iliffe's method breaks down in circumstances where it might seem to be most potentially useful, namely when an open method requires many cycles to reach periodic steady state.

References

1. H. Hausen, "Näherungsverfahen zur Berechnung des Warmeaustausches in Regeneratoren (An Approximate Method of Dimensioning Regenerative Heat Exchangers)," *Z.A.M.M.* **11**, 105–114 (1931) (RAE Library Translation, No. 98, Feb. 1946, C. K. Newcombe).
2. C. E. Iliffe, "Thermal Analysis of the Contra-flow Regenerative Heat Exchanger," *Proc. Inst. Mech. Eng.* **159**, 363–372 (1948).
3. A. N. Nahavandi, A. S. Weinstein, "A Solution to the Periodic-Flow Regenerative Heat Exchanger Problem," *Appl. Sci. Res.* **10**, 335–348 (1961).
4. A. J. Willmott, "Digital Computer Simulation of a Thermal Regenerator," *Int. J. Heat Mass Transfer* **7**, 1291–1302 (May 1964).
5. W. Nusselt, "Der Beharrungszustand im Winderhitzer (The Steady Operating Condition in the Air Draught Heater)," *Z. Ver. Deut. Ing.* **72**, 1052–1054 (1928) (RAE Library Translation No. 267, 1948, W. Shirley).
6. H. Hausen, *Heat Transfer in Counterflow, Parallel Flow and Crossflow*, English translation (1983) edited by A. J. Willmott, McGraw-Hill, New York (originally published 1976).
7. F. W. Schmidt, A. J. Willmott, *Thermal Energy Storage and Regeneration*, McGraw-Hill, New York (1981).

Chapter 6

COUNTERFLOW REGENERATORS: FINITE CONDUCTIVITY MODELS

6.1 Introduction

It was assumed in Chapter 4 that the descriptive differential equations in a period of regenerator operation could embody a lumped heat transfer coefficient, $\bar{\alpha}$, to approximate the effect of latitudinal, transverse conduction within the packing. It will be recalled that these equations can be developed by consideration of the exchange of heat between the gas flowing through a single channel of the packing and the wall of material through which the channel is driven. Further, the behavior of the regenerator is obtained by regarding the packing as a bundle of channels, identical in configuration and operation.

The relation between the rate of thermal energy transferred between gas and solid and the rate at which heat is given up/absorbed by the gas flowing through the regenerator is represented by

$$\frac{\partial t_f}{\partial y} = \frac{\bar{\alpha} A}{\dot{m}_f c_p L}(t_s - t_f) - \frac{m_f}{\dot{m}_f L}\frac{\partial t_f}{\partial \tau} \tag{6.1}$$

The same rate of heat exchange across the packing surface and the rate at which heat is stored in/recovered from the packing is described by

$$\frac{\partial t_s}{\partial \tau} = \frac{\bar{\alpha} A}{M C_s}(t_f - t_s) \tag{6.2}$$

What was not specified is just how $\bar{\alpha}$, the lumped heat transfer coefficient, can be computed. In the first part of Chapter 6, we discuss the historical background to the development of such a lumped coefficient. By approaching the problem in this way, the reader will be made aware of the assumptions built into the $\bar{\alpha}$ commonly in use today. Hausen's approximations for the so-called Φ-factor are developed for plain walls, solid cylinders and spheres with a later analysis extending this approach to hollow cylinders. A degree of unification in the theory for different geometries is discussed. A comparison between the full 3-D model and the 2-D model, using $\bar{\alpha}$, is offered together with the time-varying $\Phi(\omega)$ factor. The chapter concludes with a discussion of the potentially significant effect of axial conduction.

6.2 Lumped Heat Transfer Coefficients

The lumped heat transfer coefficient, $\bar{\alpha}$, was developed by Hausen. His background was somewhat different to that of many other workers in this field at the time. Whereas they were involved with regenerators for what today we would call the "smoke stack" industries, Hausen had worked on regenerators for refrigeration purposes for a company based in Munich. In such high-temperature regenerators, the packing is constructed of materials of low thermal conductivity with thick walls. For Hausen, the packings of his regenerators might be made of metal, certainly fabricated in many cases with relatively thin walls, that is with small semithickness w. As a consequence, Hausen could focus upon the way the *mean* solid temperature, $t_{s,m}$, as opposed to the surface solid temperature, varied with time, where

$$t_{s,m} = \frac{1}{w}\int_0^w t_s(x)\,dx \tag{6.3}$$

and where x is the direction into the packing, perpendicular to the direction, y, of gas flow.

Hausen had observed in experiments on low-temperature regenerators that the solid temperature varied *linearly* with distance and with time, except at the reversals and except at the entrances to the regenerator. In 1938/39, Hausen [1] developed a purely theoretical approach, without any calibrations based on experimental work, using the assumption that

$$\frac{\partial t_f}{\partial \tau} = \frac{\partial t_s}{\partial \tau} = constant \tag{6.4}$$

On this basis, he proceeded to integrate the following equation:

$$\frac{\partial t_s}{\partial \tau} = \kappa_s \frac{\partial^2 t_s}{\partial x^2} \tag{6.5}$$

This is subject to the boundary conditions

$$t_s = t_{s,o} \quad \text{at} \quad x = 0 \quad \text{and} \quad x = 2w \quad \text{and} \quad \frac{\partial t_s}{\partial x} = 0 \quad \text{at} \quad x = w \tag{6.6}$$

Equation (6.5) is the Fourier equation representing the unsteady state transmission of heat within a plain wall of semithickness w. The boundary conditions (6.6) merely specify that $t_{s,o}$ is the surface solid temperature at $x = 0$ and $x = 2w$ and that the temperature distribution within the wall at any instant is symmetric around the semithickness w.

The underlying assumption here is that we need only to consider a single layer of the packing at any position y along the direction of gas flow. The conclusions of assuming that $\partial t_s/\partial \tau =$ constant are then applied at all positions in the regenerator along y. Hausen assumed that although t_s varies in a negative exponential fashion at the gas entrance, $y = 0$, as a consequence of the gas temperature at this entrance, $t_{f,in}$, not varying with time, this would not have a significant effect. The linear variation of solid temperature elsewhere in the regenerator would be a dominating factor.

Upon integrating twice the right-hand side of equation (6.5), the relation

$$t_{s,o} - t_s = \frac{1}{\kappa_s} \frac{\partial t_s}{\partial \tau} \left(wx - \frac{x^2}{2} \right) \tag{6.7}$$

is obtained. Dividing throughout by w and integrating again, Hausen obtained the relation between the surface solid temperature and the mean solid temperature for which he had been seeking, namely

$$t_{s,o} - t_{s,m} = \frac{1}{\kappa_s} \frac{\partial t_{s,m}}{\partial \tau} \frac{w^2}{3} \tag{6.8}$$

Here we note that for $0 \le x \le w$, $\partial t_s/\partial \tau =$ *constant* implies that

$$\frac{\partial t_s}{\partial \tau} = \frac{\partial t_{s,m}}{\partial \tau}$$

and hence the inclusion of this in equation (6.8). Further, we recall that

$$\frac{\partial t_{s,m}}{\partial \tau} = \frac{\bar{\alpha} A}{MC_s}(t_f - t_{s,m})$$

and that, for a plain-wall configuration of the packing, $MC_s = Aw\rho_s C_s$. It follows that

$$\frac{\partial t_{s,m}}{\partial \tau} = \frac{\bar{\alpha}}{w\rho_s C_s}(t_f - t_{s,m}) \tag{6.9}$$

Noting that $\kappa_s = \lambda_s/(\rho_s C_s)$ and inserting equation (6.9) into (6.8), we obtain

$$\begin{aligned} t_{s,o} - t_{s,m} &= \frac{\rho_s C_s}{\lambda_s} \frac{\bar{\alpha}}{w\rho_s C_s} \frac{w^2}{3}(t_f - t_{s,m}) \\ &= \bar{\alpha}(t_f - t_{s,m})\frac{w}{3\lambda_s} \end{aligned} \tag{6.10}$$

Adding and subtracting t_f from the left-hand side of equation (6.10) yields

$$(t_{s,o} - t_f) + (t_f - t_{s,m}) = \bar{\alpha}(t_f - t_{s,m})\frac{w}{3\lambda_s}$$

But

$$t_{s,o} - t_f = \frac{\bar{\alpha}}{\alpha}(t_{s,m} - t_f)$$

and hence

$$\frac{\bar{\alpha}}{\alpha}(t_{s,m} - t_f) - (t_{s,m} - t_f) = \bar{\alpha}(t_f - t_{s,m})\frac{w}{3\lambda_s}$$

which yields Hausen's simple relationship:

$$\frac{1}{\bar{\alpha}} = \frac{1}{\alpha} + \frac{w}{3\lambda_s} \tag{6.11}$$

This relation was developed in 1938/39 by Hausen from the purely theoretical considerations set out above. As such, it can be applied to a packing constructed of materials of low conductivity with thick walls as well as to the packing typical of regenerators used in refrigeration plants.

Earlier work had necessitated the calibration of the equivalent of a lumped heat transfer coefficient [2, 3] using experimental data, thereby limiting the area of application to the region of the experiments. No such limitations applied to Hausen's approach.

It is important to note that the t_s used in equations (6.1) and (6.2) should be the average temperature $t_{s,m}$ when used in conjunction with the lumped coefficient $\bar{\alpha}$. It is sufficient to retain the use of t_s in the differential equations (6.1) and (6.2), for simplicity's sake, even with $\bar{\alpha}$. To this is added the proviso that in the circumstances where $\bar{\alpha}$ is used, t_s is interpreted to be the mean solid temperature defined by equation (6.3).

A particular problem, associated with this simple development of a lumped heat transfer coefficient, $\bar{\alpha}$, remains however. At the start of a period of regenerator operation, an inversion takes place in the temperature profile within the packing immediately after a reversal. The development of equation (6.11) makes no allowance for this. In general, the temperature distribution within the packing is parabolic, as indicated in equation (6.7). At a reversal, the profile is inverted, as shown in Figure 6.1.

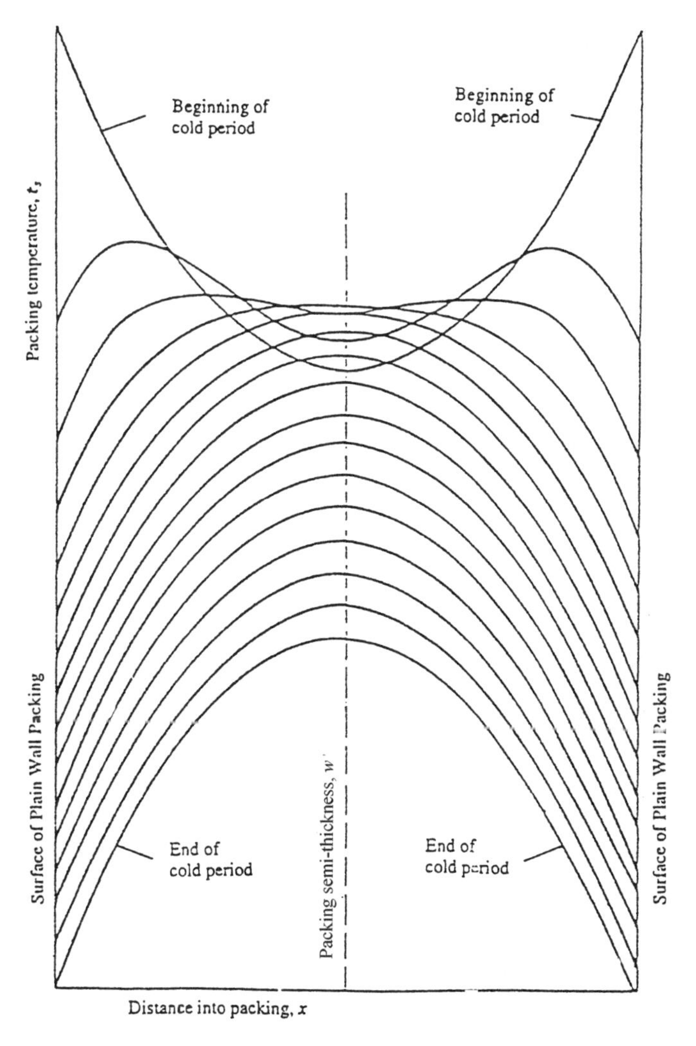

Figure 6.1: Packing temperature profile changes during a cold period of regenerator operation.

6.3 Further Development of $\bar{\alpha}$

Hausen proposed to modify equation (6.11) by the inclusion of an additional multiplying factor, Φ, with a view to representing the effect of the inversion of the temperature profile within the packing after a reversal. Equation (6.11) now becomes

$$\frac{1}{\bar{\alpha}} = \frac{1}{\alpha} + \frac{w}{\lambda_s}\Phi \tag{6.12}$$

At any position within the regenerator and at any instant, it is assumed that

$$\bar{\alpha}(t_f - t_{s,m}) = \alpha(t_f - t_{s,o})$$

from which it follows that

$$\begin{aligned} \frac{t_f - t_{s,m}}{t_f - t_{s,o}} &= \frac{\alpha}{\bar{\alpha}} = \alpha\left(\frac{1}{\alpha} + \frac{w}{\lambda_s}\Phi\right) \\ &= 1 + \frac{\alpha w}{\lambda_s}\Phi \\ &= 1 + \mathrm{Bi}\,\Phi \end{aligned} \tag{6.13}$$

Here $\mathrm{Bi} = \alpha w/\lambda_s$ is the Biot modulus. From this it follows that

$$\Phi = \frac{1}{\mathrm{Bi}}\left(\frac{t_f - t_{s,m}}{t_f - t_{s,o}} - 1\right) = \frac{1}{\mathrm{Bi}}\left(\frac{t_{s,o} - t_{s,m}}{t_f - t_{s,o}}\right) \tag{6.14}$$

The immediate observation that must be made is Φ will vary with position y and time τ as the temperature differences $t_f - t_{s,m}$ and $t_{s,o} - t_{s,m}$ vary with space and time throughout the hot and cold periods of a cycle of regenerator operation. Hausen [4] made a number of interlinking proposals whereby a single value of Φ could be used for both periods of regenerator operation:

1. He suggested that Φ be computed from the solution of equation (6.5) with boundary conditions (6.6) and

   $$\lambda_s \frac{\partial t_s}{\partial x} = \alpha(t_f - t_{s,o}) \quad \text{at} \quad x = 0 \tag{6.15}$$

   for the condition that

   $$\frac{\partial t_{s,m}}{\partial \tau} = constant$$

   in both periods of operation.

2. Hausen then suggested that Φ be computed as the chronological average of the time varying values so computed, so that

$$\Phi = \frac{1}{P}\int_0^P \frac{1}{\mathrm{Bi}}\left(\frac{t_{s,o} - t_{s,m}}{t_f - t_{s,o}}\right) d\tau \tag{6.16}$$

3. Hausen again assumed that consideration of a single layer of packing would yield a value of Φ which could be applied at all positions y in the direction of gas flow in the heat exchanger.

Hausen presented an analytical solution for equation (6.5) from which he derived the approximations commonly in use nowadays. We shall first present a numerical approach to this problem, as this may well clarify what Hausen had in mind, and will point to the improved approaches to this matter that have been developed relatively recently.

6.4 Numerical Development of the Φ-Factor

It is useful to normalize the Fourier equation and the accompanying boundary conditions. First we introduce the dimensionless variables ω and z. These are defined by

$$\omega = \frac{\kappa_s \tau}{w^2} \tag{6.17}$$

$$z = \frac{x}{w} \tag{6.18}$$

The duration, in dimensionless terms, of the period of operation is given by

$$\Omega = \frac{\kappa_s P}{w^2} \tag{6.19}$$

Equation (6.5) now takes up the simpler form:

$$\frac{\partial t_s}{\partial \omega} = \frac{\partial^2 t_s}{\partial z^2} \tag{6.20}$$

and the boundary conditions (6.6) become

$$t_s = t_{s,o} \quad \text{at} \quad z = 0 \quad \text{and} \quad z = 2 \quad \text{and} \quad \frac{\partial t_s}{\partial z} = 0 \quad \text{at} \quad z = 1 \tag{6.21}$$

The boundary condition (6.15) now takes the simpler form:

$$\frac{\partial t_s}{\partial z} = \mathrm{Bi}(t_f - t_{s,o}) \quad \text{at} \quad z = 0 \tag{6.22}$$

The requirement of Hausen that $\partial t_{s,m}/\partial \tau = constant$ is realized if we set the "heat flux term" $\mathrm{Bi}(t_f - t_{s,o}) = q$ to be constant in each period of operation. So that a heat balance is achieved between the hot and cold periods, we set the heat fluxes, q and q', to be

$$q = \frac{Q}{\Omega} \quad \text{and} \quad q' = -\frac{Q}{\Omega'} \tag{6.23}$$

where Q is an arbitrary positive constant, the same for both periods. The negative value of q' corresponds to the cooling that takes place in the cold period.

The temperatures $t_{s,i}$ are calculated at equally spaced positions within the packing in the direction z, namely at positions $\{z_i = i\Delta z \mid i = 0, 1, 2, \ldots, nm\}$ where $nm\Delta z = 1$, that is, at $x = w$. Note that z_0 corresponds to the surface of the packing. Equation (6.20) now takes the form

$$\frac{\partial t_{s,i}}{\partial \omega} = \frac{1}{\Delta z^2}\left(t_{s,i+1} - 2t_{s,i} + t_{s,i-1}\right) \tag{6.24}$$

The Binder–Schmidt method of integrating equation (6.24) consists of computing the solid temperatures, $t_{s,i,j}$, at time positions $\{\omega_j = j\Delta\omega \mid j = 0, 1, 2, \ldots, np\}$ where $np\Delta\omega = \Omega$, the duration of the period under consideration. In this method, the Euler–Cauchy method is used, namely

$$t_{s,i,j+1} = t_{s,i,j} + \Delta\omega \frac{\partial t_{s,i,j}}{\partial \omega} \tag{6.25}$$

where $\partial t_{s,i,j}/\partial \omega$ is given by equation (6.24) applied at ω_j. Sadly, this method is unstable for $p = \Delta\omega/\Delta z^2 > 0.5$, although, with a small enough time step, the integration can be undertaken. It is more common to use the implicit method of Crank and Nicolson [5], namely

$$t_{s,i,j+1} = t_{s,i,j} + \frac{\Delta\omega}{2}\left(\frac{\partial t_{s,i,j+1}}{\partial \omega} + \frac{\partial t_{s,i,j}}{\partial \omega}\right) \tag{6.26}$$

This method does not suffer from this instability problem.

The boundary conditions are incorporated in the following way. In the middle of the packing, when $z = 1$, $\partial t_{s,nm}/\partial \omega = 0$, we set $t_{s,nm+1} = t_{s,nm-1}$ whereupon equation (6.24) takes the form

$$\frac{\partial t_{s,nm}}{\partial \omega} = \frac{2}{\Delta z^2}(t_{s,nm-1} - t_{s,nm}) \tag{6.27}$$

At the surface of the packing, we set $\partial t_{s,0}/\partial z = q$, where q is the "heat flux." Now equation (6.24) takes the form

$$\frac{\partial t_{s,0}}{\partial \omega} = \frac{2}{\Delta z^2}(t_{s,1} - t_{s,0} + q\Delta z) \tag{6.28}$$

By solving equation (6.20) by one of these numerical methods, the heating and cooling of the slab of material of semithickness w can be simulated until periodic steady state is realized. At this juncture, the variation of the surface solid temperature $t_{s,o}(\omega)$ is immediately available and the variation of the mean solid temperature $t_{s,m}(\omega)$ (see equation (6.3)) can be easily calculated, for both periods of operation. From these, the time variation of $\Phi(\omega)$ can be found using a modified form of (6.14):

$$\Phi(\omega) = \frac{t_{s,o}(\omega) - t_{s,m}(\omega)}{q} \tag{6.29}$$

From this, a chronological average of Φ can be found:

$$\Phi = \frac{1}{\Omega}\int_0^{\Omega} \Phi(\omega)\, d\omega \tag{6.30}$$

This Φ is incorporated within equation (6.12) from which the value of $\bar{\alpha}$ can be computed. We shall see shortly that Hausen provides formulae by which Φ can be estimated (see equations (6.47) and (6.48)). In Table 6.1 is given a

Table 6.1: Values of Φ-factor for varying values of Ω

Ω	Φ Computed by solving numerically the Fourier equation	Φ Estimated using Hausen's formula
0.50000	0.24464	0.24444
0.75000	0.27336	0.27407
1.00000	0.28821	0.28889
1.50000	0.30313	0.30370
2.00000	0.31065	0.31111
2.50000	0.31512	0.31556
3.00000	0.31805	0.31852
3.50000	0.32018	0.32063
4.00000	0.32177	0.32222
4.50000	0.32307	0.32346
5.00000	0.32406	0.32444
5.50000	0.32488	0.32525
6.00000	0.32550	0.32593
6.50000	0.32608	0.32650
7.00000	0.32651	0.32698

set of values of Φ for different values of Ω. The first column of Φ's has been found by solving numerically equation (6.20) and then determining Φ by finding the time mean value of $\Phi(\omega)$, which, in turn, is found by computing the integral (6.30) using a Newton–Cotes quadrature formula. The second column of Φ's has been estimated using Hausen's formulae.

Figure 6.2 shows how $\Phi(\omega)$ varies with time, presented here on the normalized (0, 1) scale as ω/Ω. For $\Omega = 3$, it will be seen that $\Phi(\omega)$ quickly settles down to a steady value of 0.333. However, again for $\Omega = 3$, for $\omega/\Omega < 0.2$, $\Phi(\omega)$ rises steeply from $\omega = 0$, at the start of a period, from $\Phi(0) = -0.333$ as $\Phi(\omega)$ seeks to represent the inverting of the temperature profile within the solid immediately after a reversal. As Ω declines in value, $\Omega = 1$, $\Omega = 0.5$, and $\Omega = 0.25$, so the reversal effect occupies more and more of the period of operation, as seen in Figure 6.2. Indeed, for $\Omega = 0.25$, the curve for $\Phi(\omega)$ never "settles down."

The value of Φ embodied in the lumped heat transfer coefficient, $\bar{\alpha}$ (see equation (6.12)), is the time average value. This means that a model incorporating such an $\bar{\alpha}$ will be unable to replicate any effects of the profile inversion of the solid temperature at the start of a period of regenerator operation. This is made clear in Figure 6.3 where it will be seen that $\Phi(mean)$

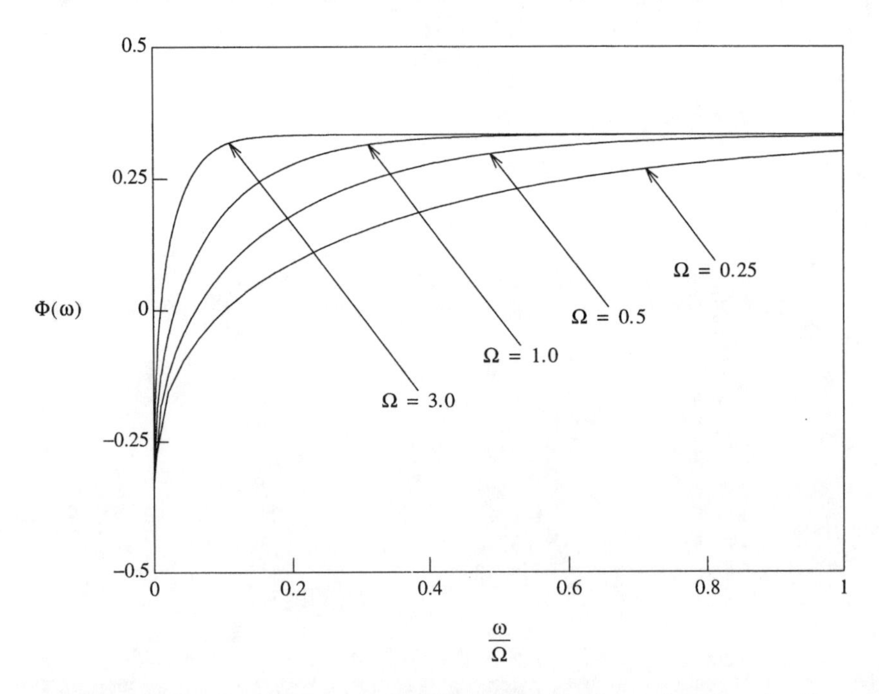

Figure 6.2: Time-varying value of $\Phi(\omega)$ displayed as a function of ω/Ω.

is a good estimate of $\Phi(\omega)$ for $\omega > 0.2$ but not so for lower values of ω. It turns out, as we shall see later, that such a model is able to replicate the time average performance of a regenerator and it is widely used.

6.5 Hausen's Development of the Φ-Factor

We begin rearranging equation (6.7) into the form

$$t_s(x) - t_{s,o} = \frac{1}{\kappa_s}\frac{\partial t_s}{\partial \tau}\left(\frac{x^2}{2} - wx\right) \tag{6.31}$$

and then adding equation (6.31) to equation (6.8). This yields

$$\begin{aligned} t_s(x) - t_{s,m} &= \frac{1}{\kappa_s}\frac{\partial t_s}{\partial \tau}\left(\frac{x^2}{2} - wx + \frac{w^2}{3}\right) \\ &= \frac{\alpha}{w\lambda_s}(t_f - t_{s,o})\left(\frac{x^2}{2} - wx + \frac{w^2}{3}\right) \end{aligned} \tag{6.32}$$

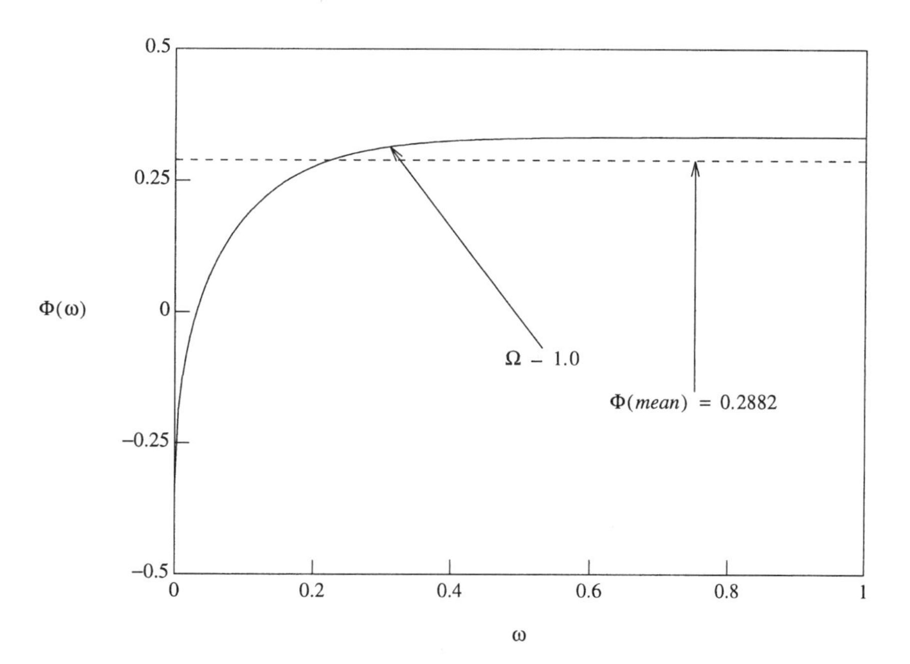

Figure 6.3: Time-varying value of $\Phi(\omega)$ displayed as a function of ω for $\Omega = 1$ together with chronological average value of Φ.

Hausen [6] then considered the general analytical solution to equation (6.5) for the case where

$$\frac{\partial t_{s,m}}{\partial \tau} = constant = K\,(\text{say})$$

namely

$$t_s(x,\tau) = A + K\tau - \frac{Kx}{\kappa_s}\left(w - \frac{x}{2}\right) + \sum_{n=1}^{\infty} B_n e^{-\beta_n^2 \kappa_s \tau} \cos \beta_n (x - w) \tag{6.33}$$

Note that A, β_n, and B_n, as used in equation (6.33) and in the equations that follow from it, are arbitrarily selected constants. The term $A + K\tau$ represents the linear variation of temperature and the term

$$\frac{Kx}{\kappa_s}\left(w - \frac{x}{2}\right)$$

represents the parabolic temperature within the packing. The summation of cosine terms seeks to represent the inversion of the temperature profile immediately after a reversal in regenerator operation. The effects of this inversion will disappear once the period of operation is sufficiently developed. That this is the case will be apparent from the consideration that

$$\lim_{\tau \to \infty} e^{-\beta_n^2 \kappa_s \tau} = 0$$

assuming that $\beta_n > 0$. (In fact, as will be seen shortly, $\beta_n = n\pi/w$, $n = 1, 2, \ldots$.) Nevertheless, even during a reversal, it is assumed that $t_{s,m}$ still varies linearly. This corresponds to the application of the $\pm q$ term in equation (6.28) throughout a complete period of operation, including when a reversal has just occurred. In Hausen's case, he realizes this necessity by requiring that the spatial average of each *cosine* term, over the semithickness, w, of the packing is zero. This is shown by

$$\frac{1}{w}\int_0^w \cos \beta_n (x - w)\, dx = 0 \tag{6.34}$$

from which it follows that

$$\beta_n = \frac{n\pi}{w} \tag{6.35}$$

Hausen then modifies equation (6.32) to allow for this effect of the reversal to yield

$$t_s(x,\tau) = t_{s,m}(\tau) + \frac{\alpha}{\lambda_s w}(t_f - t_{s,o})\left(\frac{x^2}{2} - wx + \frac{w^2}{3}\right)$$
$$+ \sum_{n=1}^{\infty} B_n e^{-(n\pi/w)^2 \kappa_s \tau} \cos\frac{n\pi x}{w} \qquad (6.36)$$

As in the numerical solution to this problem offered here, the dimensionless parameters ω and z are introduced:

$$\omega = \frac{\kappa_s \tau}{w^2} \quad \text{and} \quad z = \frac{x}{w} \quad \text{with} \quad \mathrm{Bi} = \frac{\alpha w}{\lambda_s}$$

Equation (6.36) now becomes

$$t_s(z,\omega) = t_{s,m}(\omega) + \mathrm{Bi}(t_f - t_{s,o})\left(\frac{z^2}{2} - z + \frac{1}{3}\right)$$
$$+ \sum_{n=1}^{\infty} B_n e^{-(n\pi)^2 \omega} \cos n\pi z \qquad (6.37)$$

This refers to the hot period. The corresponding equation for the cold period will be

$$t'_s(z,\omega) = t'_{s,m}(\omega) + \mathrm{Bi}'(t'_f - t'_{s,o})\left(\frac{z^2}{2} - z + \frac{1}{3}\right)$$
$$+ \sum_{n=1}^{\infty} B'_n e^{-(n\pi)^2 \omega'} \cos n\pi z \qquad (6.38)$$

At cyclic equilibrium, the heat transferred per period, q_{per}, will be the same in the hot and the cold periods. Assuming that $c_p = c'_p$, we can write

$$\mathrm{Bi}(t_f - t_{s,o})\Omega = \mathrm{Bi}'(t'_f - t'_{s,o})\Omega'$$

from which it follows that, assuming that $\kappa_s = \kappa'_s$ and $\lambda_s = \lambda'_s$,

$$t'_s(z,\omega) = t'_{s,m}(\omega) + \mathrm{Bi}(t_f - t_{s,o})\frac{\Omega}{\Omega'}\left(\frac{z^2}{2} - z + \frac{1}{3}\right)$$
$$+ \sum_{n=1}^{\infty} B'_n e^{-(n\pi)^2 \omega'} \cos n\pi z \qquad (6.39)$$

Hausen suggests that the coefficients $\{B_n, B'_n \mid n = 1, 2, \ldots\}$ be obtained from the reversal conditions, namely that

$$t_s(z,0) = t'_s(z,\Omega') \quad \text{and} \quad t_{s,m}(z,0) = t'_{s,m}(z,\Omega')$$

giving

$$\sum_{n=1}^{\infty}\left(B_n - B_n' e^{-(n\pi)^2\Omega'}\right)\cos n\pi z = \mathrm{Bi}(t_f - t_{s,o})\left(\frac{\Omega}{\Omega'} - 1\right) \tag{6.40}$$

Hausen then treats the left-hand side as a Fourier series with coefficients $B_n - B_n' e^{-(n\pi)^2\Omega'}$, which have to be found. Employing a standard method for determining such coefficients (see Hausen [6], p. 327) he deduces that

$$B_n - B_n' e^{-(n\pi)^2\Omega'} = -\frac{2}{(n\pi)^2}\,\mathrm{Bi}(t_f - t_{s,o})\frac{\Omega + \Omega'}{\Omega'} \tag{6.41}$$

He deduces similar expressions for the other reversal condition

$$t_s'(z, 0) = t_s(z, \Omega) \quad \text{and} \quad t_{s,m}'(z, 0) = t_{s,m}(z, \Omega)$$

and finds

$$B_n' - B_n e^{-(n\pi)^2\Omega} = +\frac{2}{(n\pi)^2}\,\mathrm{Bi}(t_f - t_{s,o})\frac{\Omega + \Omega'}{\Omega'} \tag{6.42}$$

Hausen then treats equations (6.41) and (6.42) as a pair of simultaneous equations in B_n and B_n'. He is then able to develop equation (6.37) in the following way:

$$t_s(z, \omega) = t_{s,m}(\omega) + \mathrm{Bi}(t_f(\omega) - t_{s,o}(\omega))\left[\left(2z^2 - z + \frac{1}{3}\right) - \frac{\Omega + \Omega'}{\Omega'}\sum_{n=1}^{\infty}\frac{2}{(n\pi)^2}\,\frac{1 - e^{-(n\pi)^2\Omega'}}{1 - e^{-(n\pi)^2(\Omega+\Omega')}}\,e^{-(n\pi)^2\Omega}\cos n\pi z\right] \tag{6.43}$$

Recalling that $t_s(0, \omega) = t_{s,o}(\omega)$ and setting $z = 0$ in equation (6.43) to yield the surface solid temperature on the left-hand side of this equation, we obtain

$$t_{s,o}(\omega) = t_{s,m}(\omega) + \mathrm{Bi}[t_f(\omega) - t_{s,o}(\omega)] \times \left(\frac{1}{3} - \frac{\Omega + \Omega'}{\Omega'}\sum_{n=1}^{\infty}\frac{2}{(n\pi)^2}\,\frac{1 - e^{-(n\pi)^2\Omega'}}{1 - e^{-(n\pi)^2(\Omega+\Omega')}}\,e^{-(n\pi)^2\omega}\right) \tag{6.44}$$

This is the analytical form of the numerical solution of equation (6.20) from which a time-varying form of $\Phi(\omega)$ could be computed. The parallel between the analytical approach of Hausen and the numerical approach offered earlier in this chapter is completed by recalling that a chronological average value of Φ is used. This can be developed from a suitably modified form, using dimensionless parameters, of equation (6.16):

$$\Phi = \frac{1}{\Omega}\int_0^{\Omega} \frac{1}{\mathrm{Bi}}\left(\frac{t_{s,o}(\omega) - t_{s,m}(\omega)}{t_f(\omega) - t_{s,o}(\omega)}\right) d\omega \tag{6.45}$$

From this, Hausen developed the form of Φ given by

$$\Phi = \frac{1}{3} - 2\left(\frac{1}{\Omega} + \frac{1}{\Omega'}\right) \times \sum_{n=1}^{\infty} \frac{1}{(n\pi)^4} \frac{\left(1 - e^{-(n\pi)^2\Omega}\right)\left(1 - e^{-(n\pi)^2\Omega'}\right)}{1 - e^{-(n\pi)^2(\Omega+\Omega')}} \tag{6.46}$$

Hausen, from these considerations, developed the approximations

$$\Phi = \frac{1}{3} - \frac{1}{45}\left(\frac{1}{\Omega} + \frac{1}{\Omega'}\right) \quad \text{where} \quad \frac{1}{\Omega} + \frac{1}{\Omega'} \leq 5 \tag{6.47}$$

and

$$\Phi = \frac{0.714}{\left[0.3 + 2\left(\frac{1}{\Omega} + \frac{1}{\Omega'}\right)\right]^{\frac{1}{2}}} \quad \text{where} \quad \frac{1}{\Omega} + \frac{1}{\Omega'} > 5 \tag{6.48}$$

These approximations are used in the generation of the second column of Φ's in Table 6.1. Relative to the figures for Φ computed numerically directly from the Fourier equation, it will be seen that Hausen's values of Φ are accurate to three significant figures, which in general is more than adequate. The reader is reminded that a time-varying $\Phi(\omega)$ may be required if the effects of the reversals need to be accommodated.

Hausen [6] developed more accurate approximations from the analytical solutions shown above. The reader is referred to Hausen's text for details. It seems clear from Table 6.1, however, that in general, the formulae (6.47) and (6.48) are more than adequate.

6.6 Formulae for Φ for Cylinders and Spheres

Hausen's work on the development of the Φ-factor for plain walls has been extended to other geometries for the packing material of regenerators. In particular, Stuke [7] developed the corresponding formulae for Φ for solid cylinders and for spheres. This work was extended still further by Razelos and Lazaridis [8], who dealt with the case of hollow cylinders.

The Fourier equation for these cases takes the form

$$\frac{\partial t_s}{\partial \tau} = \frac{\kappa_s}{r^n} \frac{\partial}{\partial r}\left(r^n \frac{\partial t_s}{\partial r}\right) \tag{6.49}$$

where $n = 1$ for the cylindrical geometry, and $n = 2$ for spheres.

At $r = 0$ at the center of the solid cylinder or at the center of the sphere, we assume that

$$\left.\frac{\partial t_s}{\partial r}\right|_{r=0} = 0$$

In the case of the hollow cylinder of radius R_O, if we assume that the gas flows down the middle of the cylinder through a channel of radius R_I and if we assume that the outer surface of the cylinder is effectively insulated, then

$$\left.\frac{\partial t_s}{\partial r}\right|_{r=R_O} = 0$$

At the heating surface of radius R_h, we set

$$\left.\frac{\partial t_s}{\partial r}\right|_{r=R_h} = \frac{\alpha}{\lambda_s}(t_f - t_s)$$

where $R_h = R$, the radius of the solid cylinder or sphere, or where $R_h = R_I$, the radius of the channel through the hollow cylinder.

The rate of change of the mean solid can be written as a modified form of equation (6.9):

$$\frac{\partial t_{s,m}}{\partial \tau} = m \frac{\alpha}{w \rho_s C_s}(t_f - t_{s,o}) \tag{6.50}$$

where the semithickness $w = R$, the radius of the solid cylinder or of the sphere, or where $w = R_O - R_I$ in the case of the hollow cylinder. The constant m takes the values shown below.

Plain wall: $m = 1$
Cylinder: $m = 2$
Sphere: $m = 3$

Stuke develops forms for the variation of solid temperature; for cylinders,

$$t_s(r,\tau) = t_{s,m}(r,\tau) + \frac{\alpha}{2\lambda_s R}[t_f(\tau) - t_{s,o}(\tau)]\left(r^2 - \frac{R^2}{2}\right) + \sum_{n=1}^{\infty} A_n e^{-\gamma_n^2 \kappa_s \tau} J_0(\gamma_n r) \tag{6.51}$$

and for spheres,

$$t_s(r,\tau) = t_{s,m}(r,\tau) + \frac{\alpha}{2\lambda_s R}[t_f(\tau) - t_{s,o}(\tau)]\left(r^2 - \frac{3R^2}{5}\right) + \sum_{n=1}^{\infty} \frac{B_n}{r} e^{-\mu_n^2 \kappa_s \tau} \sin(\mu_n r) \tag{6.52}$$

where J_0 is the zero-order Bessel function of the first kind. In these equations, A_n and B_n are constants. γ_n and μ_n are the solutions to the equations

$$J_1(\gamma_n R) = 0$$

where J_1 is the Bessel function of the first order, and

$$\mu_n R - \tan(\mu_n R) = 0$$

respectively. Hausen [6] comments that it is remarkable that the parabolic temperature profile found for these operating conditions ($\partial t_{s,m}/\partial\tau =$ *constant*) in plain walls is also found for solid cylinders and spheres. Both expressions for cylinders and spheres as they stand separate the time (τ) and space (r) variables.

The bulk heat transfer coefficient, $\bar{\alpha}$ for both solid cylinders and spheres takes the form

$$\frac{1}{\bar{\alpha}} = \frac{1}{\alpha} + \frac{R}{\lambda_s}\Phi \tag{6.53}$$

From equations (6.51) and (6.52), Stuke developed forms for Φ in the following manner: for cylinders,

$$\Phi = \frac{1}{4} - 2\left(\frac{1}{\Omega} + \frac{1}{\Omega'}\right) \times \sum_{n=1}^{\infty} \frac{\left(1 - e^{-\gamma_n^2 \kappa_s P}\right)\left(1 - e^{-\gamma_n^2 \kappa_s P'}\right)}{\gamma_n^4 \left(1 - e^{-\gamma_n^2 \kappa_s (P+P')}\right)} \tag{6.54}$$

and for spheres,

$$\Phi = \frac{1}{5} - 2\left(\frac{1}{\Omega} + \frac{1}{\Omega'}\right) \times \sum_{n=1}^{\infty} \frac{\left(1 - e^{-\mu_n^2 \kappa_s P}\right)\left(1 - e^{-\mu_n^2 \kappa_s P'}\right)}{\mu_n^4 \left(1 - e^{-\mu_n^2 \kappa_s (P+P')}\right)} \tag{6.55}$$

Stuke [7] claims, however, that the equations for Φ for plain walls, solid cylinders and spheres can be unified and suggests that the following equation can be applied as required:

$$\Phi = \frac{1}{K} - 2\left(\frac{1}{\Omega} + \frac{1}{\Omega'}\right) \times \sum_{n=1}^{\infty} \frac{1}{\Gamma^4} \frac{\left(1 - e^{-\Gamma^2 \Omega}\right)\left(1 - e^{-\Gamma^2 \Omega'}\right)}{1 - e^{-\Gamma^2(\Omega + \Omega')}} \tag{6.56}$$

where

$$\Gamma = \begin{cases} n\pi, & \text{zero of } \sin x \\ \gamma_n, & \text{zero of } J_1(x) \\ \mu_n, & \text{zero of } x - \tan x \end{cases} \qquad K = \begin{cases} 3 & \text{for plain walls} \\ 4 & \text{for solid cylinders} \\ 5 & \text{for spheres} \end{cases}$$

Hausen [6] collects together these formulae in simplified form. These are presented in Table 6.2.

For spheres and solid cylinders, we use the radius, R, as the "semithickness" from which it follows that

$$\Omega = \frac{\kappa_s P}{R^2}$$

Table 6.2: Approximations for Φ for different packing geometries

Plain walls	$\Phi = \frac{1}{3} - \frac{1}{45}\left(\frac{1}{\Omega} + \frac{1}{\Omega'}\right)$	for $\left(\frac{1}{\Omega} + \frac{1}{\Omega'}\right) \leq 5$
Solid cylinders	$\Phi = \frac{1}{4} - \frac{4}{383}\left(\frac{1}{\Omega} + \frac{1}{\Omega'}\right)$	for $\left(\frac{1}{\Omega} + \frac{1}{\Omega'}\right) \leq \frac{15}{2}$
Spheres	$\Phi = \frac{1}{5} - \frac{1}{175}\left(\frac{1}{\Omega} + \frac{1}{\Omega'}\right)$	for $\left(\frac{1}{\Omega} + \frac{1}{\Omega'}\right) \leq 10$

$$\Phi = \frac{0.714}{\left[\varepsilon + 2\left(\frac{1}{\Omega} + \frac{1}{\Omega'}\right)\right]^{\frac{1}{2}}} \quad \text{with} \begin{cases} \varepsilon = 0.3 & \text{for plain walls} \\ \varepsilon = 1.1 & \text{for solid cylinders} \\ \varepsilon = 3.0 & \text{for spheres} \end{cases}$$

for larger values of $\frac{1}{\Omega} + \frac{1}{\Omega'}$

6.7 Formula for Φ for Hollow Cylinders

Razelos and Lazaridis [8] developed a form of Φ and hence a lumped heat transfer coefficient $\bar{\alpha}$ for periodically heated and subsequently cooled hollow cylinders. They allowed the heating/cooling to take place at the inside (the more usual in practice) surface or the outer surface. The heated surface has radius R_I, the insulated, R_O.

This case is that more complicated because Φ will be a function of the *curvature* of the thickness $w = |R_O - R_I|$ as well as the thickness, w, itself. Razelos and Lazaridis developed a function for Φ in terms of Ψ, where

$$\frac{2}{\Psi} = \frac{1}{\Omega} + \frac{1}{\Omega'}, \quad \beta = \frac{R_O}{R_I}, \quad \text{and} \quad \mu = \frac{\Omega}{\Omega'} \tag{6.57}$$

and where

$$\Omega = \frac{\kappa_s P}{(R_O - R_I)^2} \quad \text{for hollow cylinders} \tag{6.58}$$

They showed that the following equation can be used:

$$\Phi = F(\beta) + \chi(\beta, \Psi, \mu) \tag{6.59}$$

where

$$F(\beta) = \frac{\beta^4(4\ln\beta - 3) + 4\beta^2 - 1}{4(\beta - 1)^3(\beta + 1)^2} \tag{6.60}$$

The form of $\chi(\beta, \Psi, \mu)$ is far more complex. Whereas for solid cylinders, the roots $\{\gamma_n \mid n = 1, 2, \ldots\}$ of the equation $J_1(\gamma_n R) = 0$ are required, for hollow cylinders, the roots $\{\lambda_n \mid n = 1, 2, \ldots\}$ of the following equation must be determined:

$$J_1\left(\frac{\lambda_n}{\beta - 1}\right) Y_1\left(\frac{\beta\lambda_n}{\beta - 1}\right) - J_1\left(\frac{\beta\lambda_n}{\beta - 1}\right) Y_1\left(\frac{\lambda_n}{\beta - 1}\right) = 0 \tag{6.61}$$

Razelos and Lazaridis define a quantity Q_n,

$$Q_n = \frac{J_1\left(\frac{\lambda_n}{\beta - 1}\right) Y_0\left(\frac{\beta\lambda_n}{\beta - 1}\right) - J_0\left(\frac{\beta\lambda_n}{\beta - 1}\right) Y_1\left(\frac{\lambda_n}{\beta - 1}\right)}{J_1\left(\frac{\lambda_n}{\beta - 1}\right) Y_0\left(\frac{\lambda_n}{\beta - 1}\right) - J_0\left(\frac{\lambda_n}{\beta - 1}\right) Y_1\left(\frac{\lambda_n}{\beta - 1}\right)} \tag{6.62}$$

Note that J_0 and Y_0 are zero-order Bessel functions of the *first* and *second* kind respectively, and J_1 and Y_1 are first-order Bessel functions, again of the *first* and *second* kind.

Given Q_n, the function $\chi(\beta, \Psi, \mu)$ is defined by the complicated function given below, which is nevertheless, similar to the Stuke [7] equation (6.56):

$$\chi(\beta, \Psi, \mu) = -\frac{4(\beta-1)}{\Psi}\sum_{n=1}^{\infty}\left(\frac{1}{\lambda_n^4}\right)\left(\frac{1}{\beta^2 Q_n^2 - 1} \times \frac{\left(1-e^{-\lambda_n^2\Psi(1+\mu)/2\mu}\right)\left(1-e^{-\lambda_n^2\Psi(1+\mu)/2}\right)}{1-e^{-\lambda_n^2\Psi(1+\mu)^2/2\mu}}\right) \tag{6.63}$$

No approximations for Φ for hollow cylinders are offered by these authors. Razelos and Lazaridis do comment, however, that for $\beta \to 1$, this equation reduces to that obtained by Hausen [6] and presented above as equation (6.46). For other values of β, Razelos and Lazaridis present values of Φ in graphical form and these are reproduced here as Figures 6.4 and 6.5. Butterfield et al. [9] propose that Hausen's formula for Φ for plain walls be used with a "generalized semithickness," Δ, where this is defined to be the packing volume divided by the surface area. Butterfield et al. give this to be

$$\Delta = \frac{2\sqrt{3}R_O^2/\pi - R_I^2}{4R_I}$$

for which Razelos and Lazaridis comment that this generates values of Φ that show a large discrepancy for $\Psi > 0.25$.

Razelos and Lazaridis rightly comment that for long periods ($\Omega \to \infty$),

$$\Phi = F(\beta) = \frac{\beta^4(4\ln\beta - 3) + 4\beta^2 - 1}{4(\beta-1)^3(\beta+1)^2} \tag{6.64}$$

What is interesting is that this $F(\beta)$ corresponds to the $1/K$ introduced by Stuke [7] in equation (6.56). We have plotted $K = 1/F(\beta)$ as a function of β in Figure 6.6. For $0 \le \beta \le 1$ this corresponds to the case where the annulus of the cylinder is insulated and the packing is heated from the outside of the cylinder, where, recall, R_I is the radius of the heated surface and R_O is the radius of the insulated surface, where $\partial t_s/\partial r = 0$.

Where $\beta = 1$, this corresponds to an infinitely thin wall but a plain wall nevertheless. Here $K = 3$ as one might expect. Indeed, Razelos and Lazaridis point out that their equation for Φ reduces Hausen's equation (6.46) for plain walls where $\beta = 1$. Where $\beta = 0$, this corresponds to the case where the inner radius of the hollow cylinder becomes zero and the hollow cylinder becomes a solid cylinder with $K = 4$ as shown in Figure 6.6, the value of K suggested by Stuke in this case.

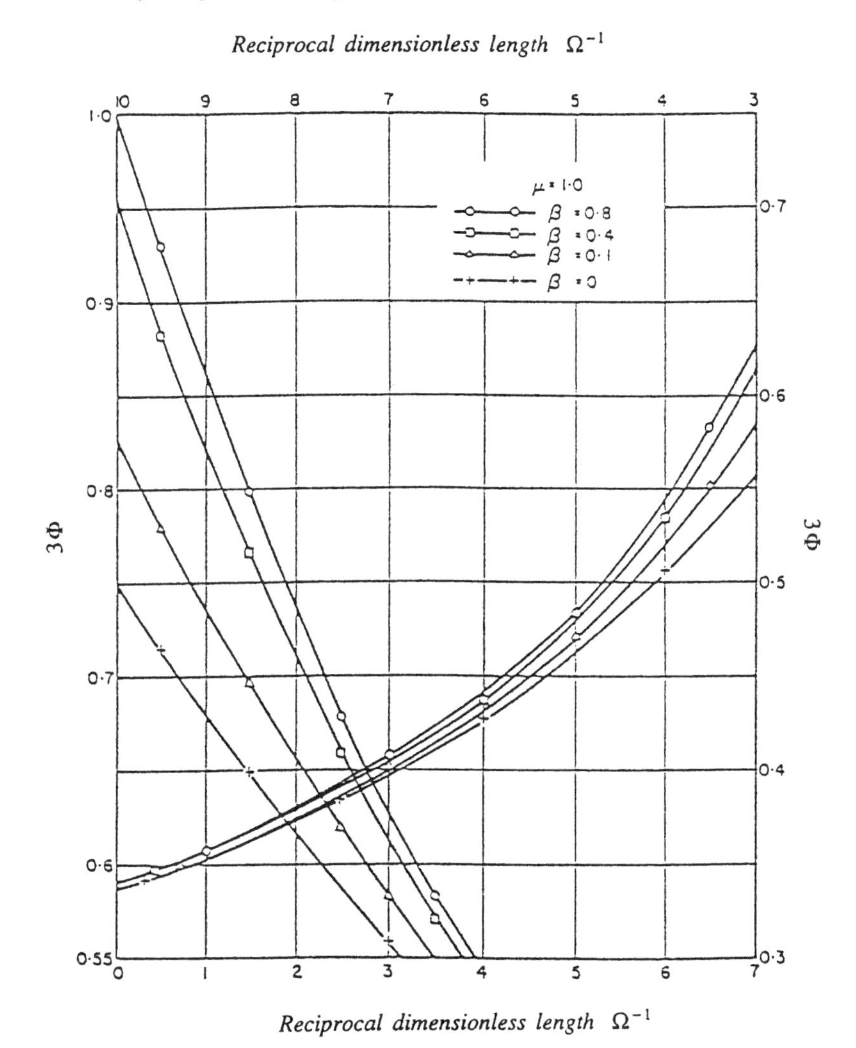

Figure 6.4: Variation of Φ-factor with dimensionless period, Ω, for different values of β for hollow cylinders, displayed as 3Φ plotted as a function of $1/\Omega$. Annulus of hollow cylinder insulated, outer surface heated/cooled. (Razelos and Lazaridis [8].)

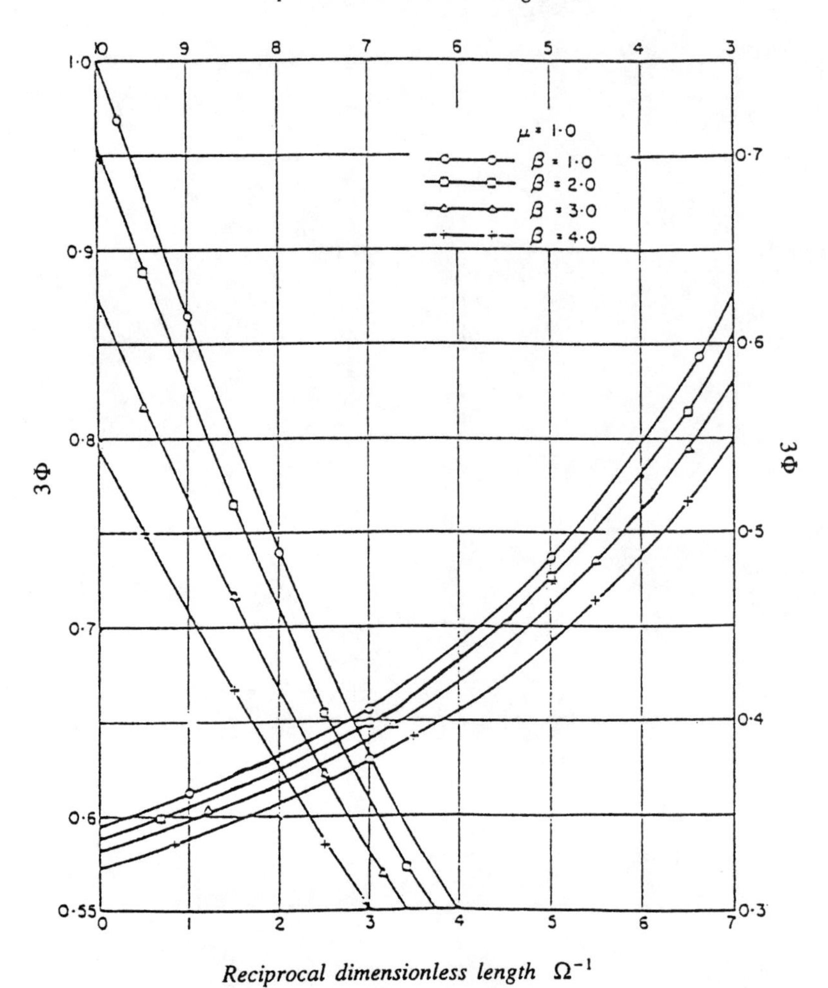

Figure 6.5: Variation of Φ-factor with dimensionless period, Ω, for different values of β for hollow cylinders. Annulus of hollow cylinder heated/cooled, outer surface insulated. (Razelos and Lazaridis [8].)

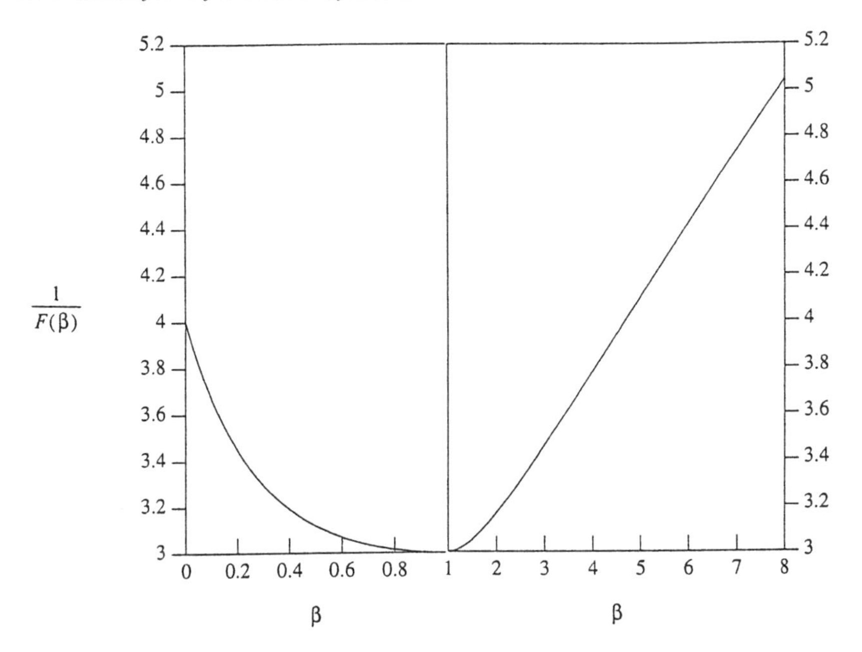

Figure 6.6: Value of K for long and equal period lengths as a function of β, where $\Phi = 1/K$ and $K = 1/F(\beta)$.

The case for $\beta > 1$ is equally remarkable. Here, the annulus of the cylinder is heated, $r = R_I$, and the outer surface is insulated, $r = R_O$. Despite the apparent complexity of the formula for K, namely

$$K = \frac{1}{F(\beta)} = \frac{4(\beta - 1)^3(\beta + 1)^2}{\beta^4(4\ln\beta - 3) + 4\beta^2 - 1}$$

K adopts an almost linear relation between $K(\beta)$ and β, certainly for $\beta > 2$. This approximation takes the form

$$K = 2.5169 + 0.3165\beta \tag{6.65}$$

which is a least squares fit of $K(\beta)$ against β for $2 < \beta < 8$. A comparison between the exact value and this linear estimate is presented in Figure 6.7. It will be seen that the exact form (6.64) should be used for $\beta < 2$ but that the approximation (6.65) offered can be used for larger values of β.

Insufficient *numerical* data is offered by Razelos and Lazaridis [8] but from the graphs given in their paper, the present author has deduced a similar approximation for hollow cylinders to that used for plain walls by Hausen, and for spheres and solid cylinders by Stuke. It is suggested that

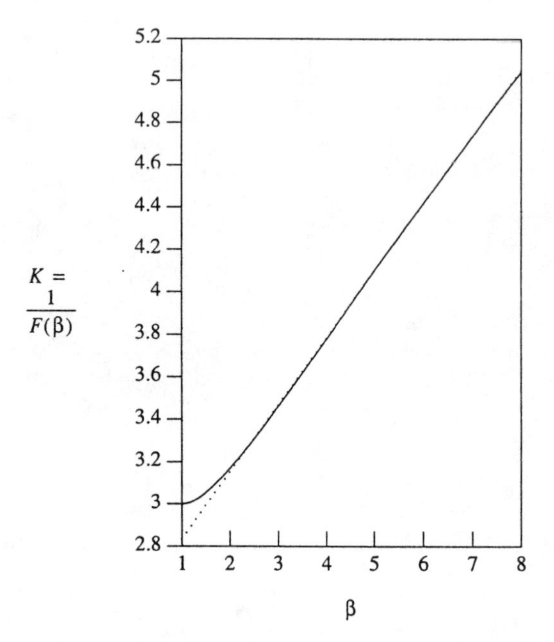

Figure 6.7: Value of K for long and equal period lengths, displayed as $1/F(\beta)$ plotted against β. Comparison between exact and linear estimate ($\beta > 1$). Linear estimate: $1/F(\beta) = 0.3165\beta + 2.5169$ (dotted).

$$\Phi_\beta = \frac{1}{K} - \frac{1}{47.267 - 2.567\,\beta}\left(\frac{1}{\Omega} + \frac{1}{\Omega'}\right) \quad \text{for} \quad \frac{1}{\Omega} + \frac{1}{\Omega'} \leq 5 \text{ (approx.)} \tag{6.66}$$

where K is given by equation (6.65).

Otherwise, with $\varepsilon \approx \frac{1}{3}[0.9 + 4.3(\beta - 1)]$,

$$\Phi_\beta = \frac{0.714}{\left[\varepsilon + 2\left(\frac{1}{\Omega} + \frac{1}{\Omega'}\right)\right]^{\frac{1}{2}}} \quad \text{for larger values of} \quad \frac{1}{\Omega} + \frac{1}{\Omega'} \tag{6.67}$$

More precise approximations should be able to be deduced by reworking the calculations suggested by Razelos and Lazaridis.

6.8 The Precise Representation of the Latitudinal Conduction in Regenerator Packing

Using the surface heat transfer coefficient, α, equation (6.1), which deals with the relation between the rate of thermal energy transferred between gas and solid and the rate at which heat is given up/absorbed by the gas flowing through the regenerator now takes the form

$$\frac{\partial t_f}{\partial y} = \frac{\alpha A}{\dot{m}_f c_p L}(t_{s,o} - t_f) - \frac{m_f}{\dot{m}_f L}\frac{\partial t_f}{\partial \tau} \tag{6.68}$$

Heat transfer within the packing, which we assume here to be a plain wall, is accommodated by the Fourier equation, namely

$$\frac{\partial t_s}{\partial \tau} = \kappa_s \frac{\partial^2 t_s}{\partial x^2} \tag{6.69}$$

At the entrance to the regenerator ($y = 0$), we assume that the gas temperature does not change in a given period of operation and takes the value $t_{f,in}$. In the middle of the wall, at all positions of y, the boundary condition

$$\left.\frac{\partial t_s}{\partial x}\right|_{x=w} = 0 \tag{6.70}$$

applies. At the surface of the packing, again at all positions y, we apply the boundary condition

$$\left.\lambda_s \frac{\partial t_s}{\partial x}\right|_{x=0} = \alpha(t_f - t_{s,o}) \tag{6.71}$$

These equations are applied to the hot and cold periods with the appropriate reversal conditions applied to take account of counterflow operation of the regenerator. This representation is called frequently the "3-D model," the three dimensions being x, y, and τ.

This system of equations can be simplified using the substitutions

$$\xi = \frac{\alpha A}{\dot{m}_f c_p L} y \qquad z = \frac{x}{w} \qquad \omega - \frac{\kappa_s}{w^2}\left(\tau - \frac{m_f}{\dot{m}_f L} y\right) \tag{6.72}$$

Note that ω is modified *slightly* in this context from that used previously (see equation (6.17)). Upon application of this introduction of the dimensionless variables ξ, z, and ω, the differential equations take the simple form

$$\frac{\partial t_f}{\partial \xi} = t_s - t_f \tag{6.73}$$

$$\frac{\partial t_s}{\partial \tau} = \frac{\partial^2 t_s}{\partial z^2} \tag{6.74}$$

The boundary conditions are as follows:

Table 6.3: Dimensionless parameters for 3-D model of a regenerator

Period	Reduced time	Reduced length	Biot modulus
Hot	Ω	Λ	Bi
Cold	Ω'	Λ'	Bi$'$

$$\left.\frac{\partial t_s}{\partial z}\right|_{z=0} = \mathrm{Bi}(t_{s,o} - t_f) \tag{6.75}$$

$$\left.\frac{\partial t_s}{\partial x}\right|_{z=1} = 0 \tag{6.76}$$

Again, at the entrance to the regenerator ($\xi = 0$), we assume that the gas temperature takes the unchanging value $t_{f,in}$, appropriate to the period under consideration. The dimensionless parameters "reduced time," Ω, and "reduced length," Λ, for this 3-D model can now be defined:

$$\Omega = \frac{\kappa_s}{w^2}\left(P - \frac{m_f}{\dot{m}_f}\right) \qquad \Lambda = \frac{\alpha A}{\dot{m}_f c_p} \tag{6.77}$$

The overall effect of the surface heat flux relative to the internal resistance to heat transfer with the packing is measured by the Biot modulus:

$$\mathrm{Bi} = \frac{\alpha w}{\lambda_s} \tag{6.78}$$

The regenerator and its mode of operation can thus be summarized for the 3-D model using just six factors, shown in Table 6.3.

6.9 Relationship Between the 3-D and 2-D Models

In the 2-D model set out in Chapter 4, two dimensions are used, namely $\bar{\xi}$ and $\bar{\eta}$, where we introduce the bar notation to denote that $\bar{\xi}$ and $\bar{\eta}$ embody the lumped heat transfer coefficient, $\bar{\alpha}$. The bar notation is also used here so that ξ (in equations (6.72)) is distinguished readily from $\bar{\xi}$ where

$$\bar{\xi} = \frac{\bar{\alpha} A}{\dot{m}_f c_p L} y \tag{6.79}$$

We note that

$$\bar{\eta} = \frac{\bar{\alpha}A}{MC_s}\left(\tau - \frac{m_f y}{\dot{m}_f L}\right) \tag{6.80}$$

These give rise to the 2-D model parameters

$$\bar{\Lambda} = \frac{\bar{\alpha}A}{\dot{m}_f c_p} \tag{6.81}$$

$$\bar{\Pi} = \frac{\bar{\alpha}A}{MC_s}\left(P - \frac{m_f}{\dot{m}_f}\right) \tag{6.82}$$

It will be seen that the 2-D model is distinct from the 3-D model, which involves the dimensions ξ, z, and ω and the corresponding descriptive parameters Λ, Ω, and Bi. In both models, the inlet temperatures and their possible time variations must be specified. Most, if not all, previous authors have restricted their considerations to unvarying inlet gas temperatures in both periods of regenerator operation.

From earlier discussions in this chapter, it will be evident that the question arises as to how adequate the 2-D model is in representing regenerative heat exchanger behavior. Moreover, it is evident that the answer revolves around the lumped heat transfer coefficient, $\bar{\alpha}$, and the Φ-factor. Because the chronological average Φ is used, as given in equation (6.30), it might be expected that the time average (in each period of operation) regenerator performance should be predicted reasonably well by the 2-D model compared with the 3-D model. However, the various forms of $\bar{\alpha}$ and Φ are developed from the assumption that the mean solid temperature varies linearly with time,

$$\frac{\partial t_{s,m}}{\partial \tau} = constant$$

In the middle of the regenerator, this assumption is generally true. At the entrance to the regenerator, if the inlet temperature does not vary with time, the solid temperature varies in a negative exponential fashion. Integrating equation (6.2) at the gas entrance, we obtain

$$t_{s,m}(0, \tau) = t_{f,in} + (t_{s,m}(0, 0) - t_{f,in})e^{-[\bar{\alpha}A/(MC_s)]\tau} \tag{6.83}$$

where $t_{s,m}(0, \tau)$ is the mean solid temperature at the entrance to the regenerator ($y = 0$) and at time τ from the start of the period of operation under consideration. At and close to the regenerator entrance, the temperature behavior is, as a consequence, nonlinear, the nonlinear effects being propagated from the entrance by the unvarying inlet gas temperature, down the length of the regenerator. Hausen [4] suggested that a factor κ/κ_0 (whose original use is not of concern here) might be used to measure the extent of the effects of these nonlinearities. It should be noted that $0 < \kappa/\kappa_0 < 1$ and that the smaller the value of κ/κ_0, the more severe the effect of these nonlinearities. Hausen restricted his considerations to the *balanced* case where $\bar{\Pi}/\bar{\Pi}' = \bar{\Lambda}/\bar{\Lambda}'$ and $\eta_{REG} = \eta'_{REG}$. In this case, Hausen indicated that κ/κ_0 can be calculated using

$$\frac{\kappa}{\kappa_0} = \frac{\eta_{REG}}{1 - \eta_{REG}} \left(\frac{1}{\bar{\Lambda}} + \frac{1}{\bar{\Lambda}'} \right) \tag{6.84}$$

Schmidt and Willmott [10] point out that Hausen suggested that, for the balanced case, κ/κ_0 can be represented as a function of the harmonic mean (suitably weighted) reduced period, $\bar{\Pi}_H$, and reduced length, $\bar{\Lambda}_H$. These are defined by

$$\frac{1}{\bar{\Pi}_H} = \frac{1}{2} \left(\frac{1}{\bar{\Pi}} + \frac{1}{\bar{\Pi}'} \right) \tag{6.85}$$

$$\frac{1}{\bar{\Lambda}_H} = \frac{1}{\bar{\Pi}_H} \left(\frac{\bar{\Pi}}{\bar{\Lambda}} + \frac{\bar{\Pi}'}{\bar{\Lambda}'} \right) \tag{6.86}$$

The variation of κ/κ_0 with $\bar{\Lambda}_H$ and $\bar{\Pi}_H$ is displayed in Figure 6.8. As one might expect, the longer the dimensionless mean period $\bar{\Pi}_H$, the greater the time available for the nonlinear behavior to penetrate the interior of the regenerator packing in the direction of gas flow. On the other hand, the larger the regenerator, the larger $\bar{\Lambda}_H$, and the nonlinearities penetrate a smaller proportion of the packing, leaving the interior of the regenerator to enjoy linear variations of gas and solid temperature.

Presented in Tables 6.4, 6.5, and 6.6 are the comparative figures for the thermal ratio, η_{REG}, computed using the 2-D model with lumped heat transfer coefficients, $\bar{\alpha}$ and $\bar{\alpha}'$, and the 3-D model described above. These relate to the Φ factors 0.3, 0.2667, and 0.2333 and to the plain-wall configuration and are updated values for the corresponding tables in the Schmidt and Willmott [10] text.

This analysis could be extended to other packing geometries but the essential principles that emerge here are applicable to all packing configurations. Although consideration is restricted to the symmetric case ($\Lambda = \Lambda'$, $\Omega = \Omega'$, and $\mathrm{Bi} = \mathrm{Bi}'$), Hausen implies that the same conclusions apply to the *balanced* and *unbalanced* cases. It will be observed that for a given value of $\bar{\Pi}_H$, the larger the value of $\bar{\Lambda}_H$ and hence the value of κ/κ_0, the closer the values of the thermal ratio, η_{REG}, computed by the 2-D and 3-D models.

The thermal ratio, η_{REG}, is a chronological average, as is the Φ-factor. Hence, when the linear temperature is dominant in regenerator operation, Hausen's proposal to use a lumped heat transfer coefficient embodying Φ is justified, certainly as far as time average (over a period of operation) tem-

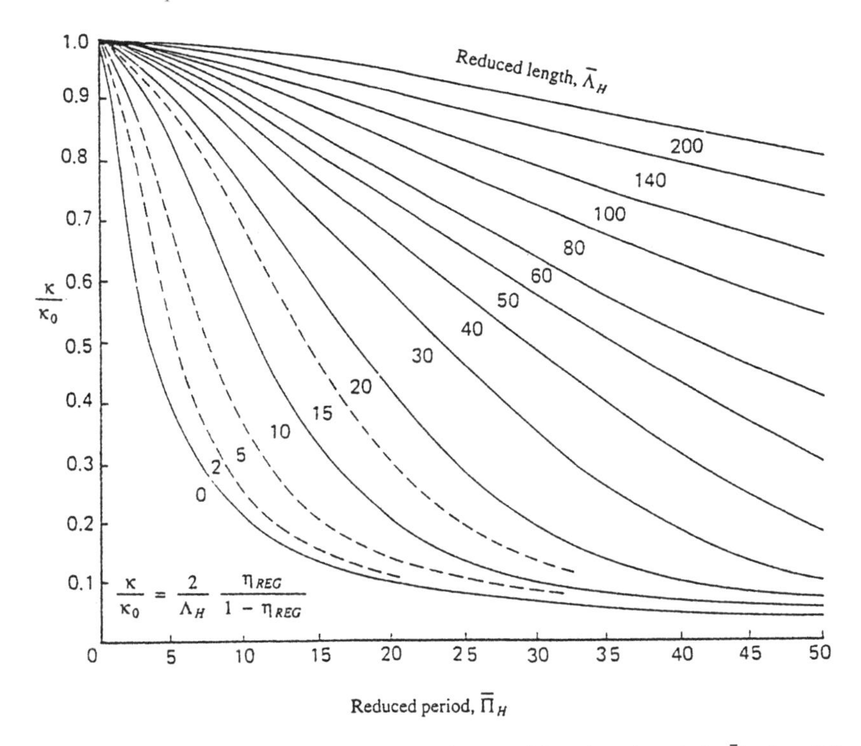

Figure 6.8: Variation of Hausen's κ/κ_0-factor with reduced period, $\bar{\Pi}_H$ for different values of reduced length, $\bar{\Lambda}_H$.

perature performance is concerned. In other words, the Φ-factor attempts to average out the effects of the inversions of the temperature profile within the solid packing, which occur at the reversals, over the whole cycle of regenerator operation. In the same way, the thermal ratios, η_{REG} and η'_{REG}, average out over each period the cumulative temperature behavior of the regenerator.

As one might expect, even for a low value of $\Phi = 0.2333$, the correspondence between the 2-D and 3-D thermal ratios is good for $\kappa/\kappa_0 > 0.8$. Low values of Φ correspond to relatively large durations of the reversal effect on the internal solid temperature profile in each period of operation.

It will be recalled that Hausen developed the approximations

$$\Phi = \frac{1}{3} - \frac{1}{45}\left(\frac{1}{\Omega} + \frac{1}{\Omega'}\right) \quad \text{where} \quad \frac{1}{\Omega} + \frac{1}{\Omega'} \leq 5 \tag{6.47}$$

and

Table 6.4: Comparison between thermal ratios, η_{REG}, as computed by the 2-D and 3-D models for $\Phi = 0.3$

$\bar{\Pi}$	$\bar{\Lambda}$	Φ	$\eta_{REG,2D}$	$\eta_{REG,3D}$
1	1	0.3000	0.3221	0.3198
1	2	0.3000	0.4911	0.4893
1	3	0.3000	0.5937	0.5924
1	4	0.3000	0.6622	0.6613
1	5	0.3000	0.7109	0.7103
1	6	0.3000	0.7474	0.7470
1	7	0.3000	0.7757	0.7754
1	8	0.3000	0.7983	0.7981
1	9	0.3000	0.8168	0.8167
1	10	0.3000	0.8322	0.8321
2	1	0.3000	0.2930	0.2863
2	2	0.3000	0.4664	0.4599
2	3	0.3000	0.5756	0.5706
2	4	0.3000	0.6490	0.6453
2	5	0.3000	0.7012	0.6985
2	6	0.3000	0.7399	0.7380
2	7	0.3000	0.7699	0.7684
2	8	0.3000	0.7936	0.7924
2	9	0.3000	0.8129	0.8120
2	10	0.3000	0.8289	0.8282
3	1	0.3000	0.2560	0.2467
3	2	0.3000	0.4305	0.4189
3	3	0.3000	0.5476	0.5374
3	4	0.3000	0.6281	0.6201
3	5	0.3000	0.6855	0.6794
3	6	0.3000	0.7280	0.7234
3	7	0.3000	0.7605	0.7570
3	8	0.3000	0.7861	0.7834
3	9	0.3000	0.8068	0.8046
3	10	0.3000	0.8238	0.8221

$$\Phi = \frac{0.714}{\left[0.3 + 2\left(\frac{1}{\Omega} + \frac{1}{\Omega'}\right)\right]^{\frac{1}{2}}} \quad \text{where} \quad \frac{1}{\Omega} + \frac{1}{\Omega'} > 5 \tag{6.48}$$

It follows that for a given 2-D symmetric configuration, the corresponding 3-D model is deduced in the manner set out below. The dimensionless time, Ω, is obtained using

Table 6.5: Comparison between thermal ratios, η_{REG}, as computed by the 2-D and 3-D models for $3\Phi = 0.8$

$\bar{\Pi}$	$\bar{\Lambda}$	Φ	$\eta_{REG,2D}$	$\eta_{REG,3D}$
1	1	0.2667	0.3221	0.3137
1	2	0.2667	0.4911	0.4842
1	3	0.2667	0.5937	0.5888
1	4	0.2667	0.6622	0.6587
1	5	0.2667	0.7109	0.7085
1	6	0.2667	0.7474	0.7457
1	7	0.2667	0.7757	0.7745
1	8	0.2667	0.7983	0.7974
1	9	0.2667	0.8168	0.8161
1	10	0.2667	0.8322	0.8316
2	1	0.2667	0.2930	0.2703
2	2	0.2667	0.4664	0.4431
2	3	0.2667	0.5756	0.5570
2	4	0.2667	0.6490	0.6351
2	5	0.2667	0.7012	0.6909
2	6	0.2667	0.7399	0.7323
2	7	0.2667	0.7699	0.7640
2	8	0.2667	0.7936	0.7891
2	9	0.2667	0.8129	0.8093
2	10	0.2667	0.8289	0.8260

Table 6.6: Comparison between thermal ratios, η_{REG}, as computed by the 2-D and 3-D models for $3\Phi = 0.7$

$\bar{\Pi}$	$\bar{\Lambda}$	Φ	$\eta_{REG,2D}$	$\eta_{REG,3D}$
1	1	0.2333	0.3221	0.3045
1	2	0.2333	0.4911	0.4758
1	3	0.2333	0.5937	0.5824
1	4	0.2333	0.6622	0.6540
1	5	0.2333	0.7109	0.7049
1	6	0.2333	0.7474	0.7429
1	7	0.2333	0.7757	0.7722
1	8	0.2333	0.7983	0.7955
1	9	0.2333	0.8168	0.8145
1	10	0.2333	0.8322	0.8303

$$\Omega = \frac{2}{15(1 - 3\Phi)} \quad \text{for} \quad \Phi \geq 0.2222 \tag{6.87}$$

$$\Omega = \frac{4}{\left(\frac{0.714}{\Phi}\right)^2 - 0.3} \quad \text{for} \quad \Phi < 0.2222 \tag{6.88}$$

From the relation

$$\frac{1}{\bar{\alpha}} = \frac{1}{\alpha} + \frac{w}{\lambda_s}\Phi \tag{6.12}$$

it can be deduced that

$$\frac{1}{\bar{\alpha}} = \frac{1}{\alpha}\left(1 + \frac{\alpha w}{\lambda_s}\Phi\right)$$

and hence

$$\Lambda = \bar{\Lambda}(1 + \text{Bi}\Phi) \tag{6.89}$$

The value of the Biot modulus, Bi, can similarly be developed from equation (6.12) using

$$\frac{1}{\bar{\Pi}} = \frac{1}{\Omega}\left(\frac{1}{\text{Bi}} + \Phi\right)$$

and hence

$$\text{Bi} = \left(\frac{\Omega}{\bar{\Pi}} - \Phi\right)^{-1} \tag{6.90}$$

6.10 Limitations of the Adequacy of the 2-D Model

Because the Φ-factor, in the lumped heat transfer coefficient, $\bar{\alpha}$, tries to average out the effects of the reversals in operation upon the temperature profile of the packing, it is not surprising that the *time variations* of the thermal performance predicted by the 2-D model are sometimes significantly different to that calculated using the 3-D model. Typically, the exit gas temperature in the cold period varies in a fairly linear manner using the 2-D model. However, there is an initial steep decline, for the 3-D model, in the exit gas temperature, as the parabolic temperature profile within the solid is inverted, before a linear variation is established. This is shown in Figure 6.9.

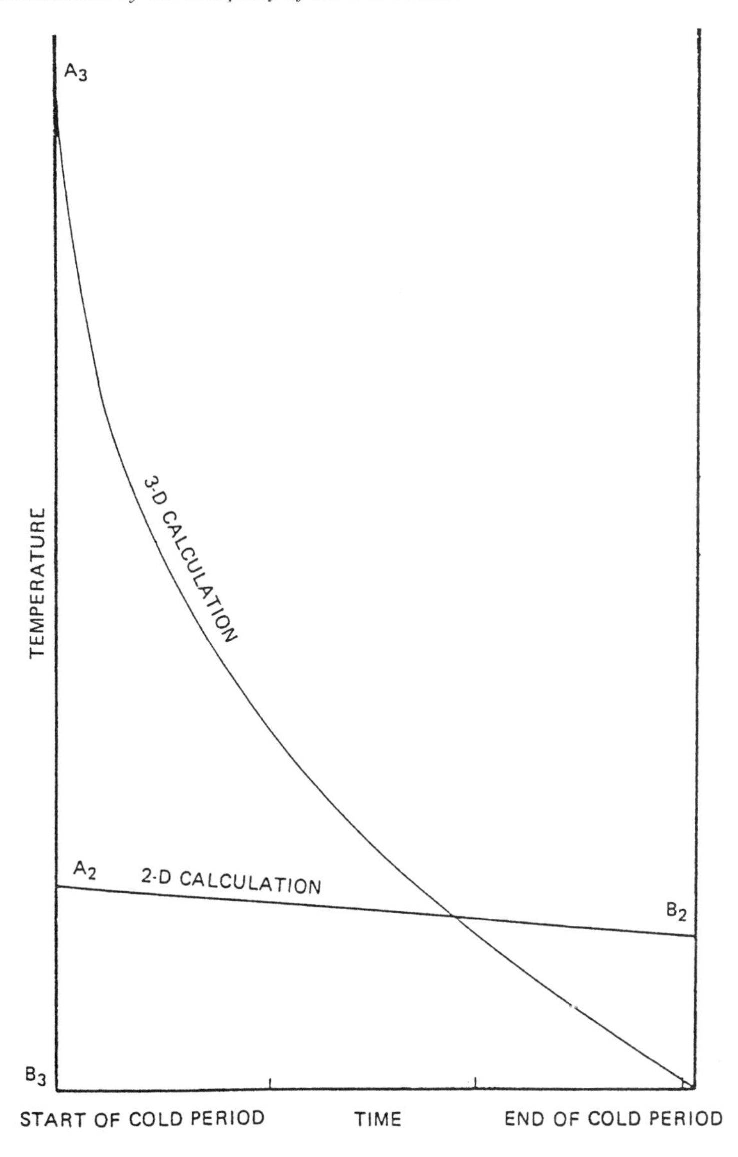

Figure 6.9: Variation of cold period exit gas temperature with time. Comparison between the 3-D and 2-D models.

A measure of this discrepancy between the models can be given by Ψ where, here,

$$\Psi = \frac{A_2 - B_2}{A_3 - B_3}$$

and where

A_2 = initial exit gas temperature (2-D model)
A_3 = initial exit gas temperature (3-D model)
B_2 = final exit gas temperature (2-D model)
B_3 = final exit gas temperature (3-D model)

For the *symmetric* case again, Ψ is displayed in Figure 6.10 as a function of reduced length, $\bar{\Lambda}$, for different values of Φ and for $\bar{\Pi} = 1$. What is clear is that for $\bar{\Lambda} > 4$ (approximately), Ψ reaches an asymptotic value that varies only with Φ. The greater the influence of a reversal, as a proportion of the duration of a period of operation, that is, the smaller the value of Φ, the greater the timewise discrepancy between the predictions of the 2-D and 3-D models. For $\bar{\Lambda} < 4$, the effect of the "κ/κ_0 nonlinear temperature behavior" is superimposed and the value of Ψ declines somewhat.

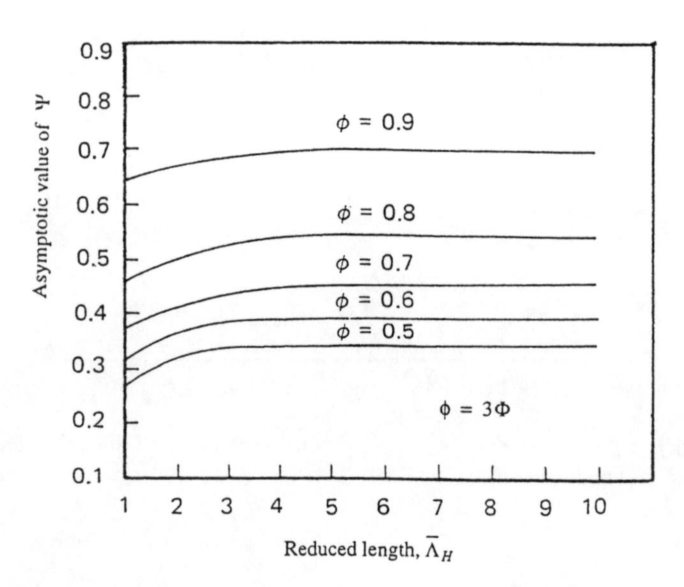

Figure 6.10: Comparison between the 3-D and 2-D models. Variation of Ψ as a function of $\bar{\Lambda}_H$ ($\bar{\Pi}_H = 1$).

What seems clear, therefore, is that if the time variations of the solid, as the temperature profile within the packing inverts after a reversal, can be built into the lumped heat transfer coefficients $\bar{\alpha}$ and $\bar{\alpha}'$, then the 2-D model might well be able to overcome, to some degree if not totally, the limitations described above. It is suggested that the time-varying value of $\Phi(\omega)$ be used within time-varying coefficients, $\bar{\alpha}(\omega)$ and $\bar{\alpha}'(\omega)$ instead of $\Phi(mean)$. What is suggested will be clear if reference is made to Figure 6.3.

6.11 Improvement of the 2-D Model by the Introduction of a Time-Varying $\Phi(\omega)$

In our numerical development of the Φ factor, it was proposed to solve the differential equation

$$\frac{\partial t_s}{\partial \omega} = \frac{\partial^2 t_s}{\partial z^2} \tag{6.20}$$

subject to the boundary condition (6.21) and the surface condition

$$\frac{\partial t_s}{\partial z} = \text{Bi}(t_f - t_{s,o}) \quad \text{at} \quad z = 0 \tag{6.22}$$

The simplest approach is that offered by Hinchcliffe and Willmott [11]. Here, $\text{Bi}(t_f - t_{s,o})$ is set equal to q for the hot period and q' for the cold, as given in equations (6.23). Equation (6.20) is then solved numerically until periodic steady state is achieved and the time variation, at successive time steps, of $\Phi(\omega)$ is found using equation (6.29). The numerical technique yields successive values of $\Phi(\omega_j)$ at positions in time $\{\omega_j = j\Delta\omega \mid j = 0, 1, 2, \ldots\}$. This time-varying $\Phi(\omega)$ is inserted into the lumped heat transfer coefficient to give $\bar{\alpha}(\omega)$ with

$$\frac{1}{\bar{\alpha}(\omega)} = \frac{1}{\alpha} + \frac{w}{\lambda_s}\Phi(\omega) \tag{6.91}$$

which is applied at all positions in the regenerator in the direction of gas flow (y) within the 2-D model. The values of $\bar{\Lambda}$, $\bar{\Lambda}'$, $\bar{\Pi}$, and $\bar{\Pi}'$ now change continuously, following the time variations of $\Phi(\omega)$. The 2-D equations are solved numerically with a chronologically varying $\Phi(\omega)$ by the finite-difference method described in Chapter 5, suitably modified. Details will be described in a later chapter in the context of generally nonlinear 2-D models.

Heggs and Carpenter [12] proposed a similar approach to approximate the thermal regenerator performance predicted by the 3-D model, using the 2-D model embodying a modified lumped heat transfer coefficient, $\bar{\alpha}\varepsilon(\omega)$. Willmott and Hinchcliffe [11] showed that

$$\frac{1}{\varepsilon(\omega)} = 1 + \mathrm{Bi}\Phi(\omega) \tag{6.92}$$

Heggs and Carpenter did not develop their $\varepsilon(\omega)$ factor employing a square-wave heat flux boundary condition such as that employed by Hausen [4] and Willmott and Hinchcliffe [11]. Instead, they specified that the fluid temperature in the boundary condition (6.22) for the solution of equation (6.20), varies linearly with time,

$$\frac{\partial t_f}{\partial \tau} = \pm R_\tau$$

where the plus sign refers to the hot period, the minus to the cold. Appropriate values of R_τ are presented graphically, having been estimated from a whole series of computational experiments using the 3-D model.

This R_τ factor is based on a consideration of *symmetric* regenerators alone, although it is suggested by Heggs and Carpenter that an average value of R_τ can be used for *balanced* regenerators. Clearly, the disadvantages of this approach lie in the facts that (1) R_τ must be interpolated from a set of graphs relating to the 3-D model of the *symmetric* regenerator only and (2) it cannot directly embody unequal periods, whereas Willmott and Hinchcliffe allow for different values q and q' (equations (6.23)) for the hot and cold periods respectively.

The difficulty with both approaches is that although both improve the accuracy of the 2-D model relative to the 3-D model, they suffer from instability. For example, for $\Phi = 0.2$, the Willmott and Hinchcliffe method is unstable. This appears to be because if

$$\mathrm{Bi}\Phi(\omega) < -1$$

at the start of a period when $\Phi(\omega) < 0$, then a negative lumped heat transfer coefficient, $\bar{\alpha}(\omega)$ is generated. Heggs and Carpenter suggest that their $\varepsilon(\omega)$ method can be used except when $\mathrm{Bi} > \Omega$, and except when Λ is small, presumably when "κ/κ_0 nonlinear effects" are manifest. Use of the Willmott and Hinchcliffe approach should be restricted to cases where $\Lambda > 4$ (approximately).

Evans [13] seeks to overcome these instability problems and to improve the accuracy of the 2-D $\Phi(\omega)$ model in the following way. He suggests the approach described below.

1. Use the 2-D model with the Hausen [4] time average Φ-factor to generate the variation with time of the fluid temperature, in the hot and cold periods, in the middle of the regenerator at periodic steady state.

2. The diffusion equation (6.20) is solved using these fluid temperatures in the boundary condition (6.22) to generate a time-varying $\Phi_E(\omega)$.
3. Insert this $\Phi_E(\omega)$ into the 2-D model and recompute the fluid temperatures as well as the solid temperatures. For Bi < 1, the solution obtained is adequate.
4. For larger values of Bi, it is proposed that stages 2 and 3 be repeated with the updated fluid temperatures. We start with an initial phi-factor $\Phi^{(1)}(\omega)$ and then generate a better $\Phi^{(2)}(\omega)$, which is then used to find an even better approximation $\Phi^{(3)}(\omega)$, and so on.

This method is superior to that of Willmott and Hinchcliffe but still suffers from instability problems when Bi > 5, when negative values of $\bar{\alpha}(\omega)$ can be generated. For Φ (time average) $= 0.2$, although the method is not unstable, it becomes inaccurate for $\Lambda > 4$. A table of values of Ψ, computed for $\bar{\Pi} = 1$ is given by Evans [13] and is reproduced here (Table 6.7). Tables 6.4, 6.5, and 6.6 were also computed by Evans.

For many applications, $\Phi > 0.26$ and it is recommended that the $\Phi_E(\omega)$ method developed by Evans be used where it is required that the *time-varying* performance within each period of operation be estimated. Where time average values only are required, it appears that the 2-D model with the Hausen Φ-factor will be sufficient except when the reduced length is small and when "κ/κ_0 nonlinear effects" may well be significant.

For such small regenerators, if the 2-D model using Φ or the time-varying $\Phi(\omega)$ factor is not sufficiently accurate, say when $\bar{\Lambda} < 4$, then the only option left is to resort to the full 3-D model. Fortunately, this situation

Table 6.7: Comparison between the values of Ψ computed as a function of $\bar{\Lambda}$ by Hinchcliffe [11] and Evans [13]

	Ψ (Hinchcliffe)	Ψ (Evans)	Ψ (Hinchcliffe)	Ψ (Evans)	Ψ (Hinchcliffe)	Ψ (Evans)
$\bar{\Lambda}$	$3\Phi = 0.9$		$3\Phi = 0.8$		$3\Phi = 0.6$	
1	1.1864	1.0022	1.3137	0.9973	Unstable	0.9108
2	1.0859	1.0058	1.1217	0.9884	Unstable	0.9152
4	1.0445	1.0182	1.0870	1.0020	Unstable	0.9125
8	1.0433	1.0296	1.0922	1.0123	Unstable	0.8853
10	1.0438	1.0308	1.0923	1.0144	Unstable	0.8640
20	1.0438	1.0408	1.0924	1.0164	Unstable	0.7201
30	1.0432	1.0486	1.0924	1.0182	Unstable	0.5391

does not arise in the majority of cases in practice. Even so, such cases may come up from time to time.

The great advantage of the chronologically varying $\Phi(\omega)$ factor is that it can be incorporated directly, if required, into nonlinear models of regenerators where the temperature dependence of the thermophysical properties involved can be represented.

6.12 The Effect of Longitudinal Thermal Conduction upon Thermal Regenerator Performance

Work in this area was developed separately but, nevertheless, extends the approach of Handley and Heggs [14] to the effect of longitudinal conduction for the single-blow problem, as described in Chapter 3. The essence of this work is to separate out transverse conduction in a direction perpendicular to gas flow from the axial or longitudinal conduction of heat in a direction parallel to gas flow. In the former case, this leads to the development of the lumped heat transfer coefficient, $\bar{\alpha}$, as given by equation (6.12). By treating axial and transverse conduction separately, we can retain the 2-D model with an additional term to represent axial conduction.

Following the analysis of equations (3.17) and (3.18) but for regenerators, the differential equations describing thermal performance of the regenerative heat exchanger take the form

$$\frac{\partial t_{s,m}}{\partial \eta} = t_f - t_{s,m} + \gamma \Lambda \frac{\partial^2 t_{s,m}}{\partial \xi^2} \tag{6.93}$$

$$\frac{\partial t_f}{\partial \xi} = t_{s,m} - t_f \tag{6.94}$$

The factor, γ, was proposed by Tipler [15] and is a measure of the potential effect of longitudinal conduction down the length of the regenerator. It is defined by

$$\gamma = \frac{\text{axial heat conduction}}{\text{heat input/extracted by the gas}} \tag{6.95}$$

The average temperature gradient down the length of the regenerator is given by

$$\frac{t_{f,in} - t_{f,x}}{L} \quad \text{(hot period)}$$

$$\frac{t'_{f,x} - t'_{f,in}}{L} \quad \text{(cold period)}$$

where $t_{f,x}$ and $t'_{f,x}$ are the time mean exit gas temperatures in the hot and cold periods respectively.

The volume of the packing of the regenerator is given by Aw where A is the available heating surface area and w is the semithickness of the plain-wall slabs that are assumed in this analysis. The effective cross-sectional area of the packing is thus Aw/L, with L equal to the length of the packing in the axial direction. The average rate of heat conduction is thus equal to

$$\frac{Aw\lambda_s(t_{f,in} - t_{f,x})}{L} \quad \text{(hot period)}$$

$$\frac{Aw\lambda_s(t'_{f,x} - t'_{f,in})}{L} \quad \text{(cold period)}$$

Now $\dot{m}_f c_p(t_{f,in} - t_{f,x})$ and $\dot{m}'_f c'_p(t'_{f,x} - t'_{f,in})$ are the average rates of heat input/extraction in the hot/cold periods respectively. These lead to the development of the factors γ and γ',

$$\gamma = \frac{Aw\lambda_s}{\dot{m}_f c_p L^2} \tag{6.96}$$

$$\gamma' = \frac{Aw\lambda_s}{\dot{m}'_f c'_p L^2} \tag{6.97}$$

These values of γ and γ' are applied within equation (6.93) for the hot and cold periods. The usual reversal conditions are applied which represent counterflow operation. The regenerator is then described by the parameters given in Table 6.8.

For the *symmetric* case with $\gamma = \gamma'$, Tipler [15] mentions the value of $\gamma = 10^{-2}$ in the case of regenerators used in gas turbines and considered the effect of axial conduction to be negligible. Willmott [16] obtained a value of $\gamma = 10^{-5}$ for a medium-size Cowper stove and considered, therefore, that longitudinal conduction effects could be ignored in this case.

Table 6.8: Parameters for 2-D model with axial conduction

Period	Reduced length	Reduced period	Conduction parameter
Hot	Λ or $\bar{\Lambda}$	Π or $\bar{\Pi}$	γ
Cold	Λ' or $\bar{\Lambda}'$	Π' or $\bar{\Pi}'$	γ'

The name *conduction parameter* was ascribed to the factor γ in the now classic paper by Bahnke and Howard [17]. They suggested that if $\Delta E/E$ is defined for the *balanced* case, where $\eta_{REG} = \eta'_{REG}$, in the following manner:

$$\frac{\Delta E}{E} = \frac{\eta_{REG}(\text{no conduction}) - \eta_{REG}(\text{conduction})}{\eta_{REG}(\text{no conduction})}$$

then the decrease in thermal effectiveness caused by axial conduction will vary directly with the conduction parameter, that is

$$\frac{\Delta E}{E} = \gamma \tag{6.98}$$

and that this "provides a very good engineering approximation" for $\Lambda > 20$ and $\gamma < 0.1$. For rapidly reversing regenerators, where $\Pi \rightarrow 0$, it can be shown that for the symmetric case, $\Delta E/E < \gamma$ for $\Lambda > 5$ using the formula of Hahnemann [18] for the thermal ratio.

Sadly, the Bahnke and Howard paper lacks a formal *mathematical* model describing regenerative heat exchanger behavior with axial conduction. In particular, the differential equation (6.93) with the term $\gamma\Lambda(\partial^2 t_{s,m}/\partial\xi^2)$ is not mentioned. From this equation, the first of Bahnke and Howard's results can be anticipated, namely $\Delta E/E$ increases in magnitude with the conduction factor γ and the reduced length Λ.

The differential equation (6.93) does not involve the cross-sectional area of the packing, except within the factor γ. Bahnke and Howard's observation follows, namely that this cross-sectional area appears to have little other influence upon the effect of axial conduction upon regenerator performance.

Finally, for $\Lambda/\Pi \geq 10$, the effect of axial conduction becomes equal to that forecast by Hahnemann for infinitely short cycle times. Readers are referred to this paper for further details. It is interesting to note that Hahnemann described γL as the *axial conduction factor*.

Of considerable interest is the contribution of J. R. Mondt in the discussion reported at the end of the paper by Bahnke and Howard. Mondt pointed out that work in the 1940s and 1950s explored the use of regenerators to improve the overall plant efficiency of various gas turbine arrangements. High efficiencies ($\eta_{REG} > 0.95$) are required, as is the case in cryogenic plant applications. In order to realize these, large heating surface areas are employed, leading to large values of reduced length Λ. However, beyond a value of $\Lambda > 40$ (approximately), further increases in heating surface area can lead to a *decrease* in thermal ratio, as a consequence of axial heat conduction. In other words, even with modest values of γ, the effect of the conduction parameter, γ, is amplified by increasing values of reduced length, Λ, as would be expected from the differential equation (6.93).

There is a problem associated with equation (6.93). The term $\gamma\Lambda$ conceals the issue that if Λ is increased by extending the value of the length, L of the regenerator, the value of γ is increased also. Hahnemann [18] overcomes this by introducing Φ_H, his *heat resistance parameter* where

$$\Phi_H = \frac{\Lambda}{\gamma} = \frac{\alpha L^2}{w\lambda_s}$$

Φ_H incorporates any increase in reduced length, Λ, effected by extending the length of the regenerator. This will be seen by comparing Figure 6.11 with 6.12. In the former, for a fixed value of γ, the thermal ratio attains an asymptotic value. This asymptote decreases with γ. In Figure 6.12, on the other hand, the factor Φ_H incorporates the effect of any increase in the length of the regenerator, L, upon Λ *and* γ and it will be seen that the thermal ratio, η_{REG} declines for $\Lambda > 40$ approximately, due to the effect

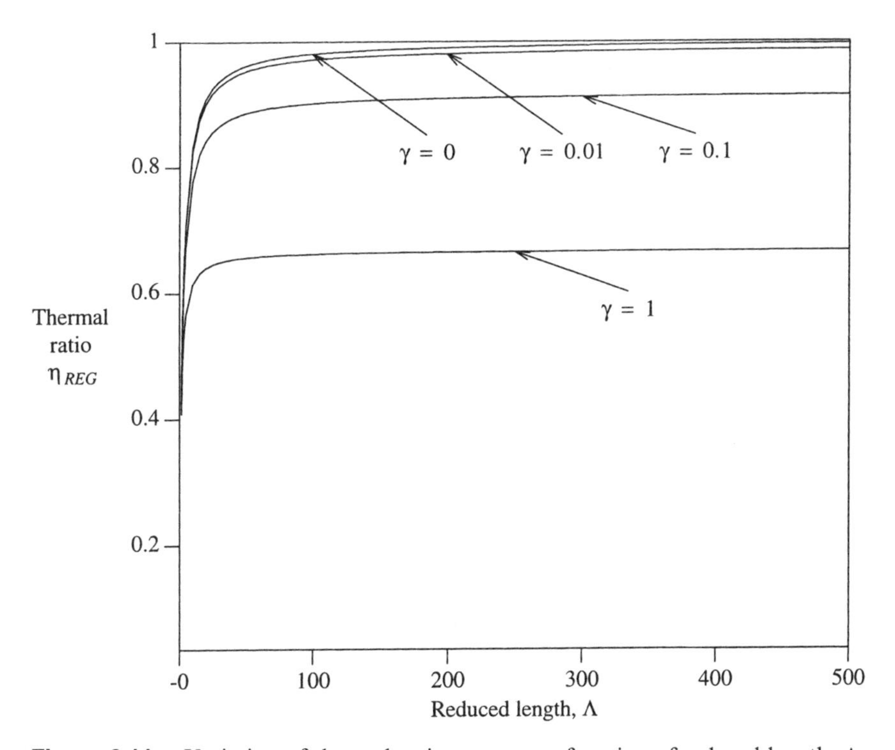

Figure 6.11: Variation of thermal ratio, η_{REG}, as a function of reduced length, Λ, for different axial conduction factors, γ, for $\Pi \to 0$ using Hahnemann's formula for the symmetric case.

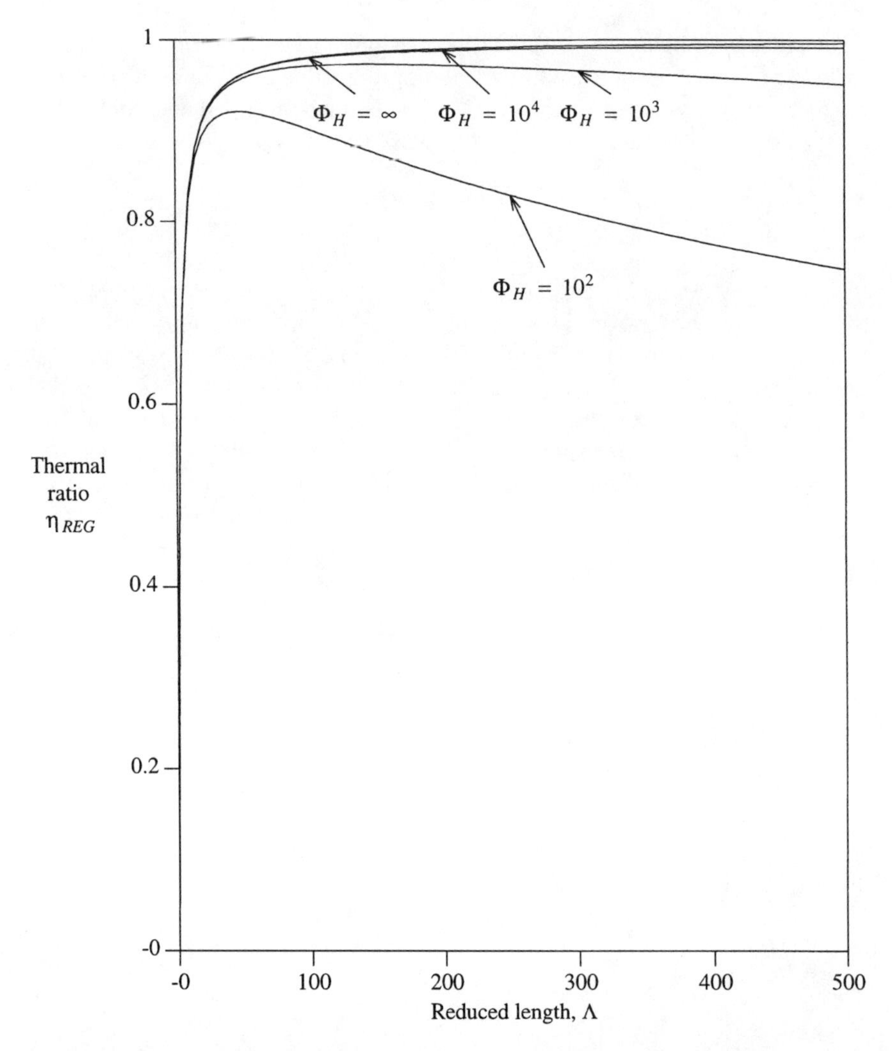

Figure 6.12: Variation of thermal ratio, η_{REG}, as a function of reduced length, Λ, for different axial conduction factors, Φ_H, for $\Pi \to 0$ using Hahnemann's formula for the symmetric case.

of longitudinal conduction. This observation was made by Hahnemann for very rapidly reversing regenerators, and Figures 6.11 and 6.12 are based on his formula for the symmetric case.

Moreover, Kulakowski [19] mentions that in order to reduce the effect of axial conduction in cryogenic and low-temperature systems, where the dimensionless length, Λ, is large, the packing of the regenerator is divided into a stack of whisker disks with as little as possible thermal contact between adjacent disks in the axial direction.

Equally interesting is the development of Mo_0, equivalent to Φ_H, the so-called *Mondt number* [20] where $Mo_0 = \Lambda/2\gamma$ for the *symmetric* case. Likewise, this focuses upon the relationship between γ and Λ on the effect of axial conduction. The graph offered in the discussion of the paper by Bahnke and Howard is equivalent to our Figure 6.12. Shah [21] offers a correlation that attempts to summarize the tables and graphs presented by Bahnke and Howard.

6.13 Concluding Remarks

This chapter has focused largely upon the effect of transverse or latitudinal conduction within the solid packing of a regenerator, and the representation of this within a lumped heat transfer coefficient, $\bar{\alpha}$. The use of a time-varying $\Phi(\omega)$ within a time-varying $\bar{\alpha}(\omega)$ is explored. Finally, the effect of axial conduction or latitudinal conduction is discussed.

References

1. H. Hausen, "Berechnung der Steintemperatur in Winderhitzen," *Arch. Eisenhüttenwes.* **10**, 473–480 (1938).
2. K. Rummel, "The Calculation of the Thermal Characteristics of Regenerators," *J. Inst. Fuel* **2**, 160–175 (Feb. 1931) (Melchett Lecture and Medal Presentation).
3. A. Schack, "Die Berechnung der Regeneratoren (The Calculation of Regenerators)," *Arch. Eisenhüttenwes.* **5/6**, 101–118 (Nov./Dec. 1943).
4. H. Hausen, "Vervollstandigte Berechnung des Warmeaustausches in Regeneratoren (Improved Calculations for Heat Transfer in Regenerators)," *Z. VDI-Beih. Verfahrenstech.* **2**, 31–43 (1942) (Iron and Steel Institute translation, June 1943).
5. J. Crank, P. Nicolson, "A Practical Method for Numerical Evaluation of Solutions of Partial Differential Equations of the Heat-Conduction Type," *Proc. Camb. Phil. Soc.* **43**, 50–67 (1947).
6. H. Hausen, *Heat Transfer in Counterflow, Parallel Flow and Crossflow*, English translation (1983) edited by A. J. Willmott, McGraw-Hill, New York (originally published 1976).
7. B. Stuke, "Berechnung des Wärmeaustausches in Regeneratoren mit zylindrischem oder kugelförmigem Füllmaterial," *Angew. Chemie* **B20**, 262–268 (1948).

8. P. Razelos, A. Lazaridis, "A Lumped Heat-Transfer Coefficient for Periodically Heated Hollow Cylinders," *Int. J. Heat Mass Transfer* **10**, 1373–1387 (1967).
9. P. Butterfield, J. S. Schofield, P. A. Young, "Hot Blast Stoves (Part II)," *J. Iron Steel Inst.* **201**, 497–508 (June 1963).
10. F. W. Schmidt, A. J. Willmott, *Thermal Energy Storage and Regeneration*, McGraw-Hill, New York (1981).
11. C. Hinchcliffe, A. J. Willmott, "Lumped Heat Transfer Coefficients for Thermal Regenerators," *Int. J. Heat Mass Transfer* **24**, 1229–1236 (Feb. 1981).
12. P. J. Heggs, K. J. Carpenter, "A Modification of the Thermal Regenerator Infinite Conduction Model to Predict the Effects of Intraconduction," *Trans. I. Chem. E.* **57**, 228–236 (1979).
13. D. J. Evans, "Non-linear Modelling of Regenerative Heat Exchangers," D.Phil. thesis, University of York (Mar. 1997).
14. D. Handley, P. J. Heggs, "The Effect of Thermal Conductivity of the Packing Material on Transient Heat Transfer in a Fixed Bed," *Int. J. Heat Mass Transfer* **12**, 549–570 (1969).
15. W. Tipler, *A Simple Theory of Heat Exchanger*, Shell Technical Report ICT/14, London (1947).
16. A. J. Willmott, "The Regenerative Heat Exchanger Computer Representation," *Int. J. Heat Mass Transfer* **12**, 997–1014 (1969).
17. G. D. Bahnke, C. P. Howard, "The Effect of Longitudinal Heat Conduction on Periodic-Flow Heat Exchanger Performance," *ASME Trans. Ser. A, J. Eng. Power* (Apr. 1964).
18. H. W. Hahnemann, "Approximate Calculation of Thermal Ratios in Heat Exchangers including Heat Conduction in Direction of Flow," *Nat. Gas Turbine Estab.* memo. M 36 (Sept. 1948).
19. B. T. Kulakowski, "Advances in Regenerator Technology," in R. K. Shah, B. T. Kulakowski, A. J. Willmott, P. J. Heggs, I. L. Maclaine-Cross, "Advances in Regenerator Design Theory and Technology," *ASME Winter Annual Meeting*, Boston (Nov. 1983).
20. J. R. Mondt, "Correlation of the Effect of Longitudinal Heat Conduction on Exchanger Performance," General Motors Research Laboratories, Warren, MI, USA GMR (405) (1964).
21. R. K. Shah, "A Correlation for Longitudinal Heat Conduction Effects in Periodic-Flow Heat Exchangers," *J. Eng. Power* **97A**, 453–454 (July 1975).

Chapter 7

INTEGRAL EQUATION METHODS FOR MODELING COUNTERFLOW REGENERATORS

7.1 Introduction

It is impossible, in a text such as this, to avoid a discussion of the methods that involve integral equations for obtaining the cyclic steady-state solution of the counterflow regenerator problem. These were developed first by Nusselt [1, 2]. The spatial temperature distributions $F(\xi)$ and $F'(\xi')$ within the solid packing at the start of the hot and cold periods respectively are defined formally within the body of this chapter. Nusselt established relations of the essential form

$$F(\xi) = f_1(F'(\xi')) \qquad F'(\xi') = f_2(F(\xi))$$

where f_1 and f_2 are functions representing the hot and cold periods and which involve the integrals embodied in Nusselt's equations. The equations above can be regarded as a pair of simultaneous equations, the solutions to which are the temperature distributions $F(\xi)$ and $F'(\xi')$. In this way, a *closed* method is developed so that the cycling of the model to dynamic equilibrium is avoided. Moreover, the calculation of the temperature distributions *within* either the hot or cold period of regenerator operation is similarly avoided: only $F(\xi)$ and $F'(\xi')$ are calculated.

In the 1920s, when Nusselt established this approach, the realization of such computational economies must have seemed a desirable proposition at a time some 30 years before digital computers became generally available. But even after the advent of electronic computing machines, these integral equations have attracted a good deal of interest from workers in the field.

Initially, different methods were proposed to solve these integral equations and one of these is set out in Chapter 5. It soon became apparent, however, that some of these methods were not as problem-free as perhaps Nusselt might have hoped. More recently, attention has focused upon two issues. Firstly, robust methods have been sought that do not break down in certain circumstances. Secondly, and remarkably enough, bearing in mind Nusselt's implicit reasons for considering the use of these integral equations, means have been sought that minimize the computational effort required to find $F(\xi)$ and $F'(\xi')$, even when computers are used! For example, the number of equations that must be handled can be reduced if $F(\xi)$ is found by solving

$$F(\xi) - f_1(f_2(F(\xi))) = 0$$

and $F'(\xi')$ is located by using

$$F'(\xi') = f_2(F(\xi))$$

What we seek to do in this chapter is to set Nusselt's equations and their solution within the framework of integral equations in general. In so doing, a means is afforded whereby the relationship between the methods can be better understood.

7.2 Initial Considerations

The integral equation methods are based on the equations developed by Nusselt [1, 2] from the differential equations (7.1) and (7.2). One of these methods, namely that of Iliffe [3] was described in Chapter 5.

$$\frac{\partial T_f}{\partial \xi} = T_s - T_f \tag{7.1}$$

$$\frac{\partial T_s}{\partial \eta} = T_f - T_s \tag{7.2}$$

The dimensionless temperature scale, T, is

$$T = \frac{t - t'_{f,in}}{t_{f,in} - t'_{f,in}} \tag{7.3}$$

It will be recalled that it is useful to introduce $F(\xi)$ and $F'(\xi')$, the spatial temperature distributions in the solid packing of the regenerator at the start of the hot and cold periods respectively. They are formally defined by

$$F(\xi) = T_s(\xi, 0) \tag{7.4}$$

$$F'(\xi') = T_s'(\xi', 0) \tag{7.5}$$

The original Nusselt [2] integral equations are greatly simplified by the introduction of the dimensionless temperature scale (7.3) and for the general case, take the form

$$F'(\xi') = e^{-\Pi'} F(\xi') + \int_{\xi'}^{\Lambda'} K'(\varepsilon - \xi') F(\varepsilon) \, d\varepsilon \tag{7.6}$$

$$1 - F(\xi) = e^{-\Pi}[1 - F'(\xi)] + \int_0^{\xi} K(\xi - \varepsilon)[1 - F'(\varepsilon)] \, d\varepsilon \tag{7.7}$$

The *kernel* of these integral equations is defined by

$$K(x) = \frac{-iJ_1(2i\,(x\Pi)^{1/2})}{(x\Pi)^{1/2}} \Pi \, e^{-x-\Pi} \tag{7.8}$$

where $iJ_1(iy)$ is a real-valued function with complex argument iy, where $i^2 = -1$ and J_1 is the Bessel function of the first type and of first order. The function $iJ_1(iy)$ is frequently called the *modified* Bessel function. It is this kernel that is the source of the many problems that successive authors have sought to overcome and that will be outlined in this chapter.

Equation (7.6) assumes that the gas flows through the packing in the cold period with the gas entering at $\xi' = \Lambda'$ and departing at $\xi' = 0$. Equation (7.7) assumes that the gas flows through the regenerator in the contraflow direction, entering at $\xi = 0$ and leaving at $\xi = \Lambda$ in the hot period. The equations are presented deliberately in this way for two reasons. Firstly, both directions ξ and ξ' are measured from the entrance of the *hot* gas to the packing and this simplifies the reversal conditions which then take the form

$$T(\xi, 0) = T'\left(\frac{\xi\Lambda'}{\Lambda}, \Pi'\right) \quad T'(\xi', 0) = T\left(\frac{\xi'\Lambda}{\Lambda'}, \Pi\right) \tag{7.9}$$

Secondly, the reader will be made aware of the two forms of the equations that appear in the literature. The other form, set out in Chapter 5, as used by Iliffe [3], sets the gas entering, in the cold period, at $\xi' = 0$ and departing at $\xi' = \Lambda'$. This is shown in equations (5.19) and (5.20). Both forms are valid, although those used in equations (7.6) and (7.7) have the advantage, as has been pointed out, that the directions ξ and ξ' are the same and have the same physical origin.

For the *symmetric* case, where

$$\frac{\Lambda'}{\Lambda} = \frac{\Pi'}{\Pi} = 1$$

the pair of equations (7.6) and (7.7) reduce to the single equation

$$F(\Lambda - \xi) + e^{-\Pi} F(\xi) + \int_0^{\xi} K(\xi - \varepsilon) F(\varepsilon) \, d\varepsilon = 1 \tag{7.10}$$

Equations (7.6) and (7.7) deal with the general nonsymmetric case where $\Lambda \neq \Lambda'$ and/or $\Pi \neq \Pi'$. Equation (7.10) exploits the symmetry, for which it can be shown that

$$F'(\Lambda - \xi) + F(\xi) = 1 \tag{7.11}$$

It turns out that, without loss of generality, it is sufficient to discuss the problems associated with the attempted solution of these integral equations in terms of the symmetric case alone, that is, in terms of equation (7.10).

These integral equations are Volterra equations of the second kind. Baker [4] in his now classic text, indicates that the numerical techniques most favored for such equations fall into two classes. There are the *quadrature* methods, such as that of Iliffe [3], in which the integral in equation (7.10)

$$\int_0^{\xi} K(\xi - \varepsilon) F(\varepsilon) \, d\varepsilon$$

is approximated by one of the Newton–Cotes methods (see Baker [4] or Delves and Mohamed [5]) and the solution of this integral equation consists of the vector $\boldsymbol{F}$ where $\boldsymbol{F} = [F_0, F_1, F_2, \ldots, F_n]^T$ and the F_k's are temperatures at the entrance and exit to the regenerator, and at equally spaced, intermediate positions. This method is described in detail in Chapter 5 with the Newton–Cotes formulae used being Simpson's rules.

On the other hand, the *series expansion* methods are those in which we seek to represent $F(\xi)$ in equation (7.10) by

$$F(\xi) = \sum_{j=0}^{n} \alpha_j \, \phi_j(\xi) \tag{7.12}$$

The solution consists of the vector $\boldsymbol{\alpha}$ where $\boldsymbol{\alpha} = [\alpha_0, \alpha_1, \alpha_2, \ldots, \alpha_n]^T$. The series expansion (7.12) embodies a set of *linearly independent* functions of degree k, $\{\phi_k(\xi) \mid k = 0, 1, 2, \ldots, n\}$. The first published paper using a series expansion is that of Nahavandi and Weinstein [6]. They employed the simple function $\phi_k(\xi) = \xi^k$.

Substitution of expansion (7.12) into equation (7.10) yields

$$\sum_{j=0}^{n} \left(\phi_j(\Lambda - \xi) + e^{-\Pi} \phi_j(\xi) + \int_0^{\xi} K(\xi - \varepsilon)\, \phi_j(\varepsilon)\, d\varepsilon \right) = 1 \tag{7.13}$$

In the *collocation* method, we apply equation (7.13) at $n + 1$ *distinct* but not necessarily equally spaced positions $\{\xi_k \mid k = 0, 1, 2, \ldots, n\}$ yielding the matrix equation

$$A\boldsymbol{\alpha} = \boldsymbol{e} \tag{7.14}$$

where $\boldsymbol{e} = [1, 1, \ldots, 1]^T$. The matrix $A = [a_{i,j}]$ is defined by

$$a_{i,j} = \phi_j(\Lambda - \xi_i) + e^{-\Pi} \phi_j(\xi_i) + \int_0^{\xi_i} K(\xi_i - \varepsilon)\, \phi_j(\varepsilon)\, d\varepsilon$$

The Nahavandi and Weinstein [6] procedure utilized this collocation approach.

Equation (7.13) can be written in more general terms as

$$\Omega(\Psi(\xi)) = 0 \tag{7.15}$$

where $\Psi(\xi)$ is an approximation of $F(\xi)$ and is given by

$$\Psi(\xi) = \sum_{j=0}^{n} \alpha_j \phi_j(\xi) \tag{7.16}$$

Equation (7.15) would be precise for $0 \leq \xi \leq \Lambda$ if $\Psi(\xi)$ were an exact solution, where it is, in fact, only an approximate solution. The method of collocation, outlined above, is one whereby the coefficients in the vector $\boldsymbol{\alpha} = [\alpha_0, \alpha_1, \alpha_2, \ldots, \alpha_n]^T$ are determined by requiring that

$$\Omega(\Psi(\xi_k)) = 0 \tag{7.17}$$

at $n + 1$ distinct collocation points $\{\xi_k \mid k = 0, 1, 2, \ldots, n\}$ all on the interval $0 \leq \xi_k \leq \Lambda$. It is assumed, implicitly, that $|\Omega(\Psi(\xi))|$ is small at all other positions on the $[0, \Lambda]$ interval.

In the Galerkin method, on the other hand, we seek to *minimize* $\Omega(\Psi(\xi))$ on the interval $0 \leq \xi \leq \Lambda$. In this approach, the vector $\boldsymbol{\alpha}$ is obtained by solving the equations

$$\int_0^\Lambda \Omega(\Psi(\xi))\, \phi_j(\xi)\, d\xi = 0 \tag{7.18}$$

for $j = 0, 1, 2, \ldots, n$.

A good deal of discussion has arisen as to the choice of functions $\{\phi_k(\xi) \mid k = 0, 1, \ldots, n\}$. It will be seen later that a number of difficulties are overcome if an appropriate choice is made.

7.3 Difficulties with the Quadrature Methods

Historically, equation (7.10),

$$F(\Lambda - \xi) + e^{-\Pi} F(\xi) + \int_0^\xi K(\xi - \varepsilon) F(\varepsilon)\, d\varepsilon = 1 \tag{7.10}$$

appeared to be so innocuous. Such an equation should be easy to solve using a quadrature method. Indeed, this is just the approach proposed by Iliffe [3] in 1948, as described in Chapter 5. Here, the integral

$$\int_0^{k\Delta\xi} K(\xi - \varepsilon)\, F(\varepsilon)\, d\varepsilon \tag{7.19}$$

in equation (7.10) is approximated by the composite Simpson's rules. It has been shown that by applying these approximations at $k = 0, 1, 2, \ldots, n$, where $n\Delta\xi = \Lambda$, along the dimensionless length of the regenerator, a set of simultaneous equations

$$(P + Q + R)\boldsymbol{F} = \boldsymbol{e} \tag{7.20}$$

is yielded where the matrix P is given by equation (5.29) and, as previously, with $Q = e^{-\Pi} I$, where I is the unit matrix. $R = [r_{i,j}]$ with $r_{j,n-j} = 1$ for $j = 0, 1, 2, \ldots, n$, and all other elements of R equal to zero. Again, $\boldsymbol{e} = [1, 1, \ldots, 1]^T$. The successful application of this technique depends upon the matrix $(P + Q + R)$ being nonsingular, that is the determinant $|P + Q + R|$ being nonzero.

Sadly, despite the simplicity of this approach to the solution of equation (7.10), it has been found that the Iliffe [3] method breaks down when the reduced length Λ is large and the ratio $\Lambda/\Pi > 10$, the so-called *long regenerator problem*, in the manner described by Willmott and Thomas [7]. The matrix $(P + Q + R)$ approaches a singular state and the matrix equation becomes difficult to solve accurately unless the number, $n + 1$, of simultaneous equations is made very large. It turns out that the collocation method, and other series expansion techniques, do not break down, once certain precautions are taken in the calculation of the matrix elements $a_{i,j}$ for equation (7.14).

Figure 7.1 shows the variation of the kernel function, $K(\Lambda - \xi)$, with ξ for $\Lambda = 30$. It will be seen immediately that K_s is very small for large s, large Λ and small Π. As a consequence, in these circumstances, many zero or almost zero elements are introduced into the bottom rows of the matrix P. One way of looking at this is to say that, unless n is large, an insufficient number of large enough elements involving K_s with small s will be introduced into the bottom rows of the matrix P causing the matrix $P + Q + R$ to approach a singular state. This effect was observed by Willmott and Thomas [7] and was described in detail by them.

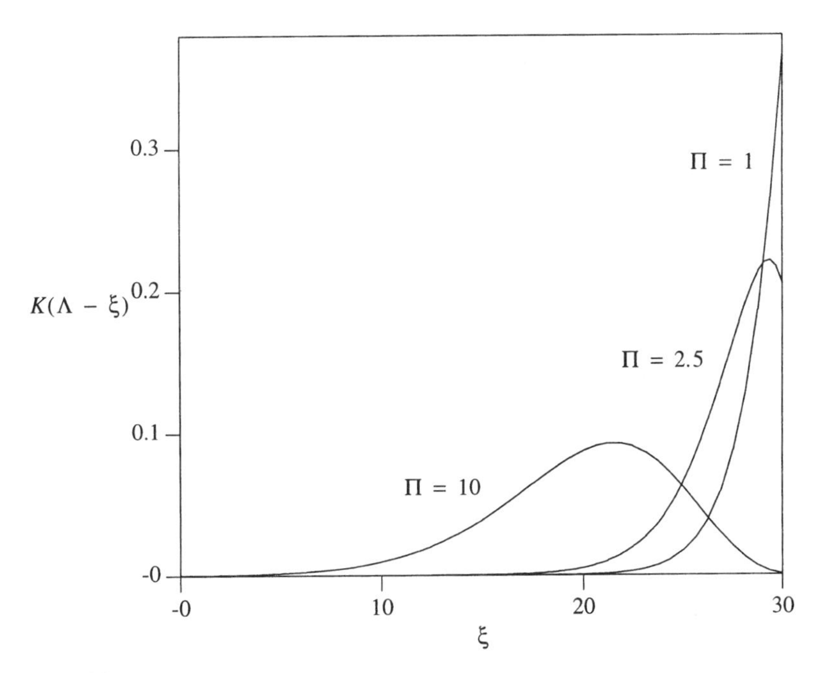

Figure 7.1: Variation of the kernel $K(\Lambda - \xi)$ with ξ for $\Lambda = 30$.

Another way of looking at the same problem is to say that it is difficult to evaluate the integral

$$\int_{0}^{k\Delta\xi} K(\xi - \varepsilon)\, F(\varepsilon)\, d\varepsilon$$

by numerical quadrature for larger values of k, if the majority of the data points involve values of K_{k-j} that are very small and *unrepresentative* of $K(\Lambda - \xi)$ over the complete range $0 \leq \xi \leq \Lambda$ under consideration. This basic flaw in the method of Iliffe [3] is implicit in any other quadrature method, using, say, a higher-order Newton–Cotes formula to approximate the integral (7.18). It makes their application difficult to all long regenerator problems. This flaw only became apparent when the method had been programmed for a computer and it was then possible to submit the method to close scrutiny over large ranges of Λ and Π.

Although this method is seemingly so straightforward, it involved Iliffe in a good deal of hand calculation (computers were simply not generally available when this work was done in the mid-1940s) and Iliffe mentions the use of "five figure logarithms to evaluate $K(\xi - \varepsilon)$ . . . and a twenty inch slide rule thereafter." Iliffe indicated that the number of hours required to solve equation (7.20) was equal to the number $n + 1$ of simultaneous equations. Workers such as Iliffe simply could only deal with the limited number of cases that were of special interest.

7.4 Difficulties with the Early Series Expansion Methods

Equation (7.10) involves the integral

$$\int_{0}^{\xi} K(\xi - \varepsilon) F(\varepsilon)\, d\varepsilon$$

and if this integral could be retained in some way, without breaking it down into a weighted sum of terms, some of which might be very small or zero, the problems associated with the quadrature techniques might be avoided. This becomes clear from consideration of Figure 7.1. The area under the $K(\Lambda - \xi)$ curve for $\Lambda = 30$ and $\Pi = 1$ is most certainly far from small, despite the fact that $K(\Lambda - \xi)$ appears to be very small for $0 \leq \xi < 20$.

It is convenient to examine this proposition by examining the method of *Nahavandi and Weinstein* [6]. It is most unlikely that Nahavandi and Weinstein were aware of the problems with the method of Iliffe outlined above. They simply developed a series expansion method in which they set $\{\phi_k(\xi) = \xi^k \mid k = 0, 1, 2, \ldots, n\}$ for equation (7.12). Equation (7.13) becomes

$$\sum_{j=0}^{n} \alpha_j \left((\Lambda - \xi)^j + e^{-\Pi} \xi^j + \int_0^{\xi} K(\xi - \varepsilon)\, \varepsilon^j \, d\varepsilon \right) = 1 \tag{7.20}$$

In this collocation scheme, equation (7.20) is applied at $n + 1$ distinct and equally spaced positions $\{\xi_k \mid k = 0, 1, 2, \ldots, n\}$ where $\xi_k = k\Delta\xi$ and $n\Delta\xi = \Lambda$. Nahavandi and Weinstein applied this scheme to the general case defined by equations (7.6) and (7.7). Without loss of generality in this context, the scheme for symmetric case embodied in equation (7.10) can be considered. In so doing, it is possible to derive the equation (7.14) but where, in this case, the matrix $A = [a_{i,j}]$ is defined by

$$a_{i,j} = (\Lambda - \xi_i)^j + e^{-\Pi} \xi_i^j + \int_0^{\xi_i} K(\xi_i - \varepsilon)\, \varepsilon^j \, d\varepsilon$$

The obvious problem is the calculation of the integral

$$\int_0^{\xi_i} K(\xi_i - \varepsilon)\, \varepsilon^j \, d\varepsilon$$

for each matrix element $a_{i,j}$. Nahavandi and Weinstein [6] did not reveal how they achieved this, but one might assume that the integral was calculated by numerical quadrature.

Nevertheless, putting this difficulty aside, the method seemed ideal to overcome the problems associated with Iliffe's [3] method. If only this problem were that simple to solve! Willmott and Duggan [8] and subsequently Baclic [9] pointed out that very large matrix elements

$$a_{i,j} = (\Lambda - \xi_i)^j + e^{-\Pi} \xi_i^j + \int_0^{\xi_i} K(\xi_i - \varepsilon)\, \varepsilon^j \, d\varepsilon$$

can arise for large values of reduced length, Λ, even for modestly large values of n, the maximum degree of the polynomial ξ^j used. These large elements can result in the process of solving the linear equations (7.14) breaking down, thereby limiting the possible range of application of the Nahavandi and Weinstein method.

Baclic [9] and Hill [10] both suggested employing linearly independent functions which attempted to avoid the generation of such large matrix elements. Baclic put forward the use of $\phi_k(\xi) = \xi^k/k!$. It is fairly easy to see, however, that for

$$\xi > (k!)^{1/k}$$

$\xi^k/k!$ rapidly increases with ξ, although not as quickly as ξ^k. It is worth noting that $(k!)^{1/k} \approx 3$ for $k = 6$ and ≈ 4.5 for $k = 10$. Large matrix elements $a_{i,j}$ can still be generated. It is easy to see with hindsight that this problem is overcome if we choose linearly independent functions $\{\phi_k(\xi) \mid k = 0, 1, 2, \ldots, n\}$ such that

$$|\phi_k(\xi)| < 1$$

for $0 \leq \xi \leq \Lambda$. Hill [10] introduced the idea of using the Chebyshev polynomials, that is, with $\phi_k(\xi) = T_k(2\xi/\Lambda - 1)$. Hill indicated that large values of reduced length, Λ, did not cause the collocation method to break down with this use of the Chebyshev polynomials. By way of example, he computed successfully the vector $\boldsymbol{\alpha}$ for the symmetric case with $\Lambda = 100$ and $\Pi = 5$ using a polynomial of degree 20. An exploration of this approach was undertaken by Willmott and Knight [11]. The Chebyshev polynomials $T_k(x)$ possess the property that $|T_k(x)| \leq 1$ and can be defined by the recurrence

$$T_{k+1}(x) = 2x\,T_k(x) - T_{k-1}(x)$$

with $T_0(x) = 1$ and $T_1(x) = x$.

The Legendre polynomials possess similar properties to those of the Chebyshev polynomials, in particular the avoiding of large matrix elements $a_{i,j}$. These Legendre polynomials, which are introduced below, have additional properties that make them most convenient to use in the solution of these integral equations.

7.5 Legendre Series Expansion Methods

One underlying property of the Legendre polynomials, $\{P_j(x) \mid j = 0, 1, 2, \ldots\}$, is that of *orthogonality*, that is,

$$\int_{-1}^{+1} P_i(x)P_j(x)\,dx = 0 \quad \text{for} \quad i \neq j \tag{7.21}$$

and

$$\int_{-1}^{+1} P_j(x)^2\,dx = \frac{2}{2j+1} \neq 0 \tag{7.22}$$

The functions $\{P_j(x) \mid j = 0, 1, 2, \ldots\}$ are polynomials of degree j of the independent variable x where $-1 \leq x \leq +1$. They possess the further additional property that they can be defined in terms of a recurrence relationship,

$$P_{j+1}(x) = \frac{2j+1}{j+1} xP_j(x) - \frac{j}{j+1} P_{j-1}(x) \tag{7.23}$$

with $P_0(x) = 1$ and $P_1(x) = x$. The polynomials possess the important property

$$|P_j(x)| \leq 1 \tag{7.24}$$

pointing to the way of avoiding excessively large matrix elements in the solution of the integral equations by a series expansion method.

The Legendre polynomials are defined in terms of the variable x on the interval $-1 \leq x \leq +1$. The regenerator model operates with the variable ξ on the interval $0 \leq \xi \leq \Lambda$. This is overcome by use of the change of variable

$$x = \frac{2\xi - \Lambda}{\Lambda}$$

It will be seen that when $\xi = 0$, then $x = -1$, whereas when $\xi = \Lambda$, then $x = +1$. In order to simplify the notation, we introduce the polynomials $\{Q_j(\xi) \mid j = 0, 1, 2, \ldots, n\}$ where

$$Q_j(\xi) = P_j\left(\frac{2\xi - \Lambda}{\Lambda}\right)$$

so that the orthogonality relation (7.21), for example, takes the form

$$\frac{2}{\Lambda} \int_0^\Lambda Q_i(\xi) Q_j(\xi)\, d\xi = 0 \quad \text{for} \quad i \neq j \tag{7.25}$$

and the series expansion (7.16) is given by

$$\Psi(\xi) = \sum_{j=0}^{n} \alpha_j Q_j(\xi) \tag{7.26}$$

There is a further property that follows from the orthogonality property (7.21) for the case where $j = 0$. Because

$$P_0(x) = Q_0(\xi) = 1$$

it follows that

$$\frac{2}{\Lambda}\int_0^{\Lambda} Q_i(\xi)\,d\xi = 0 \quad \text{for} \quad i \geq 1 \tag{7.27}$$

The recurrence relationship for the $\{Q_j(\xi) \mid j = 0, 1, 2, \ldots\}$ polynomials takes the form based on equation (7.23), namely

$$Q_{j+1}(\xi) = \frac{2j+1}{j+1}\frac{2\Lambda - \xi}{\Lambda} Q_j(\xi) - \frac{j}{j+1} Q_{j-1}(\xi) \tag{7.28}$$

with $Q_0(\xi) = 1$ and $Q_1(\xi) = (2\Lambda - \xi)/\Lambda$.

The *thermal ratio*, η_{REG}, is greatly simplified if the series solution to the integral equation takes the form given by equation (7.26). The general form of η_{REG}, for the symmetric case, is given by

$$\eta_{REG} = \frac{1}{\Pi}\int_0^{\Lambda} (F(\xi) + F'(\Lambda - \xi))\,d\xi \tag{7.29}$$

Recalling equation (7.11),

$$F'(\Lambda - \xi) + F(\xi) = 1 \tag{7.11}$$

equation (7.29) becomes

$$\eta_{REG} = \frac{1}{\Pi}\int_0^{\Lambda} (2F(\xi) - 1)\,d\xi \tag{7.30}$$

Inserting the approximation (7.26), equation (7.30) becomes

$$\eta_{REG} = \frac{1}{\Pi}\int_0^{\Lambda} \left(2\sum_{k=0}^{n} \alpha_k\, Q_k(\xi) - 1\right) d\xi \tag{7.31}$$

It will be observed that the last n terms of the summation within the integral disappear by virtue of the orthogonality of the $Q_k(\xi)$ polynomials. As a consequence, it is easy to demonstrate that the thermal ratio can be calculated using the value of α_0 alone with

$$\eta_{REG} = \frac{\Lambda}{\Pi}(2\alpha_0 - 1) \tag{7.32}$$

This relation is applicable if the coefficients $\{\alpha_j \mid j = 0, 1, 2, \ldots, n\}$ are calculated by the method of collocation, by the method of Galerkin, or by any other means. The simplicity of equation (7.32) is a function only of the properties of the Legendre polynomials. It turns out that an equally simple form emerges for the nonsymmetric case.

7.6 The Choice of Data Points

Parallels can be drawn between the method of collocation for the solution of integral equations and the method of *interpolation* for the approximation of one function by another. Again, without loss of generality, the symmetric case is considered and the function Ω is given by

$$\Omega(F(\xi)) = F(\Lambda - \xi) + e^{-\Pi} F(\xi) + \int_0^{\xi} K(\xi - \varepsilon) F(\varepsilon) \, d\varepsilon - 1 \tag{7.33}$$

If $F(\xi)$ is an exact solution of equation (7.10), then $\Omega(F(\xi)) = 0$ for $0 \leq \xi \leq \Lambda$. In the case, however, where $F(\xi)$ is *estimated*, namely

$$F(\xi) \approx \Psi(\xi) = \sum_{j=0}^{n} \alpha_j Q_j(\xi) \tag{7.34}$$

a *residual* function $R(\xi)$ is left where

$$R(\xi) = \Omega\left(\sum_{j=0}^{n} \alpha_j \phi_j(\xi)\right) \tag{7.35}$$

Similarly, if we seek to represent a function $f(\xi)$ by an approximation $p_n(\xi)$ on $[0, \Lambda]$ where $p_n(\xi)$ is a polynomial of degree n, then we can define an *error* function $e(\xi)$ equivalent to the *residual* function $R(\xi)$. Here

$$e(\xi) = f(\xi) - p_n(\xi) \tag{7.36}$$

In both interpolation and collocation, we select distinct data points $\{\xi_j \mid j = 0, 1, 2, \ldots, n\}$ and require that $e(\xi) = 0$ and $R(\xi) = 0$ at these $n + 1$ positions. No attempt is made to minimize $|e(\xi)|$ or $|R(\xi)|$ at positions ξ on the interval $[0, \Lambda]$ other than at $\{\xi_i \mid i = 0, 1, 2, \ldots, n\}$.

On the other hand, for interpolation, there is a well-known theorem (see Morris [12], Atkinson [13], or Schwarz [14]) that specifies that the maximum value of $|e(\xi)|$ on $[0, \Lambda]$ is minimized if we select the data points $\{\xi_i \mid i = 0, 1, 2, \ldots, n\}$ to be the zeros of the Chebyshev polynomial $T_{n+1}(x)$ mapped from the $[-1, 1]$ scale onto the $[0, \Lambda]$ interval. These zeros take the form

$$\xi_j = \frac{\Lambda}{2}\left[1 + \cos\left(\frac{2j + 1}{n + 1}\frac{\pi}{2}\right)\right] \tag{7.37}$$

for $j = 0, 1, 2, \ldots, n$. The theorem relies upon the function $f(\xi)$ and the interpolating function $p_n(\xi)$ both being polynomials, f of degree $n + 1$ and p_n of degree, at most n. The presence of the kernel $K(\xi - \varepsilon)$ in the integral equation precludes these assumptions. No guarantee can be given, therefore, that $|R(\xi)|$ will be minimized on $[0, \Lambda]$ if the same zeros of $T_{n+1}(x)$ be used as data points in setting up the matrix equation (7.14), although some improvement over using equally spaced data points should be expected.

It turns out, however, that from physical considerations, it is useful to employ these *Chebyshev data points* because they are relatively closely spaced around the ends of the $[0, \Lambda]$ interval, which correspond the entrances of the hot and cold gases to the regenerator. Hausen [15] pointed out that although the temperature distribution $F(\xi)$ was, in general, linear in the middle of the regenerator, most of the nonlinearities are propagated from the regenerator entrances as a consequence of the inlet gas temperatures *not* varying with time in each period of regenerator operation. The result is that use of the Chebyshev data points enables these nonlinearities to be more easily represented by the series expansion (7.16). Willmott and Duggan [8] explored this possibility and found that, indeed, modest rather than dramatic economies could be made in the number of terms required in the series expansion (they examined the Nahavandi and Weinstein [6] linearly independent functions) for an accurate solution of the integral equations, certainly as far as long regenerators are concerned, that is, where $\Lambda/\Pi > 10$ and $\Lambda > 10$.

Willmott and Knight [11] have confirmed these observations and their results are shown in Table 7.1. There, it will be observed, the value of the thermal ratio, $\eta_{REG} = \Lambda/(\Lambda + 2)$ for $\Pi = 0.1$ for $\Lambda \geq 10$, as might be expected.

It turns out, however, that for large reduced length, Λ, and *increasing* reduced period, Π, very significant economies can be effected using the Chebyshev data points. For example, with $\Lambda = 80$ and $\Pi = 5$, 30 equally spaced data points are required to achieve four-figure accuracy, whereas only

Table 7.1: Values of the thermal ratio, η_{REG}, and the number, N, of data points necessary to realize four-figure accuracy, as a function of reduced period, Π, and reduced length, Λ

	$\Pi = 0.1$				$\Pi = 1.0$				$\Pi = 5.0$			
	(a)		(b)		(a)		(b)		(a)		(b)	
Λ	η_{REG}	N	η_{REG}	N	η_{REG}	N	η_{REG}	N	η_{REG}	N	η_{REG}	N
1	0.3332	3	0.3332	3	0.3221	4	0.3221	4	0.1877	3	0.1877	3
10	0.8333	4	0.8333	4	0.8322	10	0.8322	8	0.8086	8	0.8086	7
25	0.9259	8	0.9259	6	0.9257	14	0.9257	10	0.9214	12	0.9214	10
50	0.9615	8	0.9615	8	0.9614	22	0.9614	14	0.9604	22	0.9604	13
80	0.9756	10	0.9756	10	0.9753	28	0.9753	16	0.9751	30	0.9751	16

The values in the columns (a) were evaluated with *equally spaced* collocation points; those in the columns (b) relate to the *Chebyshev* collocation points. (Symmetric case.)

16 Chebyshev collocation points are needed. This reflects the increased propagation of nonlinear temperature behavior from the regenerator entrances, consequent upon the constant inlet gas temperatures, as the period length, Π, becomes larger. One would expect larger economies to be possible for even larger values of Π. The larger values of reduced length, Λ, which Willmott and Knight [11] were able to handle using linearly independent functions, where $|\phi_j(\xi)| < 1$ is shown in this table. They used the functions

$$\phi_k(\xi) = \frac{1}{2}\left[T_k\left(\frac{2\xi}{\Lambda} - 1\right) + 1\right] \quad \text{for} \quad k = 0, 1, 2, \ldots, n \tag{7.38}$$

Their results offered in Table 7.1 and Table 7.2 are based on these functions.

The robustness of this method, as well as the collocation method based on Legendre polynomials, is demonstrated by Willmott and Knight who

Table 7.2: Values of the thermal ratio, η_{REG}, and the number, N, of Chebyshev points of collocation necessary to achieve four-figure accuracy for values Λ and Λ/Π, where Λ is the reduced length and Π is the reduced period (symmetric case)

	$\frac{\Lambda}{\Pi} = 50$		$\frac{\Lambda}{\Pi} = 20$		$\frac{\Lambda}{\Pi} = 10$	
Λ	η_{REG}	N	η_{REG}	N	η_{REG}	N
1.0	0.3333	3	0.3333	3	0.3332	2
3.0	0.6000	3	0.5999	3	0.5994	3
5.0	0.7142	3	0.7141	4	0.7134	4
7.0	0.7777	4	0.7775	3	0.7768	5
8.0	0.8000	4	0.7997	5	0.7989	6
10.0	0.8333	4	0.8330	6	0.8322	6
12.0	0.8571	4	0.8568	6	0.8559	7
16.0	0.8888	6	0.8886	7	0.8876	8
20.0	0.9090	6	0.9087	8	0.9078	8
25.0	0.9259	8	0.9256	8	0.9246	8
30.0	0.9374	8	0.9371	9	0.9362	8
40.0	0.9523	10	0.9520	10	0.9512	8
50.0	0.9614	10	0.9612	10	0.9604	10
100.0	0.9803	14	0.9802	11	0.9795	11
300.0	0.9935	17	0.9933	17	0.9930	18
500.0	0.9960	22	0.9958	21	0.9956	20
800.0	0.9974	21	0.9975	25	0.9974	27
1000.0	0.9980	28	0.9981	25	0.9973	20
2000.0	0.9989	33	0.9991	32	0.9991	32

successfully computed the temperature performance of regenerators with reduced lengths in the range

$$1 \leq \Lambda \leq 2000$$

for Λ/Π equal to 50, 20, and 10. In doing so, they used the linearly independent functions defined by equation (7.37) and the Chebyshev data points. These results are shown in Table 7.2. Without loss of generality, these relate to the symmetric case.

The ability of the method to handle such large reduced lengths, for example $\Lambda > 100$, demonstrates the robustness of the method but it is of theoretical interest only. In general, unless specific physical precautions are taken, such large reduced lengths will be accompanied by the effects of axial conductivity in the packing, as has been discussed in Chapter 6. These effects will reduce the thermal effectiveness of the regenerator.

7.7 Summary of the Collocation Method Using Legendre Polynomials and the Chebyshev Data Points

Substitution of expansion (7.26) into equation (7.10) yields

$$\sum_{j=0}^{n} \alpha_j \left(Q_j(\Lambda - \xi) + e^{-\Pi} Q_j(\xi) + \int_0^{\xi} K(\xi - \varepsilon)\, Q_j(\varepsilon)\, d\varepsilon \right) = 1 \tag{7.39}$$

In this *collocation* method, we apply equation (7.39) at $n+1$ *distinct* Chebyshev positions, defined by equation (7.37), $\{\xi_k \mid k = 0, 1, 2, \ldots, n\}$ yielding the matrix equation

$$A\boldsymbol{\alpha} = \boldsymbol{e} \tag{7.40}$$

The successive values of $\{Q_j(\xi_k) \mid j = 0, 1, 2, \ldots, n\}$ in the calculation of the matrix elements of A are obtained using the recurrence relation (7.28). Upon solving equation (7.40), the value of α_0 is extracted from $\boldsymbol{\alpha}$ and the thermal ratio, η_{REG}, is calculated using equation (7.32).

7.8 A Computational Consideration

It remains the case that the integrals

$$\int_0^{\xi_i} K(\xi_i - \varepsilon)\, Q_j(\varepsilon)\, d\varepsilon \tag{7.41}$$

must be computed in order to find the value of the matrix elements $a_{i,j}$. We have mentioned that Nahavandi and Weinstein [6] did not indicate how they calculated such integrals and it was suggested that numerical quadrature might be used. It is possible to compute the integrals (7.41) using Gaussian quadrature and the present author has encountered no difficulty in adopting such an approach.

Baclic [9] has suggested that the integral

$$\int_0^{\xi_i} K(\xi_i - \varepsilon)\, \varepsilon^j \, d\varepsilon$$

be computed by regarding the integral as the convolution of the functions $K(\xi - \varepsilon)$ and ε^j. Baclic proposed a recurrence formula whereby this integral can be obtained. Willmott and Maguire [16] examined this strategy and found that it was economical provided that the ratio Λ/Π is large (> 10) and the reduced length is not too large ($\Lambda < 100$). In other cases, Baclic's recurrence formula is very slow. In an extreme case, where $\Lambda/\Pi = 0.1$ and $\Lambda = 1000$, almost 2000 terms are required for the recurrence to realize convergence. Certainly 100 terms are required for $\Lambda = 100$ and $\Lambda/\Pi < 10$. Where appropriate, this strategy can be employed to evaluate

$$\int_0^{\xi_i} K(\xi_i - \varepsilon) Q_j(\varepsilon)\, d\varepsilon$$

simply by decomposing the Legendre polynomial into its component parts and integrating term by term. For example

$$P_2(x) = \tfrac{1}{2}(3x^2 - 1), \quad P_3(x) = \tfrac{1}{2}(5x^3 - 3x), \quad \text{and}$$
$$P_4(x) = \tfrac{1}{8}(35x^4 - 30x^2 + 3) \tag{7.42}$$

can be converted into the $Q_j(\xi)$ polynomials using the $x = (2\xi - \Lambda)/\Lambda$ change of variable.

7.9 Fast Galerkin Methods

It has been indicated that the coefficients $\{\alpha_k \mid k = 0, 1, 2, \ldots, n\}$ can be determined by the Galerkin method. Using Legendre polynomials, it is sought to minimize $\Omega(\Psi(\xi))$ on the $[0, \Lambda]$ interval and the vector $\boldsymbol{\alpha}$ is obtained by solving the equations

$$\int_0^\Lambda \Omega(\Psi(\xi))Q_j(\xi)\, d\xi = 0 \tag{7.43}$$

for $j = 0, 1, 2, \ldots, n$. Equation (7.43) can be expanded to yield, for the symmetric case,

$$\int_0^\Lambda \sum_{j=0}^{n} \left[\alpha_j \left(Q_j(\Lambda - \xi) + e^{-\Pi} Q_j(\xi) + \int_0^\xi K(\xi - \varepsilon) Q_j(\varepsilon)\, d\varepsilon \right) - 1 \right] Q_i(\xi)\, d\xi = 0 \tag{7.44}$$

for $i = 0, 1, 2, \ldots, n$. The order of integration and summation can be interchanged so that equation (7.43) becomes

$$\sum_{j=0}^{n} \alpha_j \left[\int_0^\Lambda \left(Q_j(\Lambda - \xi) + e^{-\Pi} Q_j(\xi) + \int_0^\xi K(\xi - \varepsilon) Q_j(\varepsilon)\, d\varepsilon \right) Q_i(\xi)\, d\xi \right] = \int_0^\Lambda Q_i(\xi)\, d\xi \tag{7.45}$$

for $i = 0, 1, 2, \ldots, n$. The vector $\boldsymbol{\alpha}$ is now found by solving the simultaneous linear equations

$$A\boldsymbol{\alpha} = \boldsymbol{h} \tag{7.46}$$

Here A is a square $(n + 1) \times (n + 1)$ matrix and $\boldsymbol{\alpha}$ and $\boldsymbol{h}$ are vectors of order $(n + 1)$. Willmott and Maguire [16] show that the matrix A is given by

$$a_{i,j} = \left[\int_0^\Lambda \left(Q_j(\Lambda - \xi) + e^{-\Pi} Q_j(\xi) + \int_0^\xi K(\xi - \varepsilon) Q_j(\varepsilon)\, d\varepsilon \right) Q_i(\xi)\, d\xi \right] \tag{7.47}$$

where $A = [a_{i,j}]$, for $i, j = 0, 1, 2, \ldots, n$. It can be shown that $\boldsymbol{h} = [\Lambda, 0, 0, \ldots, 0]^T$ as a consequence of the orthogonality of the functions $\{Q_k \mid k = 0, 1, 2, \ldots, .n\}$. Further, it can be verified that

$$Q_j(\Lambda - \xi) = (-1)^j Q_j(\xi) \tag{7.48}$$

It follows that the diagonal elements of the matrix A are simplified and become

$$a_{i,i} = \frac{\Lambda}{2}\frac{2}{2i+1}\left[(-1)^{-i} + e^{-\Pi}\right] + \int_0^{\Lambda}\left(\int_0^{\xi} K(\xi - \varepsilon)\, Q_i(\varepsilon)\, d\varepsilon\right) Q_i(\xi)\, d\xi \tag{7.49}$$

For the off-diagonal elements, the first terms in the matrix elements are eliminated by virtue of the orthogonality expressed by equation (7.25). In this case, for $i \neq j$,

$$a_{i,j} = \int_0^{\Lambda}\left(\int_0^{\xi} K(\xi - \varepsilon)\, Q_j(\varepsilon)\, d\varepsilon\right) Q_i(\xi)\, d\xi \tag{7.50}$$

7.10 The Nonsymmetric Case

In this section, the matrix approach to the nonsymmetric case is presented. This is an example of a method whereby the number of equations that must be handled can be reduced if $F(\xi)$ is found by solving

$$F(\xi) - f_1(f_2(F(\xi))) = 0$$

and $F'(\xi')$ is located by using

$$F'(\xi') = f_2(F(\xi))$$

as described in the introduction to this chapter. It is not proposed to describe the approach to the nonsymmetric problem outlined by Baclic and Dragutinovic [17] or that of Nahavandi and Weinstein [6], both of which do not take advantage of the economies offered by the matrix method to be described below. Instead, they treat the relations

$$F(\xi) = f_1(F'(\xi')) \qquad F'(\xi') = f_2(F(\xi))$$

as a pair of simultaneous equations, the solutions to which are the temperature distributions $F(\xi)$ and $F'(\xi')$, thereby doubling the number of equations that must be solved. Iliffe [3] treats the nonsymmetric regenerator problem in this manner also.

We begin by setting up the series approximations to $F(\xi)$ and $F'(\xi')$ as

$$F(\xi) = \sum_{j=0}^{n} \alpha_j Q_j(\xi) \tag{7.51}$$

$$F'(\xi') = \sum_{j=0}^{n} \beta_j Q_j(\xi') \tag{7.52}$$

Equation (7.6) takes the form, upon substitution of (7.52),

$$\sum_{j=0}^{n} \beta_j Q_j(\xi') = \sum_{j=0}^{n} \alpha_j \left(e^{-\Pi} Q_j(\xi') + \int_{\xi'}^{\Lambda'} K'(\varepsilon - \xi') Q_j(\varepsilon)\, d\varepsilon \right) \tag{7.53}$$

The application of the method of Galerkin to equation (7.53) yields the matrix equation

$$\Phi' \boldsymbol{\beta} = \Gamma' \boldsymbol{\alpha} \tag{7.54}$$

Here, Φ' and Γ' are $(0..n) \times (0..n)$ matrices. These are defined in the following manner:

$$\Phi' = \left[\int_{0}^{\Lambda'} Q_j(\xi') Q_i(\xi')\, d\xi' \right] \tag{7.55}$$

in which all the off-diagonal elements are zero by virtue of the orthogonality of the functions $\{Q_j(\xi') \mid j = 0, 1, 2, \ldots, n\}$. In other words, Φ' is simply a diagonal matrix. That is,

$$\Phi'_{i,i} = \frac{\Lambda'}{2} \frac{2}{2i+1} \quad \text{for} \quad i = 0, 1, 2, \ldots, n \tag{7.56}$$

The matrix Γ' enjoys a similar simplification. The off-diagonal elements take the form for $i \neq j$

$$\Gamma'_{i,j} = \int_{0}^{\Lambda'} \left(\int_{\xi'}^{\Lambda'} K'(\varepsilon - \xi') Q_j(\varepsilon)\, d\varepsilon \right) Q_i(\xi')\, d\xi' \tag{7.57}$$

and the diagonal elements for $i = j$ are

$$\Gamma'_{i,i} = \frac{e^{-\Pi'} \Lambda'}{2i+1} + \int_{0}^{\Lambda'} \left(\int_{\xi'}^{\Lambda'} K'(\varepsilon - \xi') Q_i(\varepsilon)\, d\varepsilon \right) Q_i(\xi')\, d\xi' \tag{7.58}$$

The vector $\boldsymbol{\beta}$ can be computed from $\boldsymbol{\alpha}$ using

$$\boldsymbol{\beta} = \Phi'^{-1} \Gamma' \boldsymbol{\alpha} \tag{7.59}$$

which involves only matrix multiplications because, as has been shown, Φ' is a diagonal matrix.

For the hot period, equation (7.7) is arranged somewhat into the form

$$F(\xi) = e^{-\Pi}F'(\xi) + \int_0^{\xi} K(\xi - \varepsilon)F'(\varepsilon)\,d\varepsilon + 1 - e^{-\Pi} - \int_0^{\xi} K(\xi - \varepsilon)\,d\varepsilon \tag{7.60}$$

Substituting expansions (7.51) and (7.52) into equation (7.60) yields

$$\sum_{j=0}^{n} \alpha_j Q_j(\xi) = \sum_{j=0}^{n} \beta_j \left(e^{-\Pi} Q_j(\xi) + \int_0^{\xi} K(\xi - \varepsilon) Q_j(\varepsilon)\,d\varepsilon \right) + 1 - e^{-\Pi} - \int_0^{\xi} K(\xi - \varepsilon)\,d\varepsilon \tag{7.61}$$

The Galerkin transformation is applied in a similar manner to that used for the cold period. A matrix equation

$$\boldsymbol{\Phi\alpha} = \boldsymbol{\Gamma\beta} + \boldsymbol{\zeta} \tag{7.62}$$

is obtained. The matrix Φ is a diagonal matrix, as is the corresponding matrix Φ' for the cold period. The elements of Φ take the form

$$\Phi_{i,i} = \frac{\Lambda}{2i + 1} \tag{7.63}$$

for $i = 0, 1, 2, \ldots, n$. The matrix Γ is similar in form to the cold period Γ'. The off-diagonal elements take the form for $i \neq j$

$$\Gamma_{i,j} = \int_0^{\Lambda} \left(\int_0^{\xi} K(\varepsilon - \xi) Q_j(\varepsilon)\,d\varepsilon \right) Q_i(\xi)\,d\xi \tag{7.64}$$

and the diagonal elements for $i = j$ are

$$\Gamma_{i,i} = \frac{e^{-\Pi}\Lambda}{2i + 1} + \int_0^{\Lambda} \left(\int_0^{\xi} K(\varepsilon - \xi) Q_i(\varepsilon)\,d\varepsilon \right) Q_i(\xi)\,d\xi \tag{7.65}$$

The vector $\boldsymbol{\zeta} = [\zeta_0, \zeta_1, \zeta_2, \ldots, \zeta_n]^T$ initially is of the form

$$\zeta_i = \int_0^{\Lambda} \left(1 - e^{-\Pi} - \int_0^{\xi} K(\xi - \varepsilon)\,d\varepsilon \right) Q_i(\xi)\,d\xi \tag{7.66}$$

The orthogonality of the functions $Q_j(\xi)$ leads to the following simplification:

$$\zeta_0 = \Lambda(1 - e^{-\Pi}) - \int_0^{\Lambda}\int_0^{\xi} K(\xi - \varepsilon)\, d\varepsilon\, d\xi$$

$$\zeta_i = -\int_0^{\Lambda}\left(\int_0^{\xi} K(\xi - \varepsilon) Q_i(\varepsilon)\, d\varepsilon\right) Q_i(\xi)\, d\xi \quad \text{for} \quad i > 0 \tag{7.67}$$

Equation (7.62) can now be rearranged:

$$\boldsymbol{\alpha} = \Phi^{-1}\Gamma\boldsymbol{\beta} + \Phi^{-1}\boldsymbol{\zeta} \tag{7.68}$$

Substituting for $\boldsymbol{\beta}$ using equation (7.59) gives

$$\boldsymbol{\alpha} = \Phi^{-1}\Gamma\Phi'^{-1}\Gamma'\boldsymbol{\alpha} + \Phi^{-1}\boldsymbol{\zeta} \tag{7.69}$$

Willmott and Maguire [16] denote $\Phi^{-1}\Gamma\Phi'^{-1}\Gamma'$ by M and $\Phi^{-1}\boldsymbol{\zeta}$ by $\boldsymbol{v}$. Equation (7.69) takes the simple form

$$\boldsymbol{\alpha} = M\boldsymbol{\alpha} + \boldsymbol{v}$$

or

$$\boldsymbol{\alpha} = (I - M)^{-1}\boldsymbol{v} \tag{7.70}$$

No matrix inversions are involved in calculating the elements of the matrix M because Φ and Φ' are diagonal matrices. The solution vector $\boldsymbol{\alpha}$ is obtained by solving $n + 1$ equations only (instead of $2n + 2$ equations as in the method of Nahavandi and Weinstein, for example). Once $\boldsymbol{\alpha}$ has been found, $\boldsymbol{\beta}$ is found using only matrix multiplications:

$$\boldsymbol{\beta} = \Phi'^{-1}\Gamma^{-1}\boldsymbol{\alpha} \tag{7.71}$$

This matrix method is a foundation in a unifying of open and closed methods within regenerator theory and in relating cyclic steady state to transient performance. Combined with the use of the Legendre polynomials, significant economies are offered and the method extends the work of Baclic and Dragutinovic [17], which encompassed the nonsymmetric problem. The technique described here is an application of the fast Galerkin algorithm as described by Delves and Mohamed [5].

7.11 The Calculation of the Double Integral

The problem remains as to how to calculate the double integral

$$\int_0^{\Lambda}\left(\int_0^{\xi} K(\xi - \varepsilon)\, Q_j(\varepsilon)\, d\varepsilon\right) Q_i(\xi)\, d\xi \tag{7.72}$$

This appears in equations (7.49) and (7.50) for the symmetric case and in equations (7.64) and (7.65) for the hot period of the nonsymmetric case. An equivalent form appears in equations (7.57) and (7.58) for the cold period of this nonsymmetric case. Willmott and Maguire [16] evaluated these integrals using Gaussian quadrature.

Baclic and Romie [9, 18, 19] have adopted a somewhat different approach. They converted the inner integral into algebraic form which could then be integrated. Reference has already been made to this earlier in this chapter. The approach of Baclic [9] relates to the linearly independent functions $\{\phi_j(\xi) = \xi^j/j! \mid j = 0, 1, 2, \ldots, n\}$. He used the notation

$$I_n(\xi) = (-i)^n J_n(i\xi) \tag{7.73}$$

for the modified Bessel functions of integer order. Using this notation, Baclic examined the inner integral, which he denoted by

$$A_j(\xi) = \int_0^{\xi} K(\xi - \varepsilon)\frac{\varepsilon^j}{j!}\, d\varepsilon \tag{7.74}$$

He treats the integral $A_j(\xi)$ as a convolution of $K(\xi)$ and $\xi^j/j!$ and by the application of a Laplace transform and then the inversion theorem, the algebraic form of $A_j(\xi)$ is yielded. This takes the form

$$A_j(\xi) = V_{j+1}(\Pi, \xi) - e^{-\Pi}\frac{\xi^j}{j!} \tag{7.75}$$

Baclic assumes that the approximating function for $F(\xi)$ takes the form

$$\Psi(\xi) = \sum_{j=0}^{n} \alpha_j \frac{\xi^j}{j!} \tag{7.76}$$

The residual function $R(\xi)$ takes the form of (7.77) in this case:

$$R(\xi) = \sum_{j=0}^{n} \alpha_j \left(\frac{(\Lambda - \xi)^j}{j!} + e^{-\Pi}\frac{\xi^j}{j!} + A_j(\xi) \right) - 1 \tag{7.77}$$

Upon insertion of the form of $A_j(\xi)$ developed as equation (7.75), it will be seen that the term

$$e^{-\Pi}\frac{\xi^j}{j!}$$

cancels out and one is left with

$$R(\xi) = \sum_{j=0}^{n} \alpha_j \left(\frac{(\Lambda - \xi)^j}{j!} + V_{j+1}(\Pi, \xi) \right) - 1 \tag{7.78}$$

The functions $\{V_j \mid j = 1, 2, \ldots, n\}$ are defined by Baclic [9] in the following manner, using x and y as real "dummy" variables:

$$V_m(x, y) = e^{-x-y} \sum_{n=m-1}^{\infty} \begin{bmatrix} n \\ m-1 \end{bmatrix} \left(\frac{y}{x}\right)^{\frac{n}{2}} I_n(2(xy)^{1/2}) \tag{7.79}$$

A set of equations is now developed by application of the Galerkin method:

$$\int_0^{\Lambda} (R(\xi) \frac{\xi^j}{j!} \, d\xi = 0 \quad \text{for} \quad j = 0, 1, 2, \ldots, n \tag{7.80}$$

This yields a form of

$$A\boldsymbol{\alpha} = \boldsymbol{h} \tag{7.46}$$

where $A = [a_{i,j}]$ and

$$a_{i,j} = \frac{\Lambda^{i+j+1}}{(i+j+1)!} + \sum_{m=0}^{i} (-1)^m \frac{\Lambda^{i-m}}{(i-m)!} V_{i+m+2}(\Pi, \Lambda) \tag{7.81}$$

and

$$h_i = \frac{\Lambda^{i+1}}{(i+1)!} \tag{7.82}$$

with $\boldsymbol{h} = [h_0, h_1, h_2, \ldots, h_n]^T$. Baclic claims that the functions $\{V_i(\Lambda, \Pi) \mid i = 1, 2, \ldots\}$ can be readily computed and, presumably, the infinite series in (7.79) is rapidly convergent. The elements of the matrix A can thus be evaluated without resort to numerical integration. Willmott and Maguire have indicated some limitations as to this approach and this has been discussed previously. Baclic and Dragutinovic [17] have developed this approach for the nonsymmetric problem. Their method can be readily built into the matrix method thereby requiring $(n + 1)$ rather than $(2n + 2)$ simultaneous equations to be solved.

7.12 The Thermal Ratio for Nonsymmetric Regenerators

The thermal ratio, η_{REG}, for the hot period can be written in the form

$$\eta_{REG} = \frac{\Lambda}{\Pi} \left(\frac{1}{\Lambda} \int_0^{\Lambda} F(\xi) \, d\xi - \frac{1}{\Lambda'} \int_0^{\Lambda'} F'(\xi) \, d\xi \right) \tag{7.83}$$

while the corresponding η'_{REG} for the cold period is given by

$$\eta'_{REG} = \frac{\Lambda'}{\Pi'}\left(\frac{1}{\Lambda}\int_0^{\Lambda} F(\xi)\,d\xi - \frac{1}{\Lambda'}\int_0^{\Lambda'} F'(\xi)\,d\xi\right) \tag{7.84}$$

The hot-period thermal ratio, η_{REG}, is modified upon the substitution of the series solution

$$\eta_{REG} = \frac{\Lambda}{\Pi}\left(\frac{1}{\Lambda}\int_0^{\Lambda}\sum_{i=0}^{n}\alpha_i Q_i(\xi)\,d\xi - \frac{1}{\Lambda'}\int_0^{\Lambda'}\sum_{i=0}^{n}\beta_i Q_i(\xi')\,d\xi'\right) \tag{7.85}$$

The last n terms of the sums in (7.85) disappear as a consequence of the orthogonality of the functions $\{Q_i(\xi) \mid i = 0, 1, 2, \ldots, n\}$. Recalling that $Q_0(\xi) = Q_0(\xi') = 1$, equation (7.85) reduces to the form

$$\eta_{REG} = \frac{\Lambda}{\Pi}(\alpha_0 - \beta_0) \tag{7.86}$$

with

$$\eta'_{REG} = \frac{\Lambda'}{\Pi'}(\alpha_0 - \beta_0) \tag{7.87}$$

corresponding to the cold period. These correspond to the equation (7.32) for η_{REG} for the symmetric case. Where the polynomials are not orthogonal, equation (7.32) takes the form

$$\eta_{REG} = \frac{\Lambda}{\Pi}\left(2\sum_{j=0}^{n}\alpha_j\frac{\Lambda^j}{(j+1)!} - 1\right) \tag{7.88}$$

in the case where Baclic uses $\phi_j(\xi) = \xi^j/j!$. Similar expressions for η_{REG} and η'_{REG} apply for the nonsymmetric case. The advantage of using the Legendre polynomials is immediately apparent.

7.13 The Volterra Method for Solving the Integral Equations

A rather different method for the solution of equation (7.10) was devised by Nusselt [2] in 1928. He specified, for the symmetric case, that the temperature at the start of the hot period, $F(\xi)$, could be written as

$$F(\xi) = \sum_{k=0}^{\infty}\lambda^k\phi_k(\xi) \tag{7.89}$$

where the functions $\{\phi_k(\xi) \mid k = 0, 1, 2, \ldots\}$ are not specified but are calculated at chosen values of ξ. By selecting $\{\xi_j = j\Delta\xi \mid j = 0, 1, 2, \ldots, n\}$, the solution obtained is the vector $\boldsymbol{F} = [F_0, F_1, F_2, \ldots, F_n]^T$ with the F_j's being the temperatures at the entrance and exit to the regenerator, and at equally spaced, intermediate positions.

In order to find the series (7.89), Nusselt employed Volterra's method in which the integral in equation (7.10) is first multiplied by the parameter λ which is later set equal to 1. Equation (7.10) takes the form

$$F(\Lambda - \xi) + e^{-\Pi} F(\xi) + \lambda \int_0^{\xi} K(\xi - \varepsilon) F(\varepsilon)\, d\varepsilon = 1 \tag{7.90}$$

The series (7.89) is inserted into equation (7.90) and the coefficients with the same powers of λ are equated. By this means is obtained

$$\begin{aligned}
1 &= \phi_0(\Lambda - \xi) + e^{-\Pi}\phi_0(\xi) \\
0 &= \phi_1(\Lambda - \xi) + e^{-\Pi}\phi_1(\xi) + \int_0^{\xi} \phi_0(\varepsilon) K(\xi - \varepsilon)\, d\varepsilon \\
0 &= \phi_2(\Lambda - \xi) + e^{-\Pi}\phi_2(\xi) + \int_0^{\xi} \phi_1(\varepsilon) K(\xi - \varepsilon)\, d\varepsilon \\
0 &= \phi_n(\Lambda - \xi) + e^{-\Pi}\phi_n(\xi) + \int_0^{\xi} \phi_{n-1}(\varepsilon) K(\xi - \varepsilon)\, d\varepsilon
\end{aligned} \tag{7.91}$$

A general form of equation (7.91) is considered for $s = 1, 2, \ldots, n$.

$$0 = \phi_s(\Lambda - \xi) + e^{-\Pi}\phi_s(\xi) + \int_0^{\xi} \phi_{s-1}(\varepsilon) K(\xi - \varepsilon)\, d\varepsilon \tag{7.92}$$

This equation is multiplied throughout by e^{Π} yielding

$$0 = e^{\Pi}\phi_s(\Lambda - \xi) + \phi_s(\xi) + e^{\Pi} \int_0^{\xi} \phi_{s-1}(\varepsilon) K(\xi - \varepsilon)\, d\varepsilon$$

In this equation, set $\Lambda - \xi = \xi^*$ and thus, also, $\Lambda - \xi^* = \xi$.

It follows that

$$0 = e^{\Pi}\phi_s(\xi^*) + \phi_s(\Lambda - \xi^*) + e^{\Pi}\int_0^{\Lambda-\xi^*} \phi_{s-1}(\varepsilon)K(\Lambda - \xi^* - \varepsilon)\,d\varepsilon$$

or, dropping the asterisk notation,

$$0 = e^{\Pi}\phi_s(\xi) + \phi_s(\Lambda - \xi) + e^{\Pi}\int_0^{\Lambda-\xi} \phi_{s-1}(\varepsilon)K(\Lambda - \xi - \varepsilon)\,d\varepsilon \tag{7.93}$$

If we combine equations (7.92) and (7.93) to eliminate $\phi_s(\Lambda - \xi)$, then

$$\phi_s(\xi) = \frac{\int_0^{\xi} e^{-\Pi}\phi_{s-1}(\varepsilon)K(\xi - \varepsilon)\,d\varepsilon - \int_0^{\Lambda-\xi} \phi_{s-1}(\varepsilon)K(\Lambda - \xi - \varepsilon)\,d\varepsilon}{1 - e^{-2\Pi}} \tag{7.94}$$

is obtained. For the special case where $s = 0$, we have

$$\phi_0(\xi) = \frac{1 - e^{-\Pi}}{1 - e^{-2\Pi}} \tag{7.95}$$

In other words, $\phi_0(\xi)$ is evaluated directly from equation (7.95) and then the $\phi_s(\xi)$ is obtained from $\phi_{s-1}(\xi)$ for $s = 1, 2, \ldots$ using equation (7.94). The required solution is obtained from equation (7.89) setting $\lambda = 1$, that is,

$$F(\xi) = \sum_{k=0}^{\infty} \phi_k(\xi) \tag{7.96}$$

Nusselt [2] and Iliffe [3] mention that this series is rapidly convergent. On the other hand, it is necessary to compute the integrals

$$\int_0^{\xi} \phi_{s-1}(\varepsilon)K(\xi - \varepsilon)\,d\varepsilon \quad \text{and} \quad \int_0^{\Lambda-\xi} \phi_{s-1}(\varepsilon)K(\Lambda - \xi - \varepsilon)\,d\varepsilon$$

at each stage of the process. The process is carried out to compute the functions $\{\phi_k(\xi_j) \mid k = 0, 1, 2, \ldots\}$ and hence the temperatures $\{F(\xi_j) \mid k = 0, 1, 2, \ldots\}$ at the successive positions $\{\xi_j = j\Delta\xi \mid j = 0, 1, 2, \ldots, n\}$ where $n\Delta\xi = \Lambda$. The integrals can thus be computed using the Newton–Cotes formulae in a manner similar to that used in the method of Iliffe, as described in Chapter 5.

Thus, the first integral can be written in the form

$$\int_0^{\xi} \phi_{s-1}(\varepsilon)K(\xi-\varepsilon)\,d\varepsilon = \int_0^{j\Delta\xi} \phi_{s-1}(\varepsilon)K(j\Delta\xi-\varepsilon)\,d\varepsilon$$

and integrated numerically using Simpson's rules. The second integral is evaluated in like fashion.

This means that if $n+1$ levels in the packing are considered and $s+1$ functions are found, then $2s \times n$ integrals must be evaluated. It is for this reason that techniques such as the quadrature and the series solution methods have been developed.

The present author has implemented, for a computer, this technique of Nusselt and has found that the rate of convergence is not as good as was suggested originally. In particular, the number of terms which must be considered in equation (7.96) rises rapidly with the reduced length Λ for a fixed value of Π. Moreover, the method suffers from the same problem as that of Iliffe [3] discussed previously, namely, the shape of the kernel for $\Lambda > \Pi$, especially for larger Λ, makes it difficult to evaluate the integral

$$\int_0^{k\Delta\xi} K(\xi-\varepsilon)\,F(\varepsilon)\,d\varepsilon$$

by Simpson's rules for larger values of k, if the majority of the data points involve values of $K(\xi_j)$ which are very small and *unrepresentative* of $K(\xi)$ over the complete range $0 \leq \xi \leq \Lambda$ under consideration. This is exacerbated by the need to find the *difference* between two such integrals as seen in equation (7.94). Where $\Pi > \Lambda$, the shape of the kernel "improves" and Nusselt's method appears to work better.

This method of Nusselt cannot be recommended but, nevertheless, it is described here for completeness sake since it was the first method of its kind described in the literature.

7.14 Concluding Remarks

The limiting feature of the integral equation methods is that they are restricted to coping with the linear model alone. It is far from clear how they might be adopted to deal with *nonlinear models* where, for example, temperature dependent thermophysical properties of the gas and the solid imply that the parameters ξ and η vary with space y and time τ. Provided the temperature difference between the hot- and cold-gas inlet temperatures, $t_{f,in} - t'_{f,in}$, is relatively small and/or the operating temperatures are not very low, as in cryogenic applications, then the linear model and thus the integral equation methods are applicable. Failing this, different techniques must be adopted to handle nonlinear models.

References

1. W. Nusselt, "Die Theorie des Winderhitzers (The Theory of the Air Heater)," *Z. Ver. Deut. Ing.* **71**, 85–91 (1927) (RAE Library Translation No. 269, 1948, W. Shirley).
2. W. Nusselt, "Der Beharrungszustand im Winderhitzer (The Steady Operating Condition in the Air Draught Heater)," *Z. Ver. Deut. Ing.* **72**, 1052–1054 (1928) (RAE Library Translation No. 267, 1948, W. Shirley).
3. C. E. Iliffe, "Thermal Analysis of the Contra-flow Regenerative Heat Exchanger," *Proc. Inst. Mech. Eng.* **159**, 363–372 (1948).
4. C. H. T. Baker, *The Numerical Treatment of Integral Equations*, Oxford University Press, Oxford (1977).
5. L. M. Delves, J. L. Mohamed, *Computational Methods for Integral Equations*, Cambridge University Press, Cambridge (1985).
6. A. N. Nahavandi, A. S. Weinstein, "A Solution to the Periodic-Flow Regenerative Heat Exchanger Problem," *Appl. Sci. Res.* **10**, 335–348 (1961).
7. A. J. Willmott, R. J. Thomas, "Analysis of the Long Contra-flow Regenerative Heat Exchanger," *J. Inst. Maths Applics* **14**, 267–280 (1974).
8. A. J. Willmott, R. C. Duggan, "Refined Closed Methods for the Contra-flow Thermal Regenerator Problem," *Int. J. Heat Mass Transfer* **23**, 655–662 (1980).
9. B. S. Baclic, "The Application of the Galerkin Method to the Solution of the Symmetric and Balanced Counterflow Regenerator Problem," *J. Heat Transfer* **107**, 214–221 (1985).
10. A. Hill, "Stable Closed Methods for Thermal Regenerator Simulations," D.Phil. thesis, University of York (Apr. 1988).
11. A. J. Willmott, D. P. Knight, "Improved Collocation Methods for Thermal Regenerator Simulations," *Int. J. Heat Mass Transfer* **36**(6), 1663–1670 (1993).
12. J. L. Morris, *Computational Methods in Elementary Numerical Analysis*, Wiley, New York (1983)
13. K. E. Atkinson, *An Introduction to Numerical Analysis*, 2nd edn, Wiley, New York (1988).
14. H. R. Schwarz, *Numerical Analysis. A Comprehensive Introduction*, Wiley, New York (1989). Translation of *Numerische Mathematik* (1988).
15. H. Hausen, "Vervollstandigte Berechnung des Warmeaustausches in Regeneratoren (Improved Calculations for Heat Transfer in Regenerators)," *Z. VDI-Beiheft Verfahrenstechnik* **2**, 31–43 (1942) (Iron and Steel Institute translation, June 1943).
16. A. J. Willmott, P. Z. Maguire, "Fast Galerkin Methods for Thermal Regenerator Modelling," *Int. J. Heat Mass Transfer* **37**(10), 1487–1494 (1994).
17. B. S. Baclic, G. D. Dragutinovic, "Asymmetric-Unbalanced Counterflow Thermal Regenerator Problem: Solution by the Galerkin Method and

Meaning of Dimensionless Parameters," *Int. J. Heat Mass Transfer* **34**(2), 483–498 (1991).

18. F. E. Romie, "Two Functions Used in the Analysis of Crossflow Exchangers, Regenerators and Related Equipment," *J. Heat Transfer (Tech. Note)* **109**, 518–521 (1987).
19. F. E. Romie, B. S. Baclic, "Methods for Rapid Calculation of the Operation of Asymmetric Counterflow Regenerators," *J. Heat Transfer* **110**, 785–788 (1988).

Chapter 8

NONLINEAR MODELS OF COUNTERFLOW REGENERATORS

8.1 Introduction

This chapter is concerned with the simulation of a thermal regenerator of the fixed-bed or rotary type, where the model permits the representation of the temperature dependence of the relevant thermophysical properties of the gases, between which heat is transferred in the regenerator. The model also allows the inlet temperatures and flow rates to vary with time in the case of fixed-bed regenerators or spatially across the face of the entrance of a rotary regenerator. The variation of the thermophysical properties of the packing material is similarly allowed. Such models are called *nonlinear* because the dimensionless parameters ξ and η, which characterize the simplified form of the linear model, now vary with space and time. In a nonlinear model, these parameters or the variables from which they are derived, are sometimes referred to as *nonlinear variables.*

It must be understood, however, that the term "nonlinear variable" does not imply that the functional relationship between gas specific heat, for example, and gas temperature, used in the model, is *necessarily* nonlinear. The term *nonlinear* used here is derived from the idea of a nonlinear model in which such functional relationships between, say, gas viscosity, gas specific heat, and gas thermal conductivity, and hence the surface heat transfer coefficient, on the one hand, and gas temperature, on the other, are included, as opposed to a linear model, in which they are not.

Many likely requirements of industry for regenerator predictive models include nonlinear features which are specific to the application under consideration. In certain low-temperature applications, it is important to be able to represent the temperature variation of specific heat which occurs. High-temperature regenerative heat exchangers, contrawise, can experience temperature differences of up to 1500 K between the fluid inlet and outlet in the hot and cold periods. Such a large temperature difference along the length of the heat exchanger causes significant variations in the relevant temperature-dependent thermophysical properties of the gases and solid involved in the regenerator. Hill and Willmott [1] point out that in the case of the regenerative burner, which will be discussed later, variations of the order of 10% in the value of the specific heat of air and 25% in the heat transfer coefficient proposed by Denton et al. [2] have been computed along the length of the heat exchanger. Simulations of the thermal regenerators, the *Cowper stoves*, which are used to preheat the air for iron-making and zinc- and lead-smelting blast furnaces, must permit the time variation of gas mass flow rate and the accompanying variations in the convective heat transfer coefficient.

Such nonlinear models have, as a basis, the two-dimensional model described in Chapters 4 and 5, together with the lumped heat transfer coefficient, which represents the internal resistance to heat transfer within the packing, described in Chapter 6. It is unlikely that a three-dimensional model would be required, although the possibility of using such a nonlinear 3-D model is discussed by Evans [3]. The influence of axial thermal conductivity is usually negligible and is not considered here.

The form of the lumped heat transfer coefficient, $\bar{\alpha}$, described in Chapter 6, enables the continued use of equations (5.1) and (5.2), in suitably modified form, even for models in which the heat transfer coefficients and the thermal properties of both gas and solid are allowed to vary with gas/solid temperature and therefore to vary spatially and with time in both the hot and cold periods of regenerator operation. The chronological variation of gas mass flow rate can be represented in like manner.

There is a major difference between linear and nonlinear models. The linear model described in Chapters 4 and 5 is an approximation to a wide spectrum of problems: the regenerator is specified in terms of four dimensionless parameters, Λ, Λ', Π, and Π', together with the inlet gas temperatures for the hot and cold periods, placed on a dimensionless $(0, 1)$ scale. Each problem is mapped onto these parameters so that quite different regenerator configurations with possibly vastly different physical sizes might be represented by dimensionless parameters that have similar values.

For unvarying inlet gas temperature conditions, many authors have presented graphical solutions of the differential equations, frequently in the form of the thermal effectiveness η_{REG} and η'_{REG} as functions of reduced length and reduced period.

There are no corresponding general solutions for nonlinear models, simply because there are an infinite number of such models, each tailored to the particular environment in which the regenerator is required to operate. Each model incorporates those nonlinear features that are necessary if a good enough correspondence between theory and possible practical measurements on the physical regenerator system is to be realized.

Hill and Willmott [1] recall that Kulakowski and Anielewski [4] describe three types of nonlinearity that occur in the modeling of regenerative heat exchangers.

Natural nonlinearities occur due to the temperature dependence of certain thermophysical properties. As the temperature of the gas and solid varies both spatially and chronologically within the regenerator, then so do several temperature-dependent properties, in particular $\bar{\alpha}$, c_p, C_s, and λ_s.

Constructional nonlinearities arise frequently within the context of *zoning* problems. These nonlinearities occur because several different solid materials are used in the heat-storing mass, and/or different packing geometries are used at different positions in the regenerator. The properties affected here vary only in a spatial sense, that is, down the length of the regenerator, although local space and time variations, consequent upon local temperature changes, of these variables may have to be allowed also in a representative mathematical model. The relevant properties are C_s, $\bar{\alpha}$, and λ_s.

Operational nonlinearities occur here due to the mode of operation of the regenerator. For example, in a Cowper stove, a constant fluid outlet temperature is required during the cold period. This is achieved by varying the cold flow rate with time, as will be discussed below. In this case the affected property that varies chronologically is $\dot{m}_f$.

8.2 Models and Methods

Hill and Willmott [1] as well as other authors have wisely sought to clarify the meaning of the terms *model* and *method*, and this is undertaken here. By model, is meant an idealization of the regenerator and its operation, in which are built specified assumptions. Thus, in the linear model, it is assumed that the heat transfer coefficient does not vary spatially or with time in the hot period, for example, whereas in the nonlinear model, it might be assumed that this coefficient varies with gas temperature. In both the linear and nonlinear models, it is often assumed that the inlet gas temperatures of the gases in the hot and cold period do not vary with time.

Once an idealization has been agreed and its mathematical formulation developed, it is the concern of the *methods* how the relevant differential equations are solved. Different methods must obtain the same solution for the same model. Thus the closed method of Iliffe [5] and the open method of Willmott [6] yield the same solution, at periodic steady state, for the linear model. In this chapter, a finite-difference, open method developed from that of Willmott [7] is described in some detail, followed by a closed method for the same model devised by Hill [1, 8], and comparisons drawn between the two. Needless to say, comparisons in computing speed between methods relate to the performances for the same model.

In this chapter, the means will be discussed whereby software can be developed so that, firstly, the software can be readily modified to cope with different problems. This software relates to the finite-difference representation of suitably modified forms of equations (5.1) and (5.2) and indicates how these representations can be manipulated to enable different nonlinear features to be included. Subsequently, the *closed* method of Hill will be described. Comparisons between models for certain practical problems will be offered.

It is frequently required that the surface heat transfer coefficient be computed using a correlation typically of the form

$$\mathrm{Nu} = c\mathrm{Re}^n\mathrm{Pr}^m$$

where Nu, Re and Pr are the Nusselt, Reynolds and Prandtl numbers respectively. This surface coefficient can then be embodied in a lumped heat transfer coefficient where the Hausen Φ factor is included in the representation of the resistance to heat transfer *within* the regenerator packing.

8.3 Overall Structure of Typical Software

It is most convenient if it can be arranged that the software be modular in form. It can be thus arranged, for the open, finite-difference method of solution, that the dimensionless parameters, reduced length, Λ, and reduced period, Π, be computed at any space–time position on the mesh covering the hot/cold period under consideration. Such a module isolates the particulars of an individual regenerator so that only the external part of the program is "aware" that a certain regenerator is being simulated, be it rotary or fixed bed in form. Similar arrangements apply to any closed method for nonlinear models.

The basic unit of such software is the integration over a *single time step*. This can be realized by the provision of two modules or subroutines, one that deals specifically with the start of a period, called *tstep0*, for example, and another, similarly called *tstep*, which deals with the remainder of the time steps in a period. A possible way to represent such a program structure is to use pseudocode.

It is necessary, given the solid temperatures at the start of a period, to evaluate the gas temperatures down the length of the regenerator, given also the inlet gas temperature. This is realized by *tstep0*.

The subroutine *tstep* integrates from the general time position j to the time position $j+1$. If j takes the values $0, 1, 2, \ldots, p$ where

$$p\,\Delta\tau = P$$

with P representing duration of the period under consideration and $\Delta\tau$ equal to the length of the time step, then the subroutine *doperd* ("do a period") could have the basic structure set out below:

```
SUBROUTINE doperd (parameters)

comment: first time step

   CALL tstep 0 (parameters)

comment: remaining time steps

   FOR j=0 TO p-1 DO
   CALL tstep (parameters)

END
```

The simulation of a complete cycle of regenerator operation must represent, in some way, the contraflow of the hot and cold gases through the packing of the regenerator. This is achieved by *reversing the packing temperatures* at the start of a period of operation. In this way, the same software *doperd* can be used for both periods of operation by a suitable choice of parameters when *doperd* is called. In other words, by always integrating the differential equations in the same distance direction, using *doperd*, contraflow is realized by simply reversing the packing temperatures.

This is how the subroutine *docycle* might work. The software should allow for the entrance temperature of the gas to vary with "time" within each period of operation. This can be realized by setting up a vector of inlet gas temperatures, corresponding to the time positions, $j = 0, 1, 2, \ldots, p$, just before the simulation of each period. Similarly, a vector of mass flow rates, corresponding to the same time positions, might be set up if time-varying flow rates are to be represented:

```
SUBROUTINE docycle (parameters)

Reverse the solid temperatures from the end of the last
cold period for the start of the next hot period

Set up vector of hot inlet temperatures
Set up vector of hot period gas mass flow rates

  CALL doperd (parameters)

Reverse the solid temperatures for the beginning of the
next cold period.

Set up vector of cold inlet temperatures
Set up vector of cold period gas mass flow rates

  CALL doperd (parameters)

END
```

Such software would calculate the gas and solid temperatures over space–time meshes. In the distance direction, the temperatures would be computed at levels in the regenerator denoted by $i = 0, 1, 2, \ldots, m$ where

$$m\,\Delta y = L$$

and Δy is the distance step length with L being the length of the regenerator.

The exit gas temperature at the end of the cold period is denoted by $t'_{f,x}(P)$ and the value of $t'_{f,x}(P)$ at the end of the nth cycle by ϕ_n. In an *open* method, the convergence criteria used to detect whether the model of the regenerator has reached periodic steady state might employ ε_n, where

$$\varepsilon_n = \frac{\phi_n - \phi_{n-1}}{\phi_n}$$

and cyclic state said to be realized if, say, $|\varepsilon_n| < 10^{-5}$.

The basic structure of the code, for an open method, in which the regenerator simulation is run until cyclic steady state is reached might take the form shown below.

```
WHILE |ε_n| > 10^-5 DO

{SET φ_{n-1} = φ_n
  CALL docycle (parameters)
  SET φ_n = t'_{f,x}(P')
  CALCULATE ε_n}

END
```

It is the intention in the design of such elegant top-down software construction, from the WHILE loop at the top, to the *tstep0* and *tstep* subroutines at the bottom, that it can be understood readily, so that it can be modified by a new user as the need might arise.

This form of construction of the software enables any discussion of any details about the way nonlinearities are represented to be delayed within this part of the description. On the other hand, this particular top-down construction remains intact if different methods of solution of the differential equations are to be used. Such software can be designed deliberately so that this is the case.

8.4 The Underlying Differential Equations

The basic equations take the form

$$\frac{\partial H_f}{\partial y} = \frac{\bar{\alpha} A}{\dot{m}_f L}(t_s - t_f) \tag{8.1}$$

$$\frac{\partial H_s}{\partial \tau} = \frac{\bar{\alpha} A}{M_s}(t_f - T_s) \tag{8.2}$$

where H_f is the *enthalpy* of the gas and H_s is the *enthalpy* of the packing material, defined by the standard equations

$$H_f = U_f + PV \tag{8.3}$$

$$H_s = U_s \tag{8.4}$$

Here U_f/U_s denotes the internal (thermal) energy of the gas/solid. It can be shown that for a gas, a change in enthalpy ΔH_f is related to the heat input, ΔQ, into the system by the equation

$$\Delta H_f = \Delta Q + V\,\Delta P \tag{8.5}$$

It is assumed that the pressure drop down the length of the regenerator is small and therefore

$$\Delta H_f = \Delta Q = c_p \Delta t_f \tag{8.6}$$

and similarly

$$\Delta H_s = C_s \Delta t_s \tag{8.7}$$

Here, t_f and t_s denote the gas and solid temperatures (K) respectively, while c_p is the gas specific heat (kJ/kg K) and C_s is the solid specific heat.

8.5 Integration of the Underlying Differential Equations

The 1968 method [7] is reported by Schmidt and Willmott [9] and is based on the trapezoidal method. This is applied to equations (8.1) and (8.2) above, but with certain modifications sct out below.

Consider positions $i = 0, 1, 2, \ldots, m$ down the length, L, of the regenerator, which are $\Delta y = L/m$ apart. Here, $i = 0$ corresponds to the entrance to the regenerator and $i = m$ to the exit. Consider also $j = 0, 1, 2, \ldots, p$, which are times $\Delta\tau$ apart, so that $p\,\Delta\tau = P$. Again, $j = 0$ corresponds to the start of the period under consideration, $j = p$ to its finish and P to its duration.

Applying the trapezium rule at the general position (i, j), we have

$$H_{f,i+1,j} = H_{f,i,j} + \frac{\Delta y}{2}\left(\frac{\partial H_{f,i+1,j}}{\partial y} + \frac{\partial H_{f,i,j}}{\partial y}\right) \tag{8.8}$$

where

$$\frac{\partial H_{f,i,j}}{\partial y} = \frac{\bar{\alpha}_{i,j} A}{\dot{m}_j L}(t_{s,i,j} - t_{f,i,j})$$

It follows that

$$H_{f,i+1,j} - H_{f,i,j} = \frac{\Delta y}{2}\left(\frac{\bar{\alpha}_{i+1,j} A}{\dot{m}_{f,j} L}(t_{s,i+1,j} - t_{f,i+1,j}) + \frac{\bar{\alpha}_{i,j} A}{\dot{m}_{f,j} L}(t_{s,i,j} - t_{f,i,j})\right) \tag{8.9}$$

At any stage of the integration process, it can be assumed initially that $t_{s,i+1,j}$, $t_{f,i,j}$, $t_{s,i,j}$, and $\bar{\alpha}_{i,j}$ (a function of gas temperature, $t_{f,i,j}$, as well as gas mass flow rate) are all known. We now examine the left-hand side of equation (8.9) and observe that

$$H_{f,i+1,j} - H_{f,i,j} = \int_{t_{f,i,j}}^{t_{f,i+1,j}} c_p(t_f)\,dt_f \tag{8.10}$$

The integral in equation (8.10) can also be integrated using the trapezium rule, in particular

$$\int_{t_{f,i,j}}^{t_{f,i+1,j}} c_p(t_f)\,dt_f = \frac{\Delta t_f}{2}(c_{p,i+1,j} + c_{p,i,j}) = c^*_{p,i,j}(t_{f,i+1,j} - t_{f,i,j}) \tag{8.11}$$

where

$$c^*_{p,i,j} = \tfrac{1}{2}(c_{p,i+1,j} + c_{p,i,j}) \tag{8.12}$$

Employing these relations, equation (8.9) now takes the form

$$t_{f,i+1,j} = t_{f,i,j} + \tfrac{1}{2}\left[\Delta\xi_{i+1,j}(t_{s,i+1,j} - t_{f,i+1,j}) + \Delta\xi_{i,j}(t_{s,i,j} - t_{f,i,j})\right] \tag{8.13}$$

where

$$\Delta\xi_{i+1,j} = \frac{\bar{\alpha}_{i+1,j} A \Delta y}{\dot{m}_{f,j} L c^*_{p,i,j}} = \frac{\Lambda_{i+1,j}}{m} \tag{8.14}$$

and where

$$\Delta\xi_{i,j} = \frac{\bar{\alpha}_{i,j} A \Delta y}{\dot{m}_{f,j} L c^*_{p,i,j}} = \frac{\Lambda_{i,j}}{m} \tag{8.15}$$

This is almost the same form as that used in the Willmott paper [7] with the exception of the use of $c^*_{p,i,j}$. The integration routines *tstep0* and *tstep* must exercise great care that the correct value of $c^*_{p,i,j}$ is used at each stage of the integration process.

The integration of equation (8.2) follows the same pattern. We define

$$C^*_{s,i,j} = \tfrac{1}{2}(C_{s,i,j+1} + C_{s,i,j}) \tag{8.16}$$

and develop the equation below:

$$t_{s,i,j+1} = t_{s,i,j} + \tfrac{1}{2}\left[\Delta\eta_{i,j+1}(t_{f,i,j+1} - t_{s,i,j+1}) + \Delta\eta_{i,j}(t_{f,i,j} - t_{s,i,j})\right] \tag{8.17}$$

where

$$\Delta\eta_{i,j+1} = \frac{\bar{\alpha}_{i,j+1} A \Delta\tau}{M_s C^*_{s,i,j}} = \frac{\Pi_{i,j+1}}{p} \tag{8.18}$$

and

$$\Delta\eta_{i,j} = \frac{\bar{\alpha}_{i,j} A \Delta\tau}{M_s C^*_{s,i,j}} = \frac{\Pi_{i,j}}{p} \tag{8.19}$$

Equation (8.19) becomes slightly more complicated if the effect of the gas residing in the regenerator at any instant is considered. In this case, equation (8.19) takes the form

$$\Delta\eta_{i,j} = \frac{\bar{\alpha}_{i,j} A}{M_s C^*_{s,i,j}}\left(\Delta\tau - \frac{m_f}{\dot{m}_{f,j} L}\Delta y\right) = \frac{\Pi_{i,j}}{p} \tag{8.20}$$

with equation (8.18), needless to say, adopting a similar form.

Equation (8.13) can be condensed in an analogous manner to that used for the linear case:

$$t_{f,i+1,j+1} = A1_{i,j+1}t_{f,i,j+1} + A2_{i+1,j+1}t_{s,i+1,j+1} + A3_{i,j+1}t_{s,i,j+1} \tag{8.21}$$

Here, $A1$, $A2$, and $A3$ are defined:

$$A1_{i,j} = \frac{2 - \Delta\xi_{i,j}}{2 + \Delta\xi_{i+1,j}} \tag{8.22}$$

$$A2_{i,j} = \frac{\Delta\xi_{i,j}}{2 + \Delta\xi_{i,j}} \tag{8.23}$$

$$A3_{i,j} = \frac{\Delta\xi_{i,j}}{2 + \Delta\xi_{i+1,j}} \tag{8.24}$$

Equation (8.17) can be similarly condensed:

$$t_{s,i+1,j+1} = B1_{i+1,j}t_{s,i+1,j} + B2_{i+1,j+1}t_{f,i+1,j+1} + B3_{i+1,j}t_{f,i+1,j} \tag{8.25}$$

where

$$B1_{i,j} = \frac{2 - \Delta\eta_{i,j}}{2 + \Delta\eta_{i,j+1}} \tag{8.26}$$

$$B2_{i,j} = \frac{\Delta\eta_{i,j}}{2 + \Delta\eta_{i,j}} \tag{8.27}$$

$$B3_{i,j} = \frac{\Delta\eta_{i,j}}{2 + \Delta\eta_{i,j+1}} \tag{8.28}$$

Substituting for $t_{f,i+1,j+1}$ in equation (8.25), we obtain

$$t_{s,i+1,j+1} = K1_{i+1,j}t_{s,i+1,j} + K2_{i+1,j}t_{f,i+1,j} + K3_{i,j+1}t_{s,i,j+1} + K4_{i,j+1}t_{f,i,j+1} \tag{8.29}$$

where

$$K1_{i,j} = \frac{B1_{i,j}}{1 - A2_{i,j+1}B2_{i,j+1}} \tag{8.30}$$

$$K2_{i,j} = \frac{B3_{i,j+1}}{1 - A2_{i,j+1}B2_{i,j+1}} \tag{8.31}$$

$$K3_{i,j} = \frac{A3_{i,j}B2_{i+1,j}}{1 - A2_{i+1,j}B2_{i+1,j}} \tag{8.32}$$

$$K4_{i,j} = \frac{A1_{i+1,j}B2_{i+1,j}}{1 - A2_{i+1,j}B2_{i+1,j}} \tag{8.33}$$

At first sight, this representation with its plethora of subscripts i, j may seem overawing. Notice, however, that provided all the $A1$, $A2$, $A3$, $B1$, $B2$, $B3$, $K1$, $K2$, $K3$, and $K4$ can be calculated for each space–time position, i, j, on the finite-difference mesh, equations (8.21), (8.25) and (8.29) retain their simple linear form. In practice, the coefficients are calculated *as the integration proceeds* as illustrated in the pseudocode set out below:

```
SUBROUTINE tstep (parameters)

comment: remaining time steps

   With the current value of time position j:

   {FOR i=0 (the entrance of the gas) DO

      {COMPUTE B1, B2, and B3

      CALCULATE t_s,0,j using equation (8.25)}

      FOR i=1,2,... DO

      {COMPUTE K1, K2, K3, and K4 for current space and
       time position.

      COMPUTE t_s,i+1,j+1 using equation (8.29).

      Now compute ''current values'' of A1, A2 and A3.

      COMPUTE t_f,i+1,j+1 using equation (8.21)}}

END
```

On the basis of these equations, integration can take place over the space–time mesh, as in the Willmott [7] method.

Inspection of equations (8.21) and (8.25) reveals that the quantities, $t_{f,i+1,j+1}$ and $t_{s,i+1,j+1}$, which are required in order to calculate the temperature-dependent thermophysical properties of the gas and the solid at the node position $i+1, j+1$ and thus the values of $\Delta\xi_{i+1,j+1}$ and $\Delta\eta_{i+1,j+1}$, are not immediately available: the quantities $t_{f,i+1,j+1}$ and $t_{s,i+1,j+1}$ have yet to be calculated! These equations *could* be solved as nonlinear equations by, say, the Newton–Raphson method. However, several different methods are available that can be incorporated within any software, the choice of method being determined by the user. These are now described.

1. *Point iteration (most accurate) method.* The values of $\bar{\alpha}_{i,j}^{(n+1)}$, $c_{p,i,j}^{(n+1)}$ and $C_{s,i,j}^{(n+1)}$ are first calculated either on the basis of the temperatures at the previous time step in the current cycle or at the present time step but in the previous cycle. The values of the gas and solid temperature are then estimated on this basis at this mesh point.

The values of $\bar{\alpha}_{i,j}^{(n+1)}$, $c_{p,i,j}^{(n+1)}$ and $C_{s,i,j}^{(n+1)}$ are now recomputed and, hopefully "better" values of the gas and solid temperatures obtained at this same mesh point. This process is continued iteratively until the gas and solid temperatures computed have converged. This process is called *point iteration*. A not dissimilar technique was described by Hofmann and Kappelmayer [10] in the context of the simulation of hot-blast stoves. Point iteration has become the most common approach to the problem.

2. *Previous cycle approximation*. Here, $\bar{\alpha}_{i,j}^{(n+1)}$, $c_{p,i,j}^{(n+1)}$ and $C_{s,i,j}^{(n+1)}$, for example, for the $(n+1)$th cycle in the period under consideration, are computed on the basis of $t_{f,i,j}^{(n)}$ and $t_{s,i,j}^{(n)}$ from the nth, that is the *previous* cycle simulated. This is the method outlined in the Willmott paper [7] and was ideal under circumstances where computing time was a significant issue.
3. *Previous time step approximation*. Here, $\bar{\alpha}_{i,j+1}^{(n)}$, $c_{p,i,j+1}^{(n)}$, and $c_{s,i,j+1}^{(n)}$ are computed on the basis of the values of $t_{i,j}^{(n)}$ and $T_{i,j}^{(n)}$, that is on the basis of the values of gas and solid temperature at the *previous time step*, in the same cycle. Special arrangements must be taken at the time step $j = 0$, *point iteration* being a possible option with *previous cycle* gas temperatures at $j = 0$ in the period under consideration, providing the starting values in the iteration.
4. *Quasilinear method*. Here, global values of $\bar{\alpha}$, c_p, and C_s are computed on the basis of the space–time average values of t_f and t_s for the period under consideration and for the previous cycle. These are applied uniformly over the whole space–time mesh for that period. These global values of $\bar{\alpha}$, c_p, and C_s are revised at the end of each cycle for each hot/cold period, separately. The model is cycled to periodic steady state. A closed method, based on Iliffe's technique [5] but using this quasilinear approach, was devised by Kulakowski and Anielewski [4].

8.6 Overall View of These Methods

If such facilities were offered in a piece of software, any user should be able to explore, with little difficulty, the relative suitability of these nonlinear methods for the simulation of the variety of fixed-bed or rotary regenerator configurations and operating conditions that the user wished to consider. Where the temperature difference between the hot- and cold-inlet gas temperatures is less than, say 200 K, it is likely that there will be little to choose between *point iteration* and the *quasilinear method* unless the thermal properties are changing rapidly at low temperatures, for example. Much will depend upon the balance of the accuracy of any experimental data and the applicability of the methods chosen by the user to compute the thermophysical data, including the heat transfer coefficient, to the regenerator configuration under consideration, on the one hand, and the numerical accuracy demanded of the simulation on the other.

Certainly, the *point iteration* method is the most accurate but certainly the slowest method described here. The *quasilinear* method is the least accurate but is the quickest to compute regenerator temperature performance. In between, stand the methods using previous cycle and previous time step temperatures to calculate the relevant thermophysical properties.

The computing speed offered on modern personal computers, however, is such that point iteration can be adopted immediately where it is vital to use a nonlinear model where, say, the temperature difference between the hot and cold inlet gas temperatures is large, even as much as 1000 K or greater. This is certainly the case where a limited number of simulations is required. Where a search for an optimal design of a regenerator must be undertaken so that a target performance is met, subject to specified operational and design constraints, very many simulations may well have to be completed during that search. In this case, it is suggested that the *previous time step* method be used (or even possibly the *quasilinear method* or a combination of the two) and the optimal design so obtained be used as a starting point for a refined design obtained using *point iteration.*

8.7 Time-Varying Φ-Factors Within the Lumped Heat Transfer Coefficients

In Chapter 6, it was described how a time-varying $\Phi(\omega)$ might be inserted into the lumped heat transfer coefficient to give $\bar{\alpha}(\omega)$ with

$$\frac{1}{\bar{\alpha}(\omega)} = \frac{1}{\alpha} + \frac{w}{\lambda_s}\Phi(\omega) \tag{6.91}$$

which is applied at all positions in the regenerator in the direction of gas flow (y) and at all time positions (τ) within the 2-D model. It was pointed out that the values of $\bar{\Lambda}$, $\bar{\Lambda}'$, $\bar{\Pi}$, and $\bar{\Pi}'$, within the model, now change continuously, following the time variations of $\Phi(\omega)$. The open and the closed methods for nonlinear models described in this chapter enable the 2-D equations to be solved numerically with this chronologically varying $\Phi(\omega)$-factor.

8.8 Variable Gas Flow Operation with Hot-Blast Stoves

We now come to discuss a specific application of the simulation techniques outlined above for nonlinear problems, in particular that for which the present author designed the *previous cycle* approach. A system of fixed-bed regenerators delivers a preheated gas whose temperature varies in sawtooth fashion (see Figure 8.1). During each cold period, the exit gas temperature declines with time until a reversal occurs and a fresh regenerator is brought into cold-period operation. The temperature of the gas presented by the system of regenerators then rises sharply before the inevitable decline, as the regenerator packing cools down, then starts again.

Hot-blast stoves or *Cowper*[†] stoves are fixed-bed regenerators used to preheat the hot blast (air) for the iron-making and zinc-smelting processes. It is necessary to eliminate the sawtooth effect, characteristic of the output temperatures of a system of stoves, to provide stable smelting furnace operation. This is achieved normally by running the stoves with a *bypass main* (see Figure 8.2) or in *staggered parallel* (see Figure 8.3).

In periodic steady-state operation, the hot-blast stove delivers preheated blast at a constant flow rate $\hat{\dot{m}}_f$ at a constant blast temperature $t'_{f,B}$. The flow of air is divided at a valve between the stove (in its cold period of operation) and the bypass main, or between two stoves, for *staggered parallel* operation, both stoves in their cold periods but out of phase by half a period.

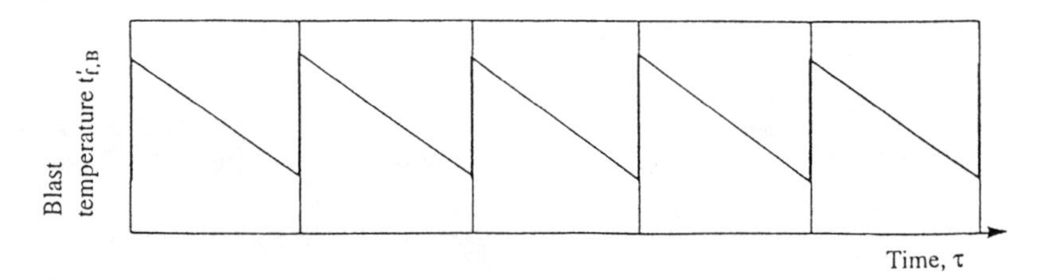

Figure 8.1: Time variation of blast temperature from Cowper stove without bypass main. (Schmidt and Willmott [9].)

[†] Green [11] reported that Cowper [12] patented this type of regenerator for use with blast furnaces in 1857. Cowper [13] discussed this invention in a paper published some three years later. Siemens [14] also developed a heat exchanger based on the regenerative principle and this was patented in 1856. However, it was in industries other than iron making, in particular glass making, and later in steel making, that Siemens's invention was exploited.

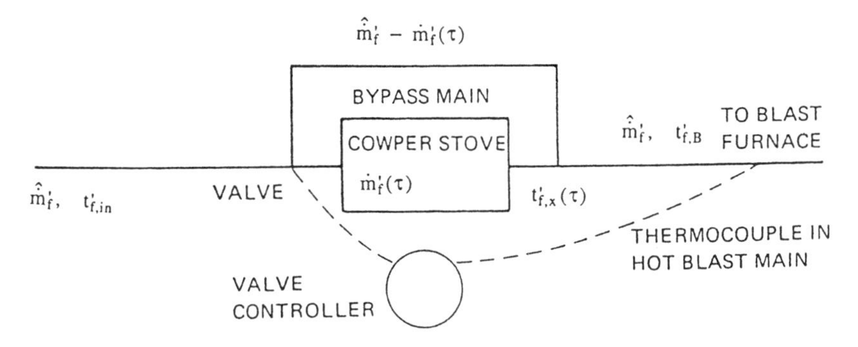

Figure 8.2: Bypass main for blast temperature control. (Schmidt and Willmott [9].)

The valve position is varied continuously by a control system in such a way that the blast temperature $t'_{f,B}$ is held constant. This imposes a steady thermal load on the system of hot-blast stoves equal to $\hat{\dot{m}}_f c'_f(t'_{f,B} - t'_{f,in})$. The time variation of $\dot{m}'_f(\tau)$, the flow rate of gas through the regenerator for bypass main operation is shown in Figure 8.4. If the system could be perfectly controlled, the instantaneous local flow rates will be determined by

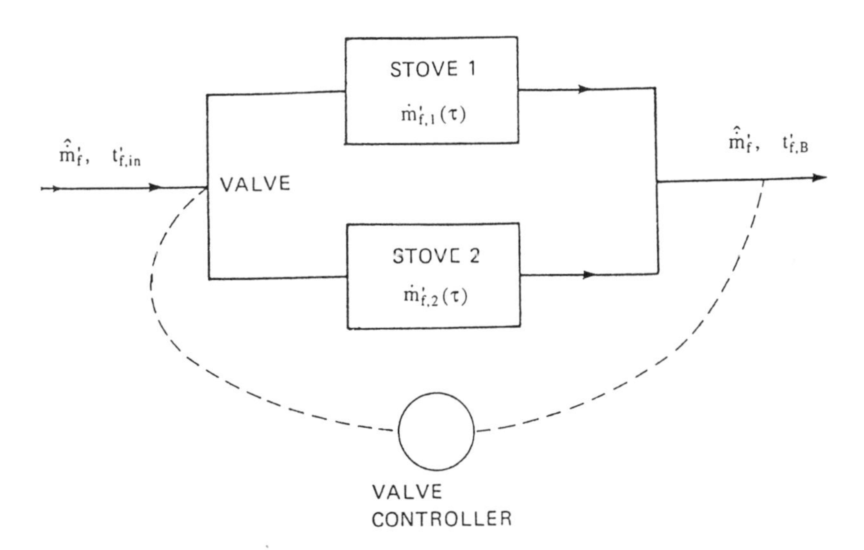

Figure 8.3: Staggered parallel arrangement for blast temperature control. (Schmidt and Willmott [9].)

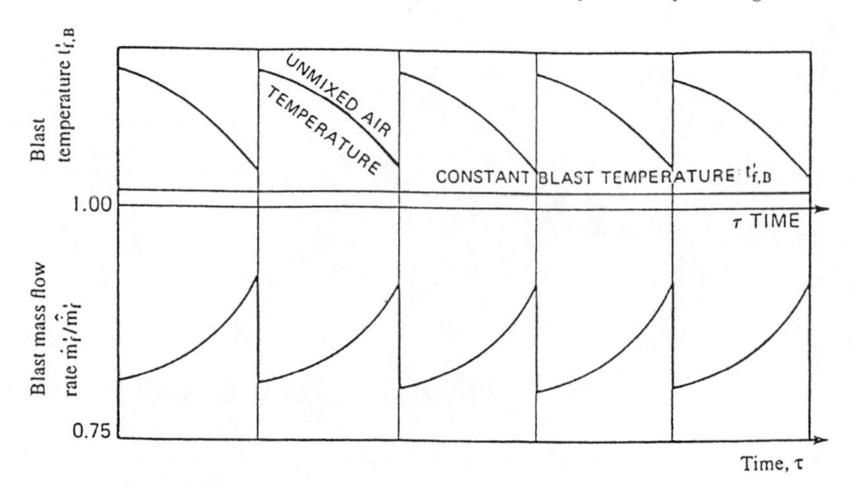

Figure 8.4: Variation of blast temperature and air flow rate: bypass main operation. (Schmidt and Willmott [9].)

Bypass main:

$$\hat{\dot{m}}'_f c'_p (t'_{f,B} - t'_{f,in}) = \dot{m}'_f(\tau) c'_p \, (t'_{f,x}(\tau) - t'_{f,in}) \tag{8.34}$$

Staggered parallel:

$$\begin{aligned} \hat{\dot{m}}'_f c'_p (t'_{f,B} - t'_{f,in}) &= \dot{m}'_{f,1}(\tau) c'_p (t'_{f,x,1}(\tau) - t'_{f,in}) \\ &\quad + \dot{m}'_{f,2}(\tau) c'_p \, (t'_{f,x,2}(\tau) - t'_{f,in}) \end{aligned} \tag{8.35}$$

where

$$\hat{\dot{m}}'_f = \dot{m}'_{f,1} + \dot{m}'_{f,2} \tag{8.36}$$

and where $\dot{m}'_{f,1}$ and $\dot{m}'_{f,2}$ are the flow rates passing through the pair of stoves, numbered respectively 1 and 2, which are operating together in staggered parallel. The values $t'_{f,x,1}(\tau)$ and $t'_{f,x,2}(\tau)$ are the corresponding exit hot-blast temperatures from the stoves prior to mixing.

The modeling of such hot-blast stove operation embodies several distinct yet intertwined procedures. Once the flow rate variation with time, $\dot{m}'_f(\tau)$, is known for the cold period, the lumped heat transfer coefficient $\bar{\alpha}'(\tau)$ and its variation with time (and temperature, if this is considered) can be computed, together with the values of $\Delta\xi_{i,j}$ and $\Delta\eta_{i,j}$ over the space–time finite-difference mesh for the cold period. The cold period can now be simulated using equations (8.21), (8.25), and (8.29).

The mechanism of controlling the time variation of flow rate, $\dot{m}_f'(\tau)$, must also be embodied in the simulation, in such a way that the mixed blast temperature, $t_{f,B}'$, is held constant. Rather different approaches are used to deal with cyclic steady state as contrasted with transient performance problems.

For stationary performance, the *previous cycle* method can be used: the variation of blast flow rate for the $(k+1)$th cycle, $\dot{m}_f'(\tau)^{(k+1)}$ is calculated on the basis of the exit air temperatures in the kth cycle. The calculation of hot-blast stove performance not only involves the calculation of the variation of the gas and solid temperatures at periodic steady state but also how the hot-blast flow rate, $\dot{m}_f'(\tau)$, changes with time during the cold period. It is necessary to impose an additional boundary condition for the problem. For bypass main operation, it is typical to set the ratio of the blast flow rate through the stove at the very end of the cold period and the total hot-blast flow rate, $\dot{m}_f'(P')/\hat{\dot{m}}_f'$, to be equal to a constant K, usually with $0.8 \leq K \leq 1.0$. In this case, the mixed blast temperature $t_{f,B}'$ is unknown and is calculated by the simulation. Alternatively, it is possible to set the value of $t_{f,B}'$ in advance and then to calculate, in a similar manner, the value of the hot-period gas flow rate, $\dot{m}_f$ which allows this value of the blast temperature, $t_{f,B}'$, to be realized with $\dot{m}_f'(P')/\hat{\dot{m}}_f' = K$ for a specified value of K. (This approach avoids the situation, if the blast temperature $t_{f,B}'$ is set too high, where a hot-blast flow rate is calculated with $\dot{m}_f'(P')/\hat{\dot{m}}_f' > 1$!) The simulation of staggered parallel follows along similar lines.

Transient performance is much more difficult to represent. Each and every cycle must be simulated during the transient phase. Either technique described above can be used *except* that the variation of blast flow rate for the current cycle, $\dot{m}_f'(\tau)$ is calculated on the basis of the exit air temperatures in the previous simulation of this *same* current cycle and this process is iterated until the values of the flow rate for this cycle, $\dot{m}_f'(\tau)$, have converged.

There is a further problem, however, when nonstationary performance of staggered parallel is considered. When a change in thermal load is imposed, each of the stoves will be at a different phase of its operation and it is thus a requirement that all three, usually four stoves be simulated separately. For steady-state calculations, provided all four stoves are identical, only one stove need be simulated.

Razelos and Benjamin [15] offer an interesting example of a Cowper stove installation together with graphical results typical of those possible using the simulation techniques described above. Figure 8.5 and Figure 8.6 contrast the hot-blast flow rates through the stoves between bypass main (sometimes called *serial*) operation and staggered parallel operation. Razelos and Benjamin present exit temperatures using the dimensionless temperature scales $T_{f,x}$ and $T_{f,x}'$ as defined by equation (4.5). The most

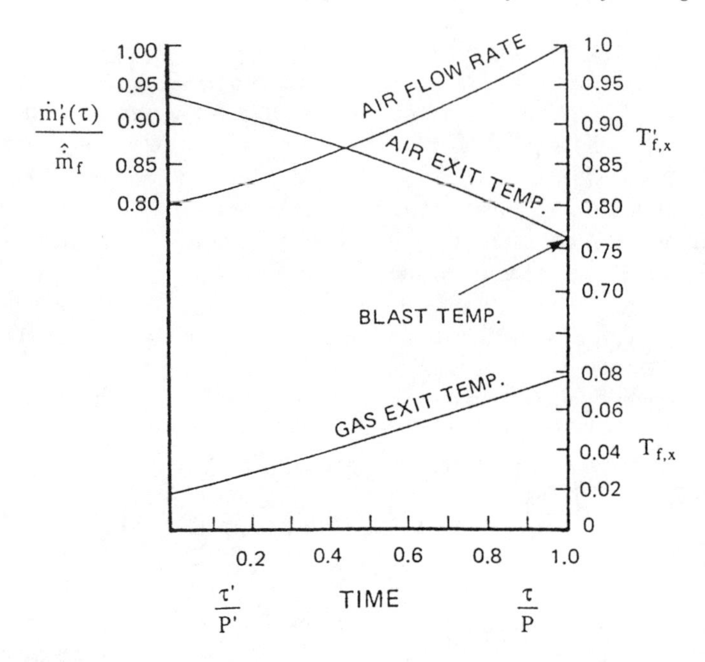

Figure 8.5: Variation of dimensionless exit blast temperature, air flow rate, and hot-period exit gas temperature: bypass main operation. (Schmidt and Willmott [9].)

important hot-blast stove dimensions and operating conditions are given in Table 8.1.

8.9 Comparison of Several Models for Different Regenerators

The design and operation of glass furnace regenerators and regenerative burners has been outlined in Chapter 4. Hill and Willmott [1] consider two such regenerators with a view to comparing the effectiveness of different models of regenerators. In particular, a glass furnace regenerator with a packing of refractory bricks and a regenerative burner, whose heat capacitance is provided by a packed bed of spheres, have been examined. The data for these regenerators is given in Table 8.2. It can readily be shown that for this data, the effect of the thermal capacity of any gas resident in the regenerator packing voids influences the reduced period, Π, for either the hot or cold periods by less than 0.35% and is thus negligible. This thermal capacity

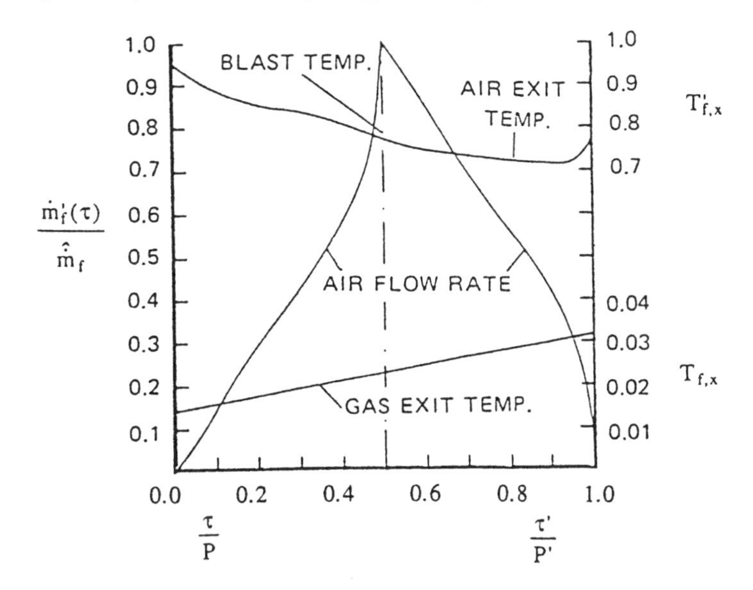

Figure 8.6: Variation of dimensionless exit blast temperature, air flow rate, and hot-period exit gas temperature: staggered parallel operation. (Schmidt and Willmott [9].)

Table 8.1: Cowper stove dimensions operating conditions [15]

Heating surface area $A = 32\,940\,\mathrm{m}^2$
Hot-period gas inlet temperature $t_{f,in} = 1400°\mathrm{C}$
Cold-blast inlet temperature $t'_{f,in} = 93°\mathrm{C}$

Required hot-blast flow rate $\hat{\dot{m}}'_{f,B} = 59\,\mathrm{m}^3/\mathrm{s}$
Target blast temperature $t'_{f,B} = 1093°\mathrm{C}$
Time on blast (cold period) $P' = 5400\,\mathrm{s}$
Maximum exit hot gas temperature $t_{f,x}(P) \leq 343°\mathrm{C}$

Hot-period gas flow rate:
Three-stove serial arrangement $\dot{m}_f = 15.23\,\mathrm{m}^3/\mathrm{s}$
Duration of hot period $P = 14\,400\,\mathrm{s}$

Four-stove staggered parallel arrangement $\dot{m}_f = 31.46\,\mathrm{m}^3/\mathrm{s}$
Duration of hot period $P = 3600\,\mathrm{s}$

Table 8.2: Data for comparison of methods of calculation

	Glass furnace		Regenerative burner	
Regenerator configuration				
Length, L (m)	4.380e+0		5.850e−1	
Area, A (m^2)	2.719e+2		8.233e+1	
Mass, M (kg)	1.990e+4		9.500e+2	
Hydraulic diameter, hd (m)	1.700e−1		—	
Semithickness, d_s (m)	2.440e−2		9.500e−3	
Free flow area, F (m^2)	2.638e+0		3.651e−1	
Solid properties				
Thermal conductivity, λ (J/(m s K)	4.000e+0		$1.859e{-}3\Theta - 2.479e{-}1$	
Specific heat, C (J/(kg K))	1.175e+3		$4.178e{-}1\Theta + 6.998e{+}2$	
Density, ρ (kg/m^3)	3.000e+3		3.644e+3	
Emissivity, ε (used for h_r, %)	38		—	
Gas composition				
	Hot	Cold	Hot	Cold
N_2 (%)	75.0	Air	72.2	Air
CO_2 (%)	14.0		8.5	
H_2O (%)	11.0		17.0	
O_2 (%)	0.0		2.3	
Operational characteristics				
	Hot	Cold	Hot	Cold
Inlet temperature, θ_{in} (K)	1678	383	1670	300
Period length, P (s)	1200	1200	300	300
Flow rate, W (kg/s)	1.75	1.42	1.122	1.246
Heat transfer coefficient correlations	Kistner's convective heat transfer coefficient [16]; Hottel charts [17] for the radiative heat transfer coefficient in the hot period		Denton et al.'s convective heat transfer coefficient [2]; Radiative effects are ignored	

does not influence reduced length, Λ. The periodic steady-state performance of these regenerators as computed using several models has been compared.

1. *Simple linear model.* The linear method of solution of Hill and Willmott [18] has been used with the hot- and cold-period dimensionless parameters, Λ, Π, Λ', and Π' computed upon the basis of the same, equal solid and gas reference temperatures, namely

$$t_{s,ref} = t_{f,ref} = \tfrac{1}{2}(t_{f,in} + t'_{f,in})$$

   Globally, within each period separately, applied values of the thermophysical properties so calculated are used in the development of the parameters Λ, Π, Λ', and Π' to be used in the hot and cold periods.
2. *Quasilinear model.* The linear solution [18] of the differential equations is obtained where Λ and Π for the hot period and Λ' and Π' for the cold period are evaluated using the Kulakowski and Anielewski [4] approach described earlier in this chapter.
3. *Spatially nonlinear model.* Here spatial, only, variations of Λ and Π are computed for each period of operation, based on the distance variation of the time mean values of the gas and solid temperatures in each period of regenerator operation, considered separately. A *closed* method of calculation for this model is described later in this chapter.
4. *The full nonlinear model.* The open method, using point iteration, of Willmott [7] was employed in which chronological as well as spatial variations in Λ and Π are allowed in each period of operation. Again, this model has been outlined previously in this chapter.

This is purely a theoretical comparison; no attempt has been made to compare theoretical and experimental results.

The solid temperature spatial distributions at the start and the end of the hot period for the *simple linear model* and the *quasilinear model* are compared with the same temperatures calculated using the *spatially nonlinear model* in Figures 8.7 and 8.8 for a glass furnace regenerator and Figures 8.9 and 8.10 for a regenerative burner. In all these figures the hot solid temperatures computed for them are shown for the *spatially nonlinear model* as bold lines, whereas those predicted by the alternative approaches, 1 and 2, are shown as dotted lines. The differences between the temperature profiles in Figures 8.7 to 8.10 are shown in Figures 8.11 to 8.14, respectively.

In the case of the glass furnace regenerator it is clearly seen from Figures 8.7 and 8.8 that the prediction of the regenerator performance at periodic steady state using the simple linear is poor, relative to the spatially nonlinear

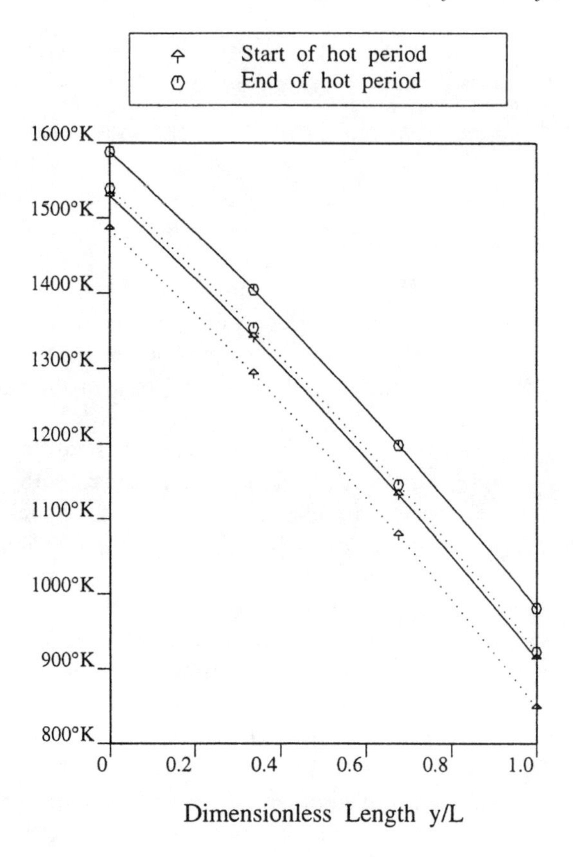

Figure 8.7: Solid temperatures at the start and end of the hot period for a glass furnace regenerator computed using the spatially nonlinear model and the simple linear model. (Hill and Willmott [1].)

model, with differences between the temperatures predicted by the simple linear model and this nonlinear model exceeding 45 K. The reason for the poor results in this case is because the reference temperatures $t_{s,ref} = t_{f,ref} = \frac{1}{2}(t_{f,in} + t'_{f,in})$ are inadequate. They are not representative of the solid and fluid temperatures in the hot period, for example, of the operation of glass furnace regenerators, which must be kept above a certain threshold to prevent the solidification of vapors carried over from the furnace. The average solid and fluid temperatures are, in fact, much higher than $\frac{1}{2}(t_{f,in} + t'_{f,in})$, resulting in a poor estimation of the relevant thermophysical properties and a weak prediction of regenerator performance by the simple linear approach.

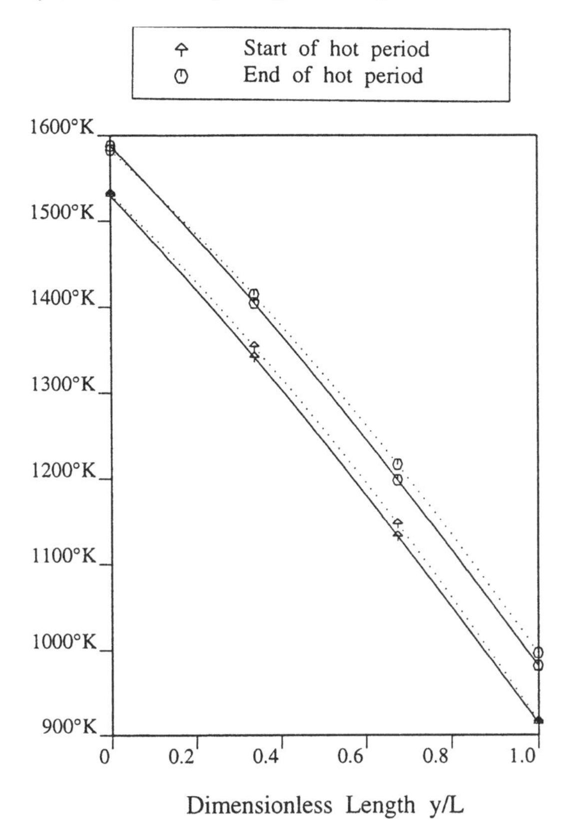

Figure 8.8: Solid temperatures at the start and end of the hot period for a glass furnace regenerator computed using the spatially nonlinear model and the quasilinear model. (Hill and Willmott [1].)

Matters are improved if the quasilinear model is used instead, as suggested in the Kulakowski and Anielewski [4] development. Figure 8.8 shows the equilibrium performance of the glass furnace regenerator estimated using this quasilinear approach but it will be seen that differences of up to 20 K are still encountered when the temperatures are compared with those produced using the spatially nonlinear model. These differences arise from the inability of the quasilinear solution to take into account the large variation in the heat transfer coefficient down the length of the regenerator, especially in the hot period when radiative effects are significant.

The *spatially nonlinear model* discussed previously gives excellent agreement with the temperatures produced by the full nonlinear model, calculated

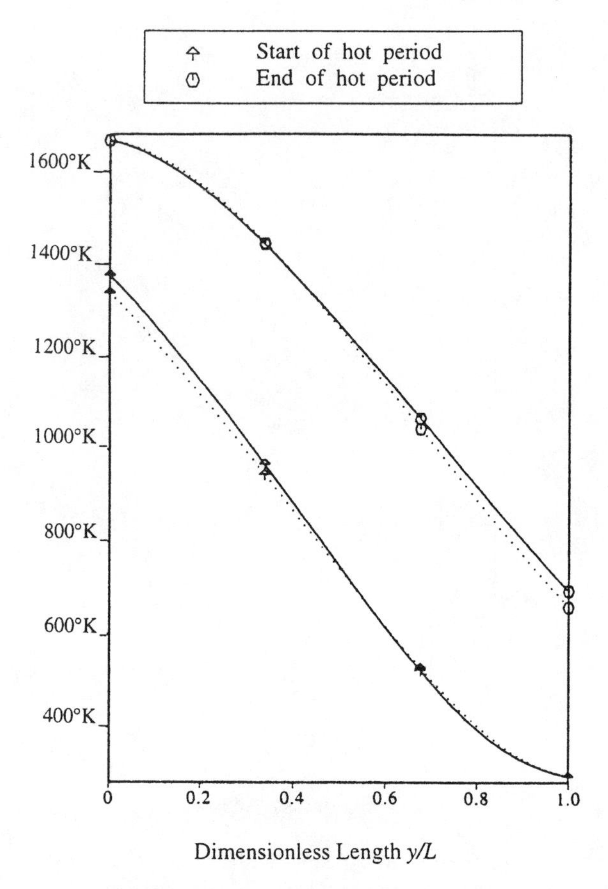

Figure 8.9: Solid temperatures at the start and end of the hot period for a regenerative burner computed using the spatially nonlinear model and the simple linear model. (Hill and Willmott [1].)

using the open scheme of Willmott [7]. Here a maximum difference of only 1 K between the solid temperatures is predicted between the two schemes.

For the regenerative burner the results obtained using the *simple linear model* are much better than those obtained for the glass furnace regenerator. This is simply because the average solid and fluid temperatures in the regenerative burner are closer to $\frac{1}{2}(t_{f,in} + t'_{f,in})$, which is used as the reference temperature. The temperatures computed using this simple linear method are different by less than 40 K compared with those predicted by the spatially nonlinear model, and, on average, these differences are half this value.

The Kulakowski and Anielewski [4] quasilinear approach gives slightly better agreement with the full nonlinear model than the simple linear

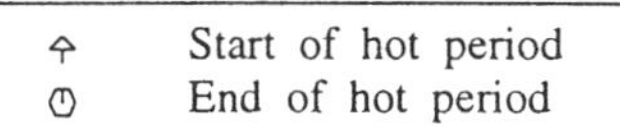

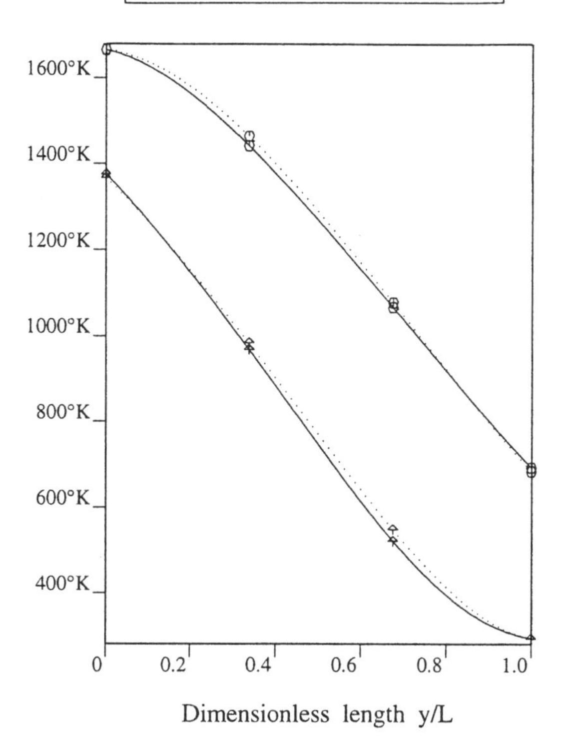

Figure 8.10: Solid temperatures at the start and end of the hot period for a regenerative burner computed using the spatially nonlinear model and the quasi-linear model. (Hill and Willmott [1].)

approach but differences of up to 30 K are still encountered between the temperature profiles produced between the quasilinear and the spatially nonlinear approaches.

Again, good agreement between the spatially nonlinear model and the full nonlinear model has been found for the regenerative burner, with a maximum difference of 4 K between the results produced by the two models. The discrepancy between the spatially nonlinear scheme and the full nonlinear model is greater for the regenerative burner than the glass furnace regenerator because the chronological variations in the solid and fluid temperatures are much greater in the case of the regenerative burner than those observed in the glass furnace regenerator. The small effect of allowing only spatial variations in the thermophysical properties

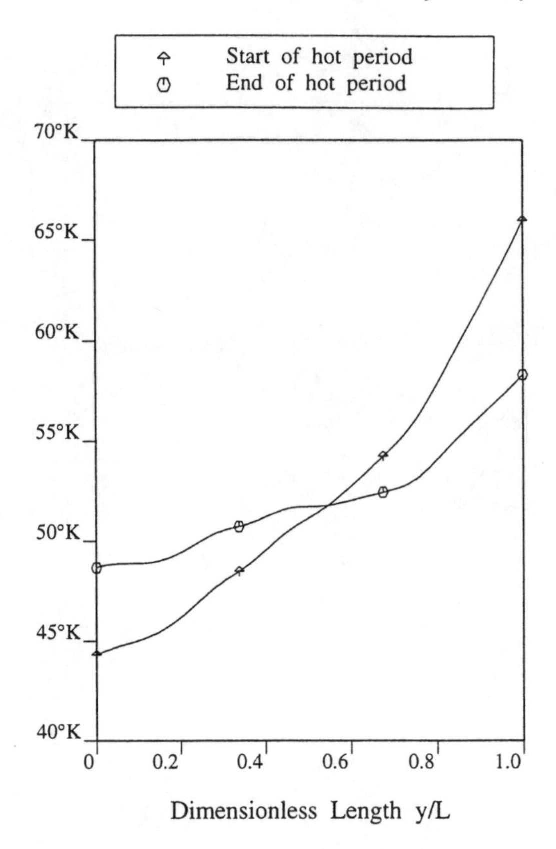

Figure 8.11: Differences between models for the solid temperatures at the start and end of the hot period for a glass furnace regenerator computed using the spatially nonlinear model and the simple linear model. (Hill and Willmott [1].)

in spatial nonlinear model can thus be detected in this instance. Should the regenerator be considered to be operated with a shorter cycle time, the chronological swing in solid temperature would become smaller and the effect of the "spatial only assumption" in this model would be diminished.

8.10 Hill Method of Analysis for the Spatially Nonlinear Model

The method adopted by Hill [1] is based on a technique developed by him for the linear model [18]. Hill acknowledges that this approach is under-

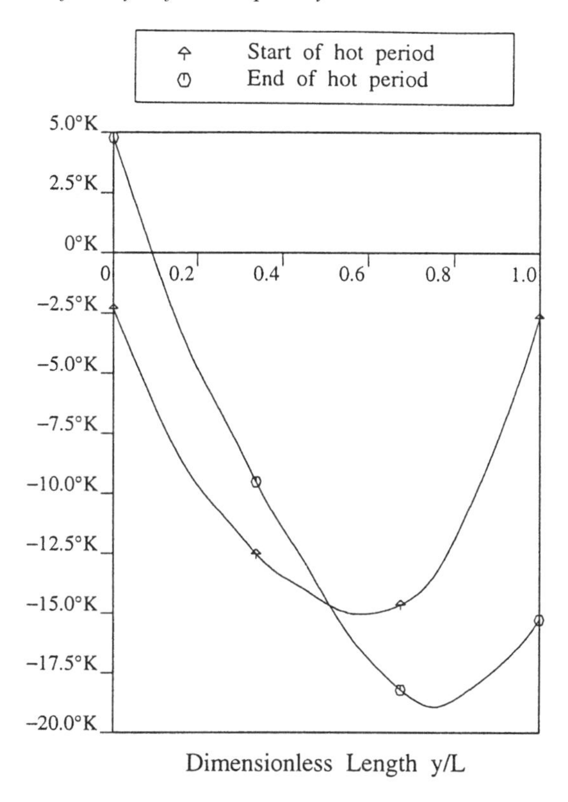

Figure 8.12: Differences between models for the solid temperatures at the start and end of the hot period for a glass furnace regenerator computed using the spatially nonlinear model and the quasilinear model. (Hill and Willmott [1].)

pinned by a method proposed by Razelos [19] in which the partial differential equations (5.1) and (5.2) are reduced to a set of $m+1$ ordinary differential equations in the time domain. This is realized by discretizing equation (5.1) and using the transformation $\{\Psi_k = e^{\eta} T_{f,k} \mid k = 0, 1, 2, \ldots, m\}$, where k is the spatial position in the regenerator. Razelos used the inaccurate and unstable Euler method to achieve this discretization, whereas Hill turned to the trapezoidal method to realize an accurate and robust method.

For the nonlinear problem, the method is not dissimilar from that of Kulakowski and Anielewski. In their approach, fixed values of the thermophysical properties, including the heat transfer coefficient, are selected for the hot and cold periods under consideration. Such values are then revised, iteratively, as in the quasilinear model.

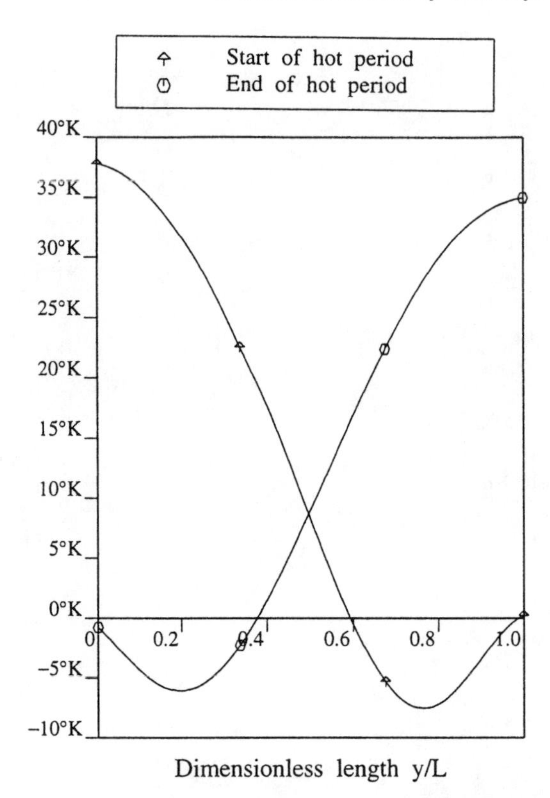

Figure 8.13: Differences between models for the solid temperatures at the start and end of the hot period for a regenerative burner computed using the spatially nonlinear model and the simple linear model. (Hill and Willmott [1].)

In this method described by Hill, it is assumed that the way in which the relevant natural parameters vary, spatially, as a function of temperature, is known. The temperatures are calculated at $m+1$ equally spaced positions in the regenerator, $\{y_k \mid k = 0, 1, 2, \ldots, m\}$ with

$$y_k = k\,\Delta y \quad (0 \leq k \leq m) \quad \text{with} \quad \Delta y = \frac{1}{m} \tag{8.37}$$

The temperature-dependent properties vary between these $(m+1)$ positions and we write

$$\Pi_k = \frac{\bar{\alpha}_k AP}{MC_{s,k}} \tag{8.38}$$

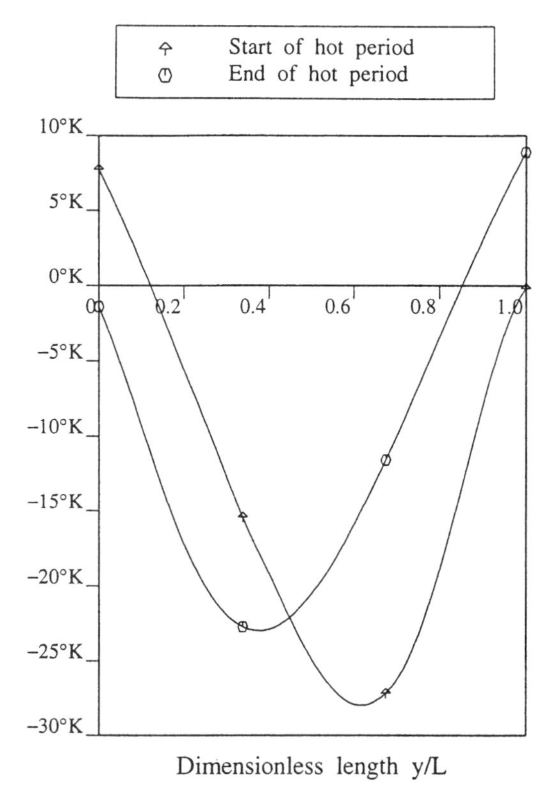

Figure 8.14: Differences between models for the solid temperatures at the start and end of the hot period for a regenerative burner computed using the spatially nonlinear model and the quasilinear model. (Hill and Willmott [1].)

$$\Lambda_k = \frac{\bar{\alpha}_k A}{\dot{m}_f c_{p,k}} \tag{8.39}$$

On the basis of the parameters Λ_k and Π_k for the hot period and Λ_k' and Π_k' for the cold period, with $k = 0, 1, 2, \ldots, m$ a closed solution to the differential equations is obtained.

At each position, k, a time mean gas temperature, $\bar{t}_{f,k}$ and a time mean solid temperature, $\bar{t}_{s,k}$ can then be computed for the hot period. In a similar fashion, $\bar{t}_{f,k}'$ and $\bar{t}_{s,k}'$ can be calculated for the cold period at the $m+1$ positions under consideration. Using these as reference temperatures, updated values of the local thermophysical properties can be computed and revised values of the parameters Λ_k and Π_k for the hot period and Λ_k' and Π_k' for the subsequent cold period, found.

A fresh closed solution can be determined and again Λ_k, Π_k, Λ_k', and Π_k' are revised. This process is allowed to continue iteratively to the convergence of the calculated values of $t_{f,k}(\tau)$ and $t_{s,k}(\tau)$ for the hot and cold periods of the equilibrium cycle.

It is important to note that this is *not* an open method of solution where the model is allowed to cycle to equilibrium. If the values of Λ_k, Π_k, Λ_k', and Π_k' were in fact known, the closed method would obtain a solution directly; an open method would still have to cycle to equilibrium. Because the values are unknown in advance, the closed method is applied repeatedly using values of Λ_k, Π_k, Λ_k', and Π_k' $(0 \le k \le m)$ which are revised in iterative fashion, as in the quasilinear method, except that here, the dimensionless parameters in both periods are allowed to vary in spatial fashion.

The spatial nonlinearities are built, in other words, into the closed method on the assumption that it is known how the thermophysical properties vary down the length of the regenerator in each period. Just how they vary is determined by this iterative application of the closed method by which, also, the temperature performance of the regenerator at cyclic equilibrium is calculated. In parallel, the spatial variations of the time average values of the thermophysical properties are located.

In the paper by Hill [1], on the basis of these parameters Λ_k, Π_k, equations (8.2) and (8.1) are presented in the form

$$\frac{\partial t_{s,k}(\tau)}{\partial \tau} = \Pi_k\left[t_{f,k}(\tau) - t_{s,k}(\tau)\right] \tag{8.40}$$

$$\frac{\partial t_{f,k}(\tau)}{\partial y} = \Lambda_k\left[t_{s,k}(\tau) - t_{f,k}(\tau)\right] \tag{8.41}$$

The trapezoidal rule is now applied, as for the linear model, to equation (8.41) to give

$$t_{f,k+1}(\tau) = a_k\, t_{f,k}(\tau) + b_k\, t_{s,k+1}(\tau) + c_k\, t_{s,k}(\tau) \qquad (0 \le k \le m-1) \tag{8.42}$$

where

$$a_k = \frac{2 - \Delta y\, \Lambda_k}{2 + \Delta y\, \Lambda_{k+1}} \qquad b_k = \frac{\Delta y\, \Lambda_{k+1}}{2 + \Delta y\, \Lambda_{k+1}} \qquad c_k = \frac{\Delta y\, \Lambda_k}{2 + \Delta y\, \Lambda_{k+1}} \tag{8.43}$$

The transformation Ψ is now introduced:

$$\Psi_k(\tau) = e^{\Pi_k \tau} t_{s,k}(\tau) \quad \text{for} \quad 0 \le k \le m \tag{8.44}$$

This yields

$$\frac{d\Psi_k(\tau)}{d\tau} = e^{\Pi_k \tau}\left(\Pi_k\, t_{s,k}(\tau) + \frac{dt_{s,k}(\tau)}{d\tau}\right) \qquad (0 \leq k \leq m) \tag{8.45}$$

which, using equation (8.40), becomes

$$\frac{d\Psi_n(\tau)}{d\tau} = e^{\Pi_k \tau}\Pi_k\, t_{f,k}(\tau) \qquad (0 \leq k \leq m) \tag{8.46}$$

Substituting for $t_{f,k}(\tau)$ using equation (8.42) gives

$$\frac{d\Psi_k(\tau)}{d\tau} = e^{\Pi_k \tau}\Pi_k\big(a_{k-1}t_{f,k-1}(\tau) + b_{k-1}t_{s,k}(\tau) + c_{k-1}t_{s,k-1}(\tau)\big)$$
$$(1 \leq k \leq m) \tag{8.47}$$

which can be written as

$$\frac{d\Psi_k(\tau)}{d\tau} = \Pi_k\, b_{k-1}\, \Psi_k(\tau) + \frac{e^{\Pi_k \tau}}{e^{\Pi_{k-1} \tau}}\Pi_k\, c_{k-1}\, \Psi_{k-1}(\tau)$$
$$+ \frac{e^{\Pi_k \tau}}{e^{\Pi_{k-1} \tau}}\frac{\Pi_k}{\Pi_{k-1}} a_{k-1} \frac{d\Psi_{k-1}(\tau)}{d\tau} \tag{8.48}$$

If we define the following:

$$\left.\begin{aligned} \zeta_{k-1} &= \Pi_k - \Pi_{k-1} \\ A_{k-1} &= \frac{\Pi_n}{\Pi_{k-1}} a_{k-1} \\ B_{k-1} &= \Pi_k\, b_{k-1} \\ C_{k-1} &= \Pi_k\, c_{k-1} \end{aligned}\right\} \quad (1 \leq k \leq m) \tag{8.49}$$

then the auxiliary equation is written as

$$\frac{d\Psi_k(\tau)}{d\tau} = B_{k-1}\, \Psi_k(\tau) + e^{\zeta_{k-1} \tau}\left(C_{k-1}\Psi_{k-1}(\tau) + A_{k-1}\frac{d\Psi_{k-1}(\tau)}{d\tau}\right)$$
$$(1 \leq k \leq m) \tag{8.50}$$

Any solution to equation (8.46) must satisfy this auxiliary equation.

The General Solution to Equation (8.46)

Using the definitions

$$\left.\begin{aligned} D_k &= (C_k + A_k\, B_{k-1}) \\ E_k &= (\zeta_k + B_{k-1} - B_k) \end{aligned}\right\} \quad (1 \leq k \leq m-1) \tag{8.51}$$

the general solution to equation (8.46) (when $t_{f,0} = 0$) can be expressed in the form

$$G_k(\tau) = e^{B_{k-1}\tau}\left(K_k + \sum_{j=0}^{k-1} K_j f_{j,k}(\tau)\right) \qquad (1 \le k \le m) \tag{8.52}$$

where $G_0(\tau) = K_0$ and the K_j values are unknown constants of integration. The function $f_{j,k}(\tau)$ is defined in the following way:

$$f_{j,k}(\tau) = \begin{cases} C_0 \int e^{(\zeta_0 - B_0)\tau}\, d\tau \quad \text{for} \quad k=1, \quad j=0 \\ D_j \int e^{E_j \tau}\, d\tau \quad \text{for} \quad j=k-1, \quad k>1 \\ \int e^{E_{k-1}\tau}\left(D_{k-1} f_{j,k-1}(\tau) + A_{n-1}\dfrac{df_{j,k-1}(\tau)}{d\tau}\right) d\tau \quad \text{for} \quad 0<j<k-2 \end{cases} \tag{8.53}$$

It is set to zero at $\tau = 0$ by introducing an appropriate constant of integration during the indefinite integration process.

8.11 Solution for Constant Inlet Temperatures

In this case we have $t_{f,0}(\tau) = t_{f,in}$ and the particular solution, $R_k(\tau)$, to equation (8.46) takes the form

$$R_k(\tau) = t_{f,in}\, e^{\Pi_n \tau} \tag{8.54}$$

The equations for the solid are now given:

$$t_{s,0}(\tau) = e^{-\Pi_0 \tau} K_0 + t_{f,in} \tag{8.55}$$

$$t_{s,k}(\tau) = e^{(B_{k-1} - \Pi_k)\tau}\left(K_k + \sum_{j=0}^{k-1} K_j f_{j,k}(\tau)\right) + t_{f,in} \qquad (1 \le k \le m) \tag{8.56}$$

Applying the contraflow reversal conditions to the equations above for the solid yields a set of $2(m+1)$ equations in the $2(m+1)$ unknowns K_j, K_j'. These can be expressed in matrix form:

$$\begin{bmatrix} I & F' \\ F & I \end{bmatrix}\begin{bmatrix} \boldsymbol{k} \\ \boldsymbol{k}' \end{bmatrix} = \begin{bmatrix} y_1 \\ y_2 \end{bmatrix} \tag{8.57}$$

where I and F are $(m+1) \times (m+1)$ matrices defined below.

The matrix I is the identity (unit) matrix and

$$F_{i,j} = \begin{cases} -e^{-\Pi_0} & \text{if } (i+j) = m \text{ and } i = m \\ -e^{B_{m-i-1}-\Pi_{m-i}} & \text{if } (i+j) = m \text{ and } i \neq m \\ -e^{B_{m-i-1}-\Pi_{m-i}} f_{j,m-i}(\tau)|_{\tau=1} & \text{if } (i+j) < m \\ 0 & \text{if } (i+j) > m \end{cases} \tag{8.58}$$

and $\boldsymbol{k}$, $\boldsymbol{y}_1$, and $\boldsymbol{y}_2$ are given by

$$\boldsymbol{k} = [K_0, K_1, K_2, \ldots, K_m]^T \tag{8.59}$$

$$y_{1,i} = t'_{f,in} - t_{f,in} = -y_{2,i} \qquad (0 \leq i \leq m) \tag{8.60}$$

Equation (8.57) can be thought of as a pair of simultaneous equations in the two unknowns $\boldsymbol{k}$ and $\boldsymbol{k}'$, which has the solution

$$\begin{bmatrix} I & F' \\ 0 & (I - FF') \end{bmatrix} \begin{bmatrix} \boldsymbol{k} \\ \boldsymbol{k}' \end{bmatrix} = \begin{bmatrix} \boldsymbol{y}_1 \\ \boldsymbol{y}_2 - F\boldsymbol{y}_1 \end{bmatrix} \tag{8.61}$$

This equation yields

$$(I - FF')\boldsymbol{k}' = \boldsymbol{y}_2 - F\boldsymbol{y}_1 \tag{8.62}$$

This is a system of only $(m + 1)$ equations but does require the computation of the product matrix FF'. The diagonal shape of the matrices allows this product matrix to be computed efficiently. Once the vector $\boldsymbol{k}'$ has been located, the vector $\boldsymbol{k}$ is easily obtained using

$$\boldsymbol{k} = \boldsymbol{y}_1 - F'\boldsymbol{k}' \tag{8.63}$$

This reduction of the solution of the *nonsymmetric* problem to an $(m + 1) \times (m + 1)$ matrix system was first realized by Hill [20] for the linear model and was subsequently applied to the nonlinear problem as described here. It should be compared with a similar method described in Chapter 7.

8.12 Obtaining the Solid and Fluid Temperatures at Periodic Steady State

If it is assumed that the values of Π_k and Λ_k are known in both periods, then the temperatures may be computed in the manner described below.

1. Compute the a_k, b_k, c_k, ζ_k, A_k, B_k, C_k, D_k, and E_k values for both periods directly from the Π_k and Λ_k values.
2. Compute the values of $f_{j,k}(\tau)$ at $\tau = 1$ for both periods.
3. Set up and solve the system of equations (8.61) to obtain the K_j, K_j' constants of integration.

4. Compute the solid temperatures at any time τ by first evaluating the $f_{j,k}(\tau)$ values and then using the equations for the solid (8.55) and (8.56).
5. Compute the gas temperatures at any time τ using equation (8.42) once the solid temperatures have been evaluated.

The values of the nonlinear variables Π_k, Λ_k are dependent on the gas and solid temperatures and these temperatures cannot be computed until the values of Π_k and Λ_k are known. In order to resolve this problem the iterative scheme described above is employed. Initial estimates of the gas and solid temperatures are made using the *linear* model and these provide the basis upon which initial values of Π_k and Λ_k can be calculated. The values of Π and Λ supplied to the linear model are evaluated using as reference temperatures the values of $t_{f,ref} = t_{s,ref} = \frac{1}{2}(t_{f,in} + t'_{f,in})$. New gas and solid temperatures are now computed, using this nonlinear *closed* method and the process repeated until convergence is attained.

The number of iterations required to reach convergence is a measure of how nonlinear the system is and is *not* related to the "length" of the regenerator, as is the case for an open method. This iterative procedure is a simple first-order process for producing iterate values of Π_k and Λ_k. Acceleration techniques, such as Aitken's δ^2 process (see Morris [21]), have been successfully applied to the scheme and convergence has been found to be very rapid in all cases considered.

At any position $y_k = k\,\Delta y$, the Π_k and Λ_k values are computed using the time-averaged gas $\bar{t}_{f,k}$ and solid $\bar{t}_{s,k}$ temperatures at that point. In the examples considered earlier in this chapter, the time-averaged temperatures were computed using Simpson's rule at the points in time $\tau = 0,\ \frac{1}{2}P,\ P$. The necessary formulae for the hot period are

$$\bar{t}_{f,k} = \tfrac{1}{6}\left[t_{f,k}(0) + 4\,t_{f,k}(\tfrac{1}{2}P) + t_{f,k}(P)\right] \tag{8.64}$$

$$\bar{t}_{s,k} = \tfrac{1}{6}\left[t_{s,k}(0) + 4\,t_{s,k}(\tfrac{1}{2}P) + t_{s,k}(P)\right] \tag{8.65}$$

Corresponding formulae were used for the cold period.

8.13 Assessment of this Closed Method for Nonlinear Problems

Hill [1] modified the open method of Willmott [7] to permit the spatial-only nonlinear model to be implemented. No differences could be detected in the temperature profiles computed by this technique and by the closed method described above. As one would expect, different methods calculated the same solutions for the same model. Hill confirmed that, as in the case of the corresponding closed method for the linear model, this method is extremely stable for large ranges in the parameters Λ and Π. Hill [1] considered $0.1 \leq \Lambda \leq 1000$ and $0.5 \leq \Pi \leq 50$. This stability remained secure for significant variations of Λ and Π down the length of the regenerator. Hill [8] extended this closed method to allow for time as well as spatial variations of Λ and Π, although this is not discussed in this text.

Finally, Hill [1] estimated that the average computing time required for the open method was *four* times greater than that required by the closed method described here. Set against this is the fact that the open method has a more flexible and accurate way of representing chronological variations of the thermophysical properties of the gases and the packing, as functions of temperature. Hill's closed method must certainly be a candidate for application in the context of the optimal design of regenerators. The reader is referred to a paper by Henry and Willmott [22].

8.14 Radiative Heat Transfer Between Gas and Solid Surface in Regenerators

In certain high-temperature regenerators, especially glass furnace regenerators, as well as Cowper stoves, the heat exchanger is fired by freshly combusted gas containing significant proportions of carbon dioxide and water vapor. The channels of the packing are sufficiently wide to avoid blockage by deposited dirt or the carryover of solid material in the hot period of operation. The radiative heat flux between CO_2 and H_2O vapor and the surface of the regenerator packing, however, is proportional to what is called the *beam length*, which in this case is the width of the channels. These factors combine to make radiative heat transfer important and, in some cases, as significant, if not larger than, the convective heat transfer between the packing surface and the gas in the hot period.

Depending upon the relative importance of convective and radiative heat transfer in the particular circumstances of regenerator construction and operation under consideration, different approximate representations of radiative heat transfer can be employed.

If t_f is the local gas temperature and $t_{s,o}$ is the corresponding surface solid temperature (K), the radiation between gas and solid is proportional to $t_f^4 - t_{s,o}^4$. Nevertheless, the underlying philosophy adapted here and in the text by Schmidt and Willmott [9] as well as by other authors, is to maintain the linearity of the model described earlier in this chapter.

The heat flux due to gas solid radiation is q_R where

$$q_R = \sigma\left(\frac{\varepsilon_s + 1}{2}\right)(\varepsilon_f t_f^4 - \alpha_f t_{s,o}^4) \tag{8.66}$$

where ε_s is the emissivity of the solid surface, ε_f is the emissivity of the gas, and α_f is the absorptivity of the gas. σ is the Stefan–Boltzmann constant. Now ε_f and α_f are functions of the gas temperature and are proportional to the percentage of CO_2 in the gas and the beam length. The radiation due to H_2O vapor is treated in a similar manner.

The value of the emissivity, ε_f, can be obtained from the Hottel charts [23] (see Figure 8.15 for carbon dioxide; a similar chart exists for water vapor), given the product of the beam length, L_0, and the proportion of

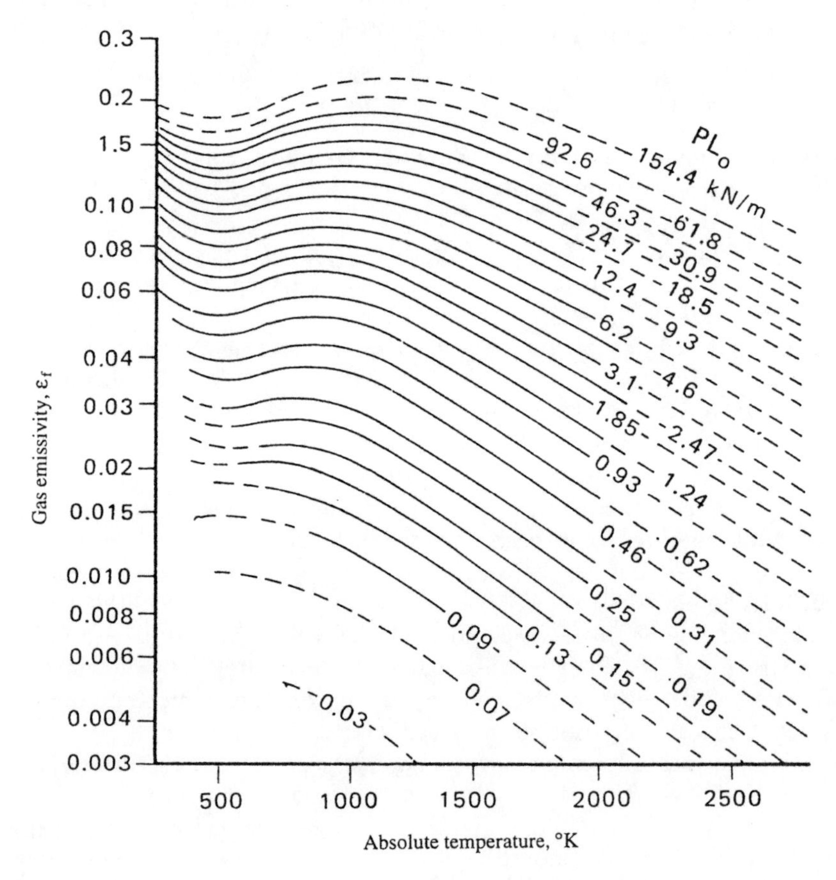

Figure 8.15: Effective emissivity of carbon dioxide [23]. (Schmidt and Willmott [9].)

CO_2, given also the gas temperature t_f K. In the circumstances where $t_f > t_{s,o}$, certainly where $t_f/t_{s,o} > 1.25$, the value of the absorptivity, α_f, can be extracted approximately from the same charts using the surface solid temperature $t_{s,o}$ in lieu of the gas temperature, t_f.

A more accurate formula is given by Hottel for α_f, namely

$$\alpha_f = \left(\frac{t_f}{t_{s,o}}\right)^{0.65} \varepsilon_f\left[t_{s,o}, L_0\left(\frac{t_{s,o}}{t_f}\right)\right] \tag{8.67}$$

where the emissivity is extracted from Figure 8.15 using the surface solid temperature, $t_{s,o}$, and the product of the beam length, L_0, and the ratio $t_{s,o}/t_f$. This and the possible inclusion of the effects of H_2O vapor are discussed by Hottel and Egbert [24].

The nonlinear form of equation (8.66) is approximated by an *equivalent radiative heat transfer coefficient*, α_R. The heat flux at any instant is given by

$$q_R = \alpha_R(t_f - t_{s,o}) \tag{8.68}$$

from which it follows that at any instant in the hot period at a particular position in the heat exchanger:

$$\alpha_R = \sigma\left(\frac{\varepsilon_s + 1}{2}\right)\left(\frac{\varepsilon_f t_f^4 - \alpha_f t_{s,o}^4}{t_f - t_{s,o}}\right) \tag{8.69}$$

The local *lumped* heat transfer coefficient $\bar{\alpha}$ is then given by

$$\frac{1}{\bar{\alpha}} = \frac{1}{\alpha + \alpha_R} + \frac{w}{\lambda_s}\Phi \tag{8.70}$$

A spatial and time average value of α_R could be estimated for the hot period and included within the dimensionless parameters Λ and Π for the simple linear model discussed above.

A more precise representation, however, can be realized within the *full nonlinear model*. Within this model, the coefficient α_R is permitted to vary timewise and spatially as a function of gas and solid temperature, using equation (8.69). In this representation, the emissivity, ε_f, and the absorptivity, α_f, are themselves functions of gas and solid temperature. The point iteration (or the previous step/cycle) methods are ideal to handle α_R in this manner.

At a position $(i+1, j+1)$ on the space–time mesh for the hot period of regenerator operation, the local value of the lumped heat transfer coefficient, $\bar{\alpha}_{i+1,j+1}$, can be estimated from the surface coefficients $(\alpha + \alpha_R)_{i+1,j+1}$ based on the previous time step (or previous cycle) values of gas and solid temperatures, t_f and t_s at this position. The surface solid temperature $t_{s,o}$ at position $(i+1, j+1)$ can itself then be estimated using

$$t_{s,o} = t_f + \frac{\bar{\alpha}}{\alpha + \alpha_R}(t_s - t_f) \tag{8.71}$$

from which a revised value of $(\alpha_R)_{i+1,j+1}$ can be obtained using equation (8.69), and thence a revised value of $\bar{\alpha}_{i+1,j+1}$. Revised values of $t_{f,i+1,j+1}$ and $t_{s,i+1,j+1}$ can now be recomputed within the point iteration scheme described previously together with "improved" values of the other temperature-dependent thermophysical properties at position $(i + 1, j + 1)$.

The accuracy of this approach depends upon the reliability of equation (8.71) and the form of the Hausen Φ-factor employed. Short of using a full 3-D nonlinear model (see Evans [3]) in difficult cases where the inversion of the temperature profile *within* the solid is especially significant for shorter cycles and/or thicker packing bricks, it may be sufficient to employ the time-varying factor $\Phi(\omega)$ discussed earlier in this chapter as well as in Chapter 6. In this way, improved values of the surface solid temperature using equation (8.71) might be yielded. Little or no work, it seems, has been reported in the literature despite the importance of radiative heat transfer in certain high-temperature regenerators.

8.15 Convective Heat Transfer Between Gas and Solid Surface

In most cases, this is far more important than the effects of radiative heat transfer. Elsewhere in this book, the use of dimensionless parameters, in particular the *reduced length*, Λ, of a bed submitted to a single blow, as well as of a packing in a period of regenerator operation, has been introduced. The dimensionless length, Ω, of a period together with the *reduced period*, Π, of regenerative heat exchanger operation have been discussed with the *Biot modulus*, Bi. The point has been made that these dimensionless parameters enable the *same* model and hence the *same* calculated dimensionless temperatures, T, to refer to different regenerators, possibly very different in physical size and in operating conditions but represented by the same dimensionless parameters.

Hausen [25] points out that the processes that take place in the convective heat transfer between a gas and a solid surface are extremely complex. They depend upon the velocities of flow and the physical properties of the gases (or liquids). Were the flow gas *laminar*, the flow might be considered to be made up of thin layers of gas, almost motionless at the solid surface but increasing in velocity with distance from that surface. Under such circumstances, heat transfer will be facilitated by thermal conduction in a direction perpendicular to the direction of gas flow and perpendicular to the solid surface. The concept of *convection* arises as a consequence of the flowing gas carrying away or delivering heat, thereby maintaining the temperature difference between the gas and the solid surface, which *enforces* the necessary transfer of heat.

The situation becomes far more complicated once *turbulent flow* develops. If the gas velocity increases, a *critical velocity* is reached, below which laminar flow continues, but beyond which a rapid mixing motion takes place. Here the transfer of heat is facilitated by the rapid motion of gas particles in a direction perpendicular as well as parallel to the main direction of gas flow. It is thought that a thin layer of gas remains attached to the solid surface, through which heat is still conducted. But, if the gas velocity increases, the flow of gas becomes more turbulent and the thickness of this attached layer decreases and its significance becomes less.

In general, it is not possible to develop solutions to the differential equations for viscous fluid flow and thermal conduction in the fluid, except in special cases. Even where a solution seems attainable, it refers only to fully developed laminar flow at positions remote from the entrance of the fluid to the channels of the packing. As a consequence, any knowledge that we have of the heat transfer coefficients, α, is based on the results of experimental work for different packings. The need for such experimental work is illustrated by the fact that turbulent flow and thus the enhanced heat transfer is promoted by varying the "roughness" of the solid surface.

It is not proposed to describe how such experimental work has been carried out. Be it sufficient to say, as Hausen [25] points out, that laboratory work can be supplemented by the observation of regenerative heat exchangers constructed for practical use. A particular problem remains, however: measurements of heat transfer coefficients and pressure drops will relate to particular circumstances, for example to the size of the channels through which the gas flows, the composition of the gas and the operating temperatures. It turns out that correlations derived from such measurements can be applied to other operating conditions, for example, from laboratory apparatus to large-scale industrial plant. This is realized by use of the *similarity principle*.

Previously, it has been discussed how certain dimensionless parameters enable the same *model* to refer to different practical regenerators. In the same way, other dimensionless parameters enable the same *experimentally determined correlations* to be applied to different operating conditions so that values of the heat transfer coefficient and of the pressure drop can be calculated for low and high temperatures, atmospheric or higher pressures and for a variety of gas compositions. Other experimental data is required, however, in particular how gas thermal conductivity, gas viscosity, and gas specific heat each vary with gas temperature and composition.

8.16 Dimensionless Parameters for Convective Heat Transfer

A dimensionless surface heat transfer rate can be obtained by dividing the heat flux expressed in terms of the heat transfer coefficient, $\alpha(T_f - T_{s,o})$, by the heat flux generated by the transfer of heat by conduction through the gas, $\lambda_f(T_f - T_{s,o})/\delta$, where δ is the *characteristic width* of the channel. This yields the Nusselt number, Nu, namely

$$\mathrm{Nu} = \frac{\alpha\delta}{\lambda_f}$$

Hausen [25] suggests that the Reynolds number, Re, can be interpreted as the ratio of the inertial forces within the fluid to the frictional forces as the gas passes through the channels. This parameter characterizes the field of flow of the gas and, as might be expected, influences the rate of heat transfer between gas and solid. The Reynolds number is defined by the relation

$$\mathrm{Re} = \frac{v\delta}{\nu_f}$$

where v is the mean velocity of flow of the gas through the channels and ν_f is the kinematic viscosity of the gas, where $\nu_f = \eta_f/\rho_f$, with η_f being the dynamic viscosity of the gas and ρ_f its density. In some circumstances, it is necessary, in addition, to consider the thermal effects within the gas which arise as a consequence of there being a temperature profile within the gas across the channel width. In other words, some representation might be made of the temperature dependence of the specific heat, c_p, the viscosity and the thermal conductivity of the gas and this is realized in the Prandtl number, Pr, where

$$\mathrm{Pr} = \frac{\eta_f c_p}{\lambda_f}$$

Many correlations for the heat transfer coefficient are represented in the form

$$\mathrm{Nu} = c\,\mathrm{Re}^n\,\mathrm{Pr}^m$$

where c, n, and m are determined by the relevant experimental work. A number of other related parameters are also used by different authors including the Stanton number, St, and the Colburn factor, j, where

$$\mathrm{St} = \frac{\mathrm{Nu}}{\mathrm{Re}\,\mathrm{Pr}} \quad \text{and} \quad j = \mathrm{St}\,\mathrm{Pr}^{2/3} = \frac{\mathrm{Nu}}{\mathrm{Re}\,\mathrm{Pr}^{1/3}}$$

Schmidt and Willmott [9] point out that an extensive survey of all heat transfer correlations available in 1965 was presented by Barker [26].

Schmidt and Willmott provide details of the correlations suitable for packed beds including those made up of spheres (as used in regenerative burners), Raschig rings and Berl saddles. These rings and saddles were designed to provide a higher porosity and hence a lower pressure drop per unit length of bed than those properties for spheres. It is not intended, therefore, to replicate this extensive data in this text.

8.17 Data for Certain Temperature-Dependent Thermophysical Properties

The surface heat transfer coefficient is computed frequently, as has been pointed out, using a correlation typically of the form

$$\mathrm{Nu} = c\,\mathrm{Re}^n\,\mathrm{Pr}^m$$

where Nu, Re and Pr are the Nusselt, Reynolds and Prandtl Numbers respectively. These dimensionless groups are functions of certain temperature dependent properties, namely gas thermal conductivity, λ_f, gas kinematic viscosity, ν_f, and gas specific heat, c_p, where c_p is of course used elsewhere in the model as well as within the Prandtl number. Any software should be able to provide some means of computing these temperature-dependent properties. Evans [27] has suggested that Tables 8.3–8.5 might be used, which are based on figures published by Yaws [28]. Earlier data is available from the same author [29]. A reader might wish to consider data published elsewhere, for example, that from the National Bureau of Standards [30] or in the UK Steam Tables [31].

1. *Dynamic viscosity*. Table 8.3 gives the coefficients for computing the dynamic gas viscosity, η_f, as a function of gas temperature (K), t_f. The value of η_f is given in *micropoise*.

Table 8.3: Dynamic viscosity, η_f

	A	B	C	$t_{f,max}$ (K)	$t_{f,min}$ (K)
N_2	42.606	0.475	-9.88×10^{-5}	1500	150
O_2	44.224	0.562	-1.13×10^{-4}	1500	150
CO_2	11.336	0.499	-1.0876×10^{-4}	1500	195
SO_2	−11.103	0.502	-1.08×10^{-4}	1000	200
H_2O	−36.826	0.429	-1.62×10^{-5}	1073	280

$\eta_f = A + Bt_f + Ct_f^2$.

2. *Gas specific heat*. Table 8.4 gives the coefficients for computing the gas specific heat, c_p, as a function of gas temperature (K), t_f. The value of c_p is given in J/(mol K).
3. *Gas thermal conductivity*. Table 8.5 gives the coefficients for computing the gas thermal conductivity, λ_s, as a function of gas temperature (K), t_f. The value of λ_s is given in W/(m K).

The problem remains of how to compute these physical properties for gas mixtures. Typically, a combusted fuel gas with excess air will consist of a mixture made up of V_{CO_2}, V_{H_2O}, V_{N_2}, and V_{O_2} where V_j is the proportion by volume in the mixture of component j. Sulfur dioxide may be present but, hopefully, in small quantities!

In the case of dynamic viscosity, the value of $\eta_{f,MIXTURE}$ can be computed using

$$\eta_{f,MIXTURE} = \frac{\sum_{j=1}^{4} \eta_{f,j} V_j M_j^{1/2}}{\sum_{j=1}^{4} V_j M_j^{1/2}}$$

where $\eta_{f,j}$ is the dynamic viscosity of component j calculated at the temperature required, and M_j is the molecular weight of component j, for example, $M_{CO_2} = 44$. For thermal conductivity, it is possible to use

$$\lambda_{f,MIXTURE} = \frac{\sum_{j=1}^{4} \lambda_{f,j} V_j M_j^{1/3}}{\sum_{j=1}^{4} V_j M_j^{1/3}}$$

where the specific heat is quoted in J/(mol K), as in Yaws [28], a weighted mean based on the proportions of the components by volume is sufficient. Where the specific heat is specified in say, kJ/kg K, a weighted mean based on mass fractions of the components should be used.

For the cold period of regenerator operation, where the working gas is typically air, it is possible to use data for these parameters published in the literature for air specifically. Alternatively, it is possible to treat air as a gas mixture where, for example, $V_{CO_2} = 0.03\%$, $V_{H_2O} = 1.97\%$, $V_{N_2} = 77.5\%$, and $V_{O_2} = 20.5\%$.

Table 8.4: Specific heat, c_p

	A	B ($\times 10^{-3}$)	C ($\times 10^{-5}$)	D ($\times 10^{-9}$)	E ($\times 10^{-13}$)	$t_{f,min}$ (K)	$t_{f,max}$ (K)
N_2	29.342	−3.5395	1.0076	−4.3116	2.5935	50	1500
O_2	29.526	−8.8999	3.8083	−32.629	88.607	50	1500
CO_2	27.437	42.315	−1.9555	3.9968	−2.9872	50	5000
SO_2	29.637	34.735	0.9290	−29.885	109.37	100	1500
H_2O	33.933	−8.4186	2.9906	−17.825	36.934	100	1500

$c_p = A + Bt_f + Ct_f^2 + Dt_f^3 + Et_f^4$.

Table 8.5: Gas thermal conductivity, λ_s

	$A\ (\times 10^{-3})$	$B\ (\times 10^{-5})$	$C\ (\times 10^{-8})$	$t_{f,min}$ (K)	$t_{f,max}$ (K)
N_2	3.09	7.5930	−1.1014	78	1500
O_2	1.21	8.6157	−1.3346	80	1500
CO_2	−11.83	10.174	−2.2242	195	1500
SO_2	3.94	4.4847	0.21066	198	1000
H_2O	0.53	4.7093	4.9551	275	1073

$\lambda_s = A + Bt_f + Ct_f^2$.

References

1. A. Hill, A. J. Willmott, "Modelling the Temperature Dependence of Thermophysical Properties in a Closed Method for Regenerative Heat Exchanger Simulations," *Proc. Inst. Mech. Engrs Pt A: J. Power Energy* **205**, 195–206 (1991).
2. W. H. Denton, C. H. Robinson, R. S. Tibbs, "The Heat Transfer and Pressure Loss in Fluid Flow through Randomly Packed Spheres," *Harwell Report (UKAEA) AERE-R 4346* (May 1963).
3. D. J. Evans, "Non-linear Modelling of Regenerative Heat Exchangers," D.Phil. thesis, University of York (Mar. 1997).
4. B. Kulakowski, J. Anielewski, "Application of the Closed Methods of Computer Simulation to Non-linear Regenerator Problems," *Arch. Automat. Telemech.* **24**, 43–63 (1979).
5. C. E. Iliffe, "Thermal Analysis of the Contra-flow Regenerative Heat Exchanger," *Proc. Inst. Mech. Eng.* **159**, 363–372 (1948).
6. A. J. Willmott, "Digital Computer Simulation of a Thermal Regenerator," *Int. J. Heat Mass Transfer* **7**, 1291–1302 (May 1964).
7. A. J. Willmott, "Simulation of a Thermal Regenerator under Conditions of Variable Mass Flow," *Int. J. Heat Mass Transfer* **11**, 1105–1116 (1968).
8. A. Hill, "Stable Closed Methods for Thermal Regenerator Simulations," D.Phil. thesis, University of York (Apr. 1988).
9. F. W. Schmidt, A. J. Willmott, *Thermal Energy Storage and Regeneration*, McGraw-Hill, New York (1981).
10. E. E. Hofmann, A. Kappelmayer, "Mathematical Model of a Hot Blast Stove with External Combustion Chamber," *Proc. Conf. Mathematical Methods in Metallurgical Process Development*, London, pp. 115–132 (1970).
11. D. R. Green, "The Evolution of the Cowper Stove," *J. Hist. Metall. Soc.* **9**(2), 41–48 (1975).
12. E. A. Cowper, *Patent No. 1404*, 19 May 1857.
13. E. A. Cowper, "On Some Regenerative Hot Blast Stoves," *Proc. IMechE*, 54–73 (1860).
14. F. Siemens, *Patent No. 2861*, 2 Dec. 1856.

15. P. Razelos, M. K. Benjamin, "Computer Model of Thermal Regenerators with Variable Mass Flow Rates," *Int. J. Heat Mass Transfer* **21**, 735–741 (Oct. 1977).
16. H. Kistner, "Großversuche an einer zu Studienzwecken gebauten Regenerativ-Kammer (Part II: Determination of Heat Transfer Coefficients and Pressure Drops in Double Chequered and In-line Arrangements of Double Grating Bricks,") *Arch. Eisenhüttenwes.* **3**, 751–768 (1930).
17. H. C. Hottel, A. F. Sarofim, *Radiative Transfer*, McGraw-Hill, New York (1967).
18. A. Hill, A. J. Willmott, "A Robust Method for Regenerative Heat Exchanger Calculations," *Int. J. Heat Mass Transfer* **30**, 241–249 (1987).
19. P. Razelos, "An Analytic Solution to the Electric Analog Simulation of the Regenerative Heat Exchanger with Time-Varying Fluid Inlet Temperatures," *Warme- und Stoffubertragung* **12**, 59–71 (1979).
20. A. Hill, A. J. Willmott, "Accurate and Rapid Thermal Regenerator Calculations," *Int. J. Heat Mass Transfer* **32**, 465–476 (1989).
21. J. L. Morris, *Computational Methods in Elementary Numerical Analysis*, Wiley, New York (1983).
22. M. P. Henry, A. J. Willmott, "A Declarative Language for the Thermal Design of Regenerative Heat Exchangers," *Int. J. Heat Mass Transfer* **33**, 703–723 (1990).
23. H. C. Hottel, R. B. Egbert, "The Radiation of Furnace Gases," *Trans. ASME* **63**, 297–307 (1941).
24. H. C. Hottel, R. B. Egbert, "Radiant Heat Transmission from Water Vapour," *Trans. AIChE* **38**, 531–568 (1942).
25. H. Hausen, *Heat Transfer in Counterflow, Parallel Flow and Crossflow*, English translation (1983) edited by A. J. Willmott, McGraw-Hill, New York (originally published 1976).
26. J. J. Barker, "Heat Transfer in Packed Beds," *Ind. Eng. Chem.* **57**, 43–51 (Apr. 1965).
27. D. J. Evans, Private communication (1999).
28. C. L. Yaws, *Chemical Properties Handbook*, McGraw-Hill, New York (1999).
29. C. L. Yaws, *Physical Properties: A Guide to the Physical, Thermodynamic and Transport Properties for Industrially Important Chemical Compounds*, McGraw-Hill, New York (1977).
30. "Tables of Thermal Properties of Gases," *National Bureau of Standards Circ. 564* (Nov. 1955).
31. *UK Steam Tables in S.I. Units*, Edward Arnold, London (1970, reprinted 1975).

Chapter 9

TRANSIENT RESPONSE OF COUNTERFLOW REGENERATORS

9.1 Introduction

Previously, the periodic steady-state and transient performance of thermal regenerators has been studied separately. It has not been clear how the two are related. Indeed, *closed* methods for the calculation of regenerators have been devised whereby the cyclic equilibrium performance can be computed without simulating any transients prior to equilibrium being reached. The relationship between *open* and *closed* methods for solving the descriptive differential equations has been understood only in the most general terms. Willmott et al. [1] have developed a matrix formulation of the linear simulations of the operation of regenerators. This approach unifies the different numerical methods for the solution of the relevant equations. It clarifies the relationship between the open and closed approaches to this problem.

How does a thermal regenerator, initially at cyclic steady state, approach a new equilibrium following a step change in gas inlet temperature? It will be shown in this chapter that by representing mathematically a cycle of regenerator operation in matrix form, an overall view of transient operation is found by examining the dominant eigenvalue of the relevant matrix. Such a matrix treatment proves to be quite general and an immediate mathematical relationship between dynamic equilibrium and transient performance is thereby obtained. In this way, the relationship between the open and closed methods is established.

The regenerator can be regarded as a forced oscillation device, where oscillations in temperature of the heat-storing packing are imposed by the alternate washing of the solid surface by the hot and cold gases. In counterflow regenerators, the hot and cold gases pass alternately through the same passages in the packing but in opposite directions. The temporary storage of thermal energy in the packing of a regenerator implies necessarily that the regenerator possesses an inertia. Consequently, changes in the operating conditions do not result immediately in changes in the level and amplitude of the cyclic oscillations in gas and solid temperature. Instead, following an increase in cold-gas flow rate and/or a decrease in the hot-side inlet gas temperature, for example, the inertia of the regenerator restrains the response of the system and cyclic steady state is realized only gradually.

It has been indicated that if a cycle of regenerative heat exchanger operation is represented mathematically in matrix form, a means is provided whereby an overall view of transient operation can be found by examining the dominant (largest in modulus) eigenvalue of the relevant matrix. In this chapter, this and earlier theoretical work on this transient performance is presented, and attempts to parameterize the nature and the scale of the behavior of a regenerator under nonequilibrium conditions are discussed.

For the most part, considerations are restricted to the linear model outlined in Chapter 5, namely

$$\text{Fluid:}\quad \frac{\partial T_f}{\partial \xi} = T_s - T_f \tag{9.1}$$

$$\text{Solid:}\quad \frac{\partial T_s}{\partial \eta} = T_f - T_s \tag{9.2}$$

The dimensionless parameters *reduced length*, Λ, Λ', and *reduced period*, Π, Π', prove to be as significant in describing transient performance as they are in representing behavior at cyclic equilibrium.

There have been two different approaches to the transient performance problem. The first approach, applied by London and co-workers [2–4] to rotary regenerators, and by Willmott and Burns [5–7] to fixed-bed and rotary configurations, has sought to parameterize the response of a regenerator initially at periodic steady state to one or more step changes in operation. This early work has now been set on the platform afforded by the matrix approach to regenerator linear simulations developed by

Willmott and Scott [1]. "What is the nature and scale of the transient response to step changes in operation?" and "How long does a regenerator take to regain cyclic steady state following a step change in an operating parameter?" are typical questions which this first approach seeks to answer. It is this treatment which is presented in this text.

The other approach, typified by papers by Beets and Elshout [8] and by Strausz [9], has been to devise feedback control strategies to enable systems of hot-blast stoves for iron making, typically operating in staggered parallel mode, to deliver time-varying thermal loads or to deliver constant thermal loads under conditions where the operating conditions vary with time.

One possible strategy is to control a Cowper stove system using a target cycle time of operation. An increase in thermal load, as a consequence of an increase in the flow rate and/or the target exit temperature of the cold blast, will require a corresponding increase in the heat input to the system, an increase realized by raising the flow rate and/or the inlet temperature of the combustion products passing through the stove in the hot period. Despite any application of such an increase in heat input to the system, the inertia of the regenerator precludes the immediate satisfaction of the new thermal demand.

Provided the step increase in load is not too large, the new load can be achieved, in the short term, using a reduced cycle time. A control strategy for a system of hot-blast stoves under such circumstances will regulate the hot-period gas flow rate (and/or the inlet temperature) and the cycle time in such a way that during transient conditions, the new thermal load is satisfied and that once periodic steady state has been realized again, the target cycle time is restored. In other words, the hot-period gas flow rate is regulated in such a way that any temporary short fall in the heat supply to the system during transient conditions is made up in the longer term. Needless to say, large step changes in thermal load cannot be accommodated in this way and this is discussed by Zuidema [10]. The computer simulation of such systems of regenerators is discussed in Chapter 8.

This and other such feedback control methodologies are applicable to industrial regenerators and require little or no information about the transient performance of such regenerators. They require only a record of measurements of present and past operation and performance of the hot-blast stove system. From such measurements, simple heat balance calculations can be undertaken and the required hot-gas flow rate in the next cycle estimated. Any error associated with this estimate can be corrected, at least partially, at the next regenerator reversal when the control strategy is applied.

9.2 Response to a Step Change in Operating Conditions

London et al. [4] illustrated the responses to changes in operating conditions by defining the following dimensionless parameters, $\varepsilon f_1(\tau)$ and $\varepsilon f_2(\tau)$, which are given by

$$\varepsilon f_1(\tau) = \frac{t_{f,x}(\tau) - t_{f,x}(0)}{t_{f,x}(\infty) - t_{f,x}(0)} \tag{9.3}$$

$$\varepsilon f_2(\tau) = \frac{t'_{f,x}(\tau) - t'_{f,x}(0)}{t'_{f,x}(\infty) - t'_{f,x}(0)} \tag{9.4}$$

where $0 \leq \tau \leq \infty$. At the equilibrium immediately prior to a step change, at $\tau = 0$, we note that

$$\varepsilon f_1(0) = \varepsilon f_2(0) = 0$$

whereas when the new equilibrium is established

$$\varepsilon f_1(\infty) = \varepsilon f_2(\infty) = 1$$

London restricted his considerations to rotary regenerators where the exit temperatures of the gas, $t_{f,x}$, used in equations (9.3) and (9.4), is an average over the face of the heat-storing mass exposed currently to the gas stream under consideration. Because this type of regenerator operates continuously, London was able to represent $\varepsilon f_1(\tau)$ and $\varepsilon f_2(\tau)$ as continuous functions of time, τ.

The approach of London [4] ignores two factors:

1. The thermal state of each sector of the rotating packing will be different at the moment when a step change is effected. One sector might be at the start of its cold period, another midway through its cold period, and another midway through its hot period, for example.
2. The response on the cold side to step changes in operation on the hot side (and vice versa) will be delayed by a time determined by the speed of rotation of the packing.

London was able to ignore these factors by considering only cases where $\Lambda/\Pi = \Lambda'/\Pi' > 100$. In these cases, the angular velocity of the rotor is great relative to the gas flow rates.

This treatment can be extended, however, to fixed-bed regenerators. Step changes are regarded as taking place only at the reversals. The responses $\varepsilon f_1(\tau)$ and $\varepsilon f_2(\tau)$ are measured in terms of the chronological average exit gas temperatures over successive periods of regenerator operation. The current values of $t_{f,x}$ and $t'_{f,x}$ for the nth cycle following a step change are calculated in the following way:

Hot period, nth cycle:

$$t_{f,x}(n(P+P')-P') = \frac{1}{P}\int_{(n-1)(P+P')}^{n(P+P')-P'} t_{f,x}(\tau)\,d\tau \tag{9.5}$$

Cold period, nth cycle:

$$t'_{f,x}(n(P+P')) = \frac{1}{P'}\int_{n(P+P')-P'}^{n(P+P')} t'_{f,x}(\tau)\,d\tau \tag{9.6}$$

These values of $t_{f,x}$ and $t'_{f,x}$ are inserted into equations (9.3) and (9.4) to yield $\varepsilon f_1(n(P+P')-P')$ and $\varepsilon f_2(n(P+P'))$ respectively. Notice that $\tau = n(P+P')-P'$ relates to the end of the hot period and $\tau = n(P+P')$ to the end of the cold period of the nth cycle. These times are employed for fixed-bed regenerator systems so that εf_1 and εf_2 are able to represent the full extent of the effect of a previous step change in the current cycle. This chronological average has as its equivalent the spatial average (across the exit face of the packing) used for rotary regenerators. As close a correspondence as possible between fixed-bed and rotary regenerators is achieved thereby.

Except in the case where a nonlinear model must be used, when a step change in inlet gas temperature is made, the dimensionless parameters Λ, Λ', Π, and Π' are unaffected and the thermal ratios, η_{REG} and η'_{REG}, remain unchanged. Here, this enables $t_{f,x}(\infty)$ and $t'_{f,x}(\infty)$ to be calculated prior to the step change in the hot period, for example, inlet gas temperature from $t_{f,in}$ to $t^*_{f,in}$:

$$\eta_{REG} = \frac{t_{f,in} - t_{f,x}(0)}{t_{f,in} - t'_{f,in}} = \frac{t^*_{f,in} - t_{f,x}(\infty)}{t^*_{f,in} - t'_{f,in}} \tag{9.7}$$

$$\eta'_{REG} = \frac{t'_{f,x}(0) - t'_{f,in}}{t_{f,in} - t'_{f,in}} = \frac{t_{f,x}(\infty) - t'_{f,in}}{t^*_{f,in} - t'_{f,in}} \tag{9.8}$$

It follows that the responses $\varepsilon f_1(\tau)$ and $\varepsilon f_2(\tau)$ can be calculated immediately as $t_{f,x}$ and $t'_{f,x}$ are computed.

When a nonlinear model must be used, some or all of the parameters Λ, Λ', Π, and Π' are modified as the effect of a step change is gradually felt. Similarly, when step changes are made to one or both of the flow rates, these parameters are again altered. In general, therefore, the temperatures $t_{f,x}(\infty)$ and $t'_{f,x}(\infty)$ are known only when periodic steady state has been realized again. In both these cases, therefore, the values of $\varepsilon f_1(\tau)$ and $\varepsilon f_2(\tau)$ can be computed only retrospectively.

London [4] exhibited the response of a regenerator to step changes in operation by displaying the variation of $\varepsilon f_1(\tau)$ and $\varepsilon f_2(\tau)$ with time τ. Green [11], on the other hand, compared the effect of a step change under different circumstances in terms of the time taken to reestablish periodic steady state. Both approaches are applied here and the dependence of $\varepsilon f_1(\tau)$ and $\varepsilon f_2(\tau)$ on the parameters describing the system is examined. These parameters are the reduced lengths and reduced periods, Λ, Λ', Π, and Π', both before and after the step change. In the linear case, these completely describe the regenerator both before and during the period of transient behavior.

In the case of the *linear* model, it is implicit that the responses εf_1 and εf_2 are independent of the size of the step change in inlet gas temperature imposed. This is not so for the *nonlinear* model, nor is it the case where step changes in gas flow rate are considered. Here the parameters Λ, Λ', Π, and Π' after the step change *will* depend on the magnitude of the step changes considered, as will, as a consequence, the responses εf_1 and εf_2.

9.3 Step Changes in Inlet Gas Temperature

Symmetric Case

All earlier work was restricted to *symmetric* regenerators where $\Lambda = \Lambda'$ and $\Pi = \Pi'$. In Figure 9.1 are displayed the responses $\varepsilon f_1(\eta)$ and $\varepsilon f_2(\eta)$ to a step change in hot-inlet gas temperature. These responses are represented as functions of the dimensionless time parameter η for a reduced length, $\Lambda = 10$. It will be observed that the responses are independent of reduced period, Π, for $\Lambda/\Pi = 100$, 200, and 400, thereby confirming the observations of London et al. [4]. Willmott and Burns investigated cases outside the ranges examined by London and have shown that, for example $\Lambda = 28$, $\varepsilon f_1(\eta)$ and $\varepsilon f_2(\eta)$ do exhibit some dependence on reduced period Π for $\Pi = 16$, 8, 4, and 2. Nevertheless, the dimensionless time to reestablish equilibrium, Θ, remains pretty well independent of reduced period. The responses are displayed in Figure 9.2.

The key thing to note is that εf_1 represents the *same side* response, that is, in the period of operation in which the step change in inlet temperature is made, namely here the hot-side response. εf_2 represents, on the other hand, the *opposite side* response in the opposite period of operation, here the cold period. The linear nature of the model permits consideration of step changes of inlet gas temperature in the hot period alone. Should the step change be made in the cold-inlet gas temperature, then εf_1 represents the cold-period response and εf_2 the hot-period response.

In this work of London et al. [4], Willmott and Burns [5], and others, the transient performance of regenerators has been examined by carrying out

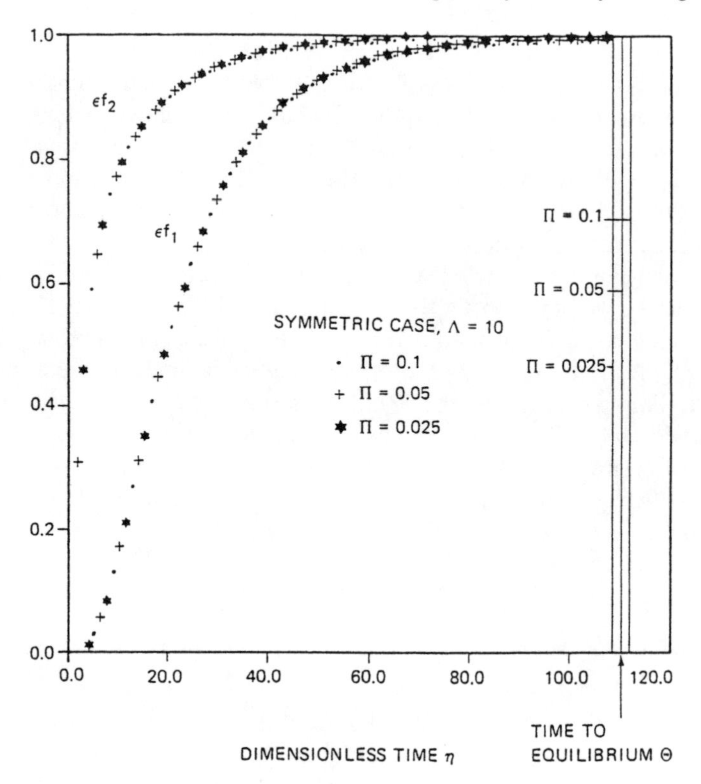

Figure 9.1: Responses εf_1 and εf_2 to step changes in hot-inlet gas temperature for $\Lambda/\Pi > 100$. (Schmidt and Willmott [16].)

computational experiments using the open methods of regenerator simulation. These are equivalent to carrying out experiments on a physical system, such as those reported by Cutland [12]. The attempts made to parameterize the results obtained have been based on analyses of the computed responses and carried out in a conventional manner. We now set out a matrix analysis of this problem in which the transient responses are described in terms of the properties of the relevant matrix.

The Matrix Analysis

In essence, a cycle of regenerator operation can be represented by a matrix equation

$$\boldsymbol{T}^{(n+1)} = \boldsymbol{M}\boldsymbol{T}^{(n)} + \boldsymbol{\alpha} \tag{9.9}$$

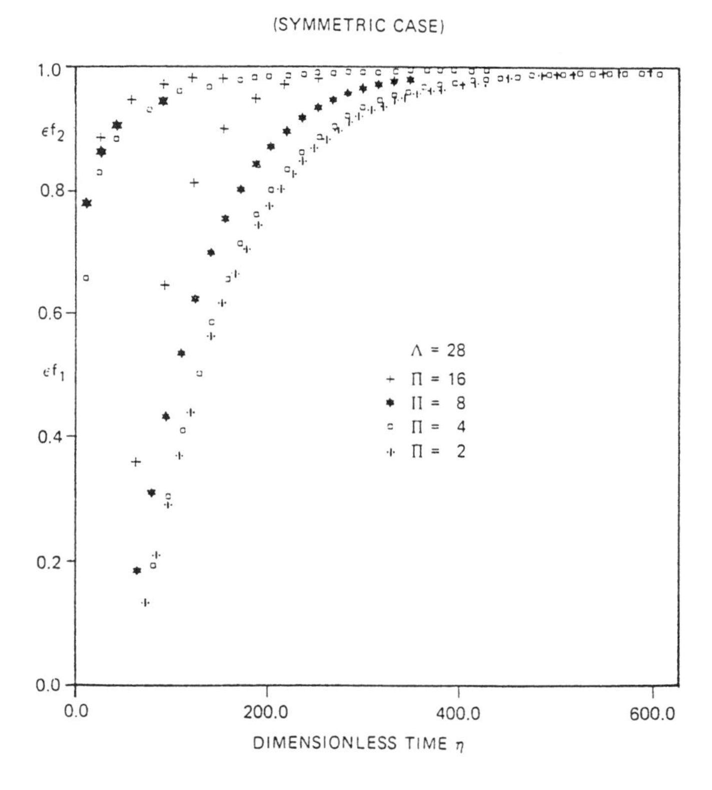

Figure 9.2: Dependence of the responses of εf_1 and εf_2 to step changes in hot-inlet gas temperature on Π for $\Lambda/\Pi > 100$. (Schmidt and Willmott [16].)

where the matrix M itself is a product of two often, but not necessarily, triangular matrices, Γ and Γ', where Γ represents the hot period and Γ' the cold period of regenerator operation. The cycle here is considered to be a cold period followed by a hot period. The vector $\boldsymbol{T}^{(n)}$ represents the spatial temperature distribution in the packing of the regenerator at the start of the nth cycle, at the end of the previous hot period. Using this notation, $T^{(\infty)}$ is the corresponding temperature distribution at periodic steady state.

This equation is based on the linear model of the regenerator. The solid temperatures, T_s, in the vector $\boldsymbol{T}$ are *dimensionless temperatures* defined by equation (9.10) onto a [0, 1] temperature scale:

$$T_s = \frac{t_s - t'_{f,in}}{t_{f,in} - t'_{f,in}} \tag{9.10}$$

The vector $\boldsymbol{\alpha}$ embodies the fact that the dimensionless inlet gas temperatures are 1 in the hot period and 0 in the cold period.

From this simple equation (9.9) flows a whole series of computational and practical conclusions. For periodic steady state, equation (9.9) becomes

$$\boldsymbol{T}^{(\infty)} = M\boldsymbol{T}^{(\infty)} + \boldsymbol{\alpha} \tag{9.11}$$

In addition, we note that for the nth cycle, the temperature distribution at the end of the cold period is obtained using

$$\boldsymbol{T}'^{(n)} = \Gamma'\boldsymbol{T}^{(n)} \tag{9.12}$$

In the *open* methods, the model is cycled from an arbitrary solid temperature distribution, $\boldsymbol{T}^{(0)}$. It can be shown readily that

$$\boldsymbol{T}^{(n)} = (I + M + M^2 + \cdots + M^{n-1})\boldsymbol{\alpha} + M^n\boldsymbol{T}^{(0)} \tag{9.13}$$

This equation throws fresh light on the convergence of this open method toward dynamic equilibrium. Equation (9.13) can be regarded as an iterative process and this will converge provided that

$$\lim_{n\to\infty} M^n = 0 \tag{9.14}$$

In that case, for n large enough, any effect of the initial temperature distribution, $\boldsymbol{T}^{(0)}$, upon $\boldsymbol{T}^{(n)}$ will be lost. Equally the series $I + M + M^2 + \ldots + M^{n-1}$ will converge, in which case

$$\lim_{n\to\infty} \boldsymbol{T}^{(n)} = (I - M)^{-1}\boldsymbol{\alpha} \tag{9.15}$$

It will be shown later that this provides an algebraic connection between the closed and open methods of regenerator simulation.

For the general, real matrix M, $\lim_{r\to\infty} M^r = 0$, provided that

$$|\lambda_1| < 1 \tag{9.16}$$

where λ_1 is the dominant eigenvalue (largest in modulus) of the matrix M. Further, the rate of convergence will be determined by the absolute size of λ_1.

9.4 The Effect of a Step Change in Inlet Gas Temperature

Without loss of generality, for the *linear* problem, we need only consider initial equilibrium operating conditions where $T_{f,in} = T'_{f,in} = 0$. The regenerator simulation is then submitted to a unit step change in inlet gas temperature, in the *hot* period, say. The new operating conditions are now $T_{f,in} = 1$ and $T'_{f,in} = 0$. The solid temperature profile is $\boldsymbol{T}^{(0)} = \boldsymbol{0}$ prior to the step change. The conditions at the new cyclic equilibrium are represented by

$$\boldsymbol{T}^{(\infty)} = M\boldsymbol{T}^{(\infty)} + \boldsymbol{\alpha} \tag{9.17}$$

During the transient phase, the temperature profile at the end of the nth cycle is

$$\boldsymbol{T}^{(n)} = M\boldsymbol{T}^{(n-1)} + \boldsymbol{\alpha} \tag{9.18}$$

Subtracting equation (9.18) from (9.17), following the analysis of Varga [13], we find:

$$\boldsymbol{\varepsilon}^{(n)} = M\boldsymbol{\varepsilon}^{(n-1)} = M^n\boldsymbol{\varepsilon}^{(0)} = M^n\boldsymbol{T}^{(\infty)} \tag{9.19}$$

where $\boldsymbol{\varepsilon}^{(n)} = \boldsymbol{T}^{(\infty)} - \boldsymbol{T}^{(n)}$ and $\boldsymbol{\varepsilon}^{(0)} = \boldsymbol{T}^{(\infty)} - \boldsymbol{T}^{(0)} = \boldsymbol{T}^{(\infty)}$ because $\boldsymbol{T}^{(0)} = \boldsymbol{0}$. Clearly $\boldsymbol{\varepsilon}^{(n)}$ tends to $\boldsymbol{0}$ as n increases provided that the iterative process converges. How *quickly* the process converges, that is how fast the regenerator approaches cyclic equilibrium, will depend on the matrix M as indicated by equation (9.19). Within M are embodied, mathematically, the constructional and operational features of the regenerator under consideration.

In this analysis, it is assumed that the eigenvalues $\lambda_1, \lambda_2, \ldots, \lambda_{N+1}$ of the matrix M are real and distinct, and that, as a consequence, the corresponding eigenvectors $\boldsymbol{x}_1, \boldsymbol{x}_2, \ldots, \boldsymbol{x}_{N+1}$ are linearly independent and span the space of all vectors of order $N+1$. In this case, we can expand $\boldsymbol{T}^{(\infty)}$ as a linear combination of the eigenvectors:

$$\boldsymbol{T}^{(\infty)} = \sum_{j=1}^{N+1} \theta_j \boldsymbol{x}_j \tag{9.20}$$

where $\{\theta_j \mid j = 1, 2, \ldots, N+1\}$ are determined by the values of the elements of the vector $\boldsymbol{T}^{(\infty)}$ and the way chosen to scale the eigenvectors (e.g., by arranging that $\boldsymbol{x}_j^T\boldsymbol{x}_j = 1$).

Equation (9.19) now takes the form

$$\boldsymbol{\varepsilon}^{(n)} = \sum_{j=1}^{N+1} \lambda_j^n \theta_j \boldsymbol{x}_j = \lambda_1^n \left[\theta_1 \boldsymbol{x}_1 + \sum_{j=2}^{N+1} \left(\frac{\lambda_j}{\lambda_1} \right)^n \theta_j \boldsymbol{x}_j \right] \tag{9.21}$$

and

$$\boldsymbol{\varepsilon}^{(n)} \approx \lambda_1^n \theta_1 \boldsymbol{x}_1 \quad \text{as} \quad n \to \infty \tag{9.22}$$

because $|\lambda_j/\lambda_1| < 1$ for $j = 2, 3, \ldots, N+1$.

Moreover, $\boldsymbol{\varepsilon}^{(n)} \to \boldsymbol{0}$ as $n \to \infty$ if $|\lambda_1| < 1$ and thus the process will converge provided that $|\lambda_1| < 1$. From a physical point of view, it is known that

$\boldsymbol{T}^{(n)}$ approaches $\boldsymbol{T}^{(\infty)}$ monotonically and that no oscillations occur in $\boldsymbol{T}^{(n)}$ from one cycle to the next as would be implied by a negative eigenvalue. The value of λ_j^n would change sign as n changes from even to odd if $\lambda_j < 0$. This implies that all the eigenvalues should be positive and lie in the [0, 1] range. The convergence will be slower the closer λ_1 becomes to 1. Willmott and Scott [14] have computed the eigenvalues of M for the symmetric case where the reduced period $\Pi = 1$ for different values of reduced length Λ. These are shown in Table 9.1.

These values were calculated using a formulation of the matrix M based on the method of Iliffe [15], although Willmott et al. [1] have shown elsewhere that a matrix M developed from an entirely different method of solving the differential equations (see Schmidt and Willmott [16]), yields the same eigenvalues. This is to be expected because the model, independent of the numerical technique employed, *does* simulate the dynamics of the regenerator.

The key things to note from Table 9.1 are:

1. All the calculated eigenvalues lie in the [0, 1] range, as one might expect from the physical considerations discussed above.
2. The dominant eigenvalue λ_1 increases with reduced length, Λ. This coincides with the observations of Willmott and Burns [5] and others [4, 7, 11, 17], which have been discussed at the start of this section, that for a fixed value of reduced period, Π, the rate of convergence to cyclic equilibrium *decreases* with reduced length.

That the reduced length Λ is an effective measure of the inertia of a symmetric regenerator is illustrated in Figure 9.3, where $\varepsilon f_1(\eta)$ and the time to equilibrium, Θ, are displayed, for a fixed value of reduced period $\Pi = 4$, for different reduced lengths $\Lambda = 32$, 24, 16, and 8. The initial lag in the response, which increases with Λ, is typical of a distributed parameter system, of which a thermal regenerator is an example.

Table 9.1: Variation of the eigenvalues of M with $\Pi = 1$

Reduced length Λ	λ_1	λ_2	λ_3	λ_4	λ_5	λ_6	λ_7	λ_8	λ_9	λ_{10}	λ_{11}
1	0.283	0.156	0.141	0.138	0.137	0.136	0.134	0.132	0.129	0.126	0.124
5	0.705	0.386	0.250	0.193	0.167	0.157	0.141	0.131	0.120	0.107	0.098
10	0.878	0.642	0.449	0.325	0.246	0.208	0.171	0.149	0.130	0.109	0.094
15	0.935	0.782	0.610	0.463	0.347	0.274	0.214	0.179	0.156	0.132	0.113

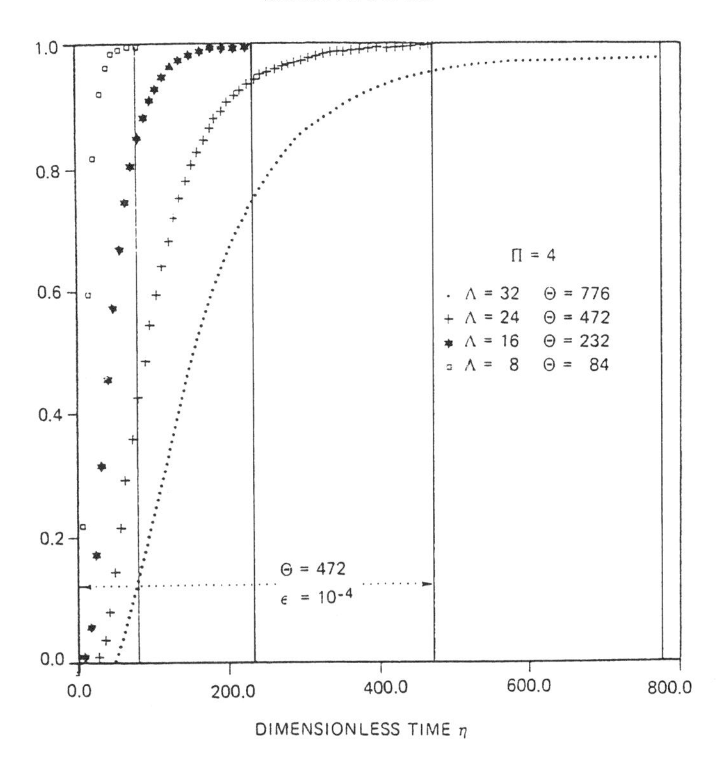

Figure 9.3: Dependence of the response εf_1 and the time to periodic steady state, Θ, upon Λ. Vertical lines denote when periodic steady state is restored. (Schmidt and Willmott [16].)

This reflection of the inertia of a regenerator in the total dimensionless time, Θ, needed to reestablish the periodic steady state following a step change in operating conditions must be considered further. In a computer simulation of a regenerative heat exchanger, it is important to provide a suitable measure of convergence to equilibrium that might be adopted. This must take account of the different rates of convergence between the same and opposite side responses, and between different regenerator configurations. Willmott [18] proposed the pseudothermal ratios for the nth cycle, namely

$$Z_n = \frac{T_{f,in} - t_{f,x}(\Pi)}{t_{f,in} - t'_{f,in}} \tag{9.23}$$

$$Z'_n = \frac{t'_{f,x}(\Pi') - t'_{f,in}}{t_{f,in} - t'_{f,in}} \tag{9.24}$$

where $t_{f,x}(\Pi)$ and $t'_{f,x}(\Pi')$ are the exit gas temperatures at the end of the hot and cold periods respectively. He suggested that, provided that

$$|Z_n - Z_{n-1}| < \varepsilon \quad \text{or} \quad |Z'_n - Z'_{n-1}| < \varepsilon \tag{9.25}$$

where ε is a small number, say 10^{-4} or 10^{-5}, a simulation could be said to have attained a new periodic steady state in the nth cycle. If the true value of Z_∞ at equilibrium is α, say, this presumes that if

$$|Z_n - Z_{n-1}| < \varepsilon \quad \text{then} \quad |Z_n - \alpha| < \varepsilon \quad \text{also.} \tag{9.26}$$

If a regenerator possesses a large thermal inertia, however, the convergence of Z_n to α will be slow, possibly very slow. In this case, the assumption (9.26) may not be valid, so that although $|Z_n - Z_{n-1}| < \varepsilon$, the simulation may still be some way from equilibrium with $|Z_n - \alpha| > \varepsilon$.

An estimate α^* of α can be obtained employing Aitkin's formula (see Henrici [19], pp. 70–74) as embodied in

$$\alpha^* = \frac{Z_n Z_{n-2} - Z_{n-1}^2}{Z_n - 2Z_{n-1} + Z_{n-2}} \tag{9.27}$$

Willmott and Burns [5] found that for regenerators with a large thermal inertia, provided that, using equation (9.27),

$$|Z_n - \alpha| \approx |Z_n - \alpha^*| < \varepsilon \tag{9.28}$$

the simulation can be regarded safely as having attained cyclic steady state. Just when should the criterion (9.28) be used? Schmidt and Willmott [16] have shown that if $K_Z < 0.5$, then it is sufficient to use the criterion embodied in equation (9.25), but where $0.5 \leq K_Z < 1.0$, it is necessary to use criterion (9.28) where

$$K_Z = \frac{Z_n - Z_{n-1}}{Z_{n-1} - Z_{n-2}}$$

Employing these revised convergence criteria, the times Θ to equilibrium have been determined and are displayed in Figure 9.3. A value of $\varepsilon = 10^{-4}$ was employed in the convergence criteria. On the assumption that cycle time has little influence on Θ, Willmott and Burns developed the general formula

$$\Theta = 0.622\Lambda^2 + 4.144\Lambda + 6.464 \tag{9.29}$$

relating time to equilibrium, Θ, to reduced length, Λ, with $\varepsilon = 10^{-4}$ and for $\Lambda < 40$. For another stopping criterion $\varepsilon = E$, the value of Θ_E, the time to cyclic steady state using $\varepsilon = E$, can be estimated using

$$\Theta_E = \frac{\Theta \log_e E}{\log_e \varepsilon} \tag{9.30}$$

For example, if $E = 10^{-2}$, then

$$\Theta_E = \frac{\Theta \log_e(10^{-2})}{\log_e(10^{-4})} = \frac{\Theta}{2} \tag{9.31}$$

Example 9.1 Using the convergence criteria with $\varepsilon = 10^{-4}$, estimate the number of cycles required to restore periodic steady state following a step change in inlet gas temperature for the symmetric regenerator configuration where $\Lambda = \Lambda' = 10$ and $\Pi = \Pi' = 0.5$.

Using equation (9.29) we obtain

$$\begin{aligned}\Theta &= 0.622(10^2) + 4.144(10) + 6.464 \\ &= 62.2 + 41.4 + 6.446 = 110.046\end{aligned}$$

The total cycle time is $\Pi + \Pi' = 1.0$, from which it follows that approximately 110 cycles are required before cyclic steady state is restored.

Example 9.2 Again estimate the number of cycles to periodic steady state required for a reduced length $\Lambda = \Lambda' = 10$ but with a reduced period $\Pi = \Pi' = 2.0$ and with $\varepsilon = 10^{-2}$.

The total time to equilibrium, using equation (9.31) is

$$\Theta_E = \frac{\Theta \log_e(10^{-2})}{\log_e(10^{-4})} = \frac{\Theta}{2} = \frac{110.046}{2} = 55.023$$

The cycle time is $\Pi + \Pi' = 4.0$ and therefore $55.023/4 \approx 14$ cycles are needed.

Equation (9.29) embodies the observations of Green [11] and Burns and Willmott [5] that for small values of reduced period, Π, the total time to equilibrium, Θ, can be made up of many short cycles or a few longer cycles and that this total time Θ varies only with reduced length, Λ. This concept, it has been shown [14], for a constant value of reduced length, Λ, can be represented by

$$\frac{\Pi}{\log_e \lambda_1} = constant = \kappa \tag{9.32}$$

The parameter $\Pi/(\log_e \lambda_1)$ has been called the *dimensionless time constant*, κ, for a given regenerator configuration. Typical values of this time constant are displayed in Table 9.2.

9.5 Limitation of the Matrix Method

The development of the *dimensionless time constant*, κ, points to a limitation of the matrix method as described here. Recalling equation (9.22),

$$\boldsymbol{\varepsilon}^{(n)} \approx \lambda_1^n \theta_1 \mathbf{x}_1 \quad \text{as} \quad n \to \infty \tag{9.22}$$

a norm $\|\cdot\|$ of $\boldsymbol{\varepsilon}^{(n)}$ is considered, for example

$$\|\boldsymbol{\varepsilon}^{(n)}\| = \left(\frac{1}{m+1}\sum_{j=0}^{m} \varepsilon_j^2\right)^{1/2} \tag{9.33}$$

It can be seen that

Table 9.2: Variation of the dimensionless time constant

Reduced length Λ	Reduced period Π	Dominant eigenvalue λ_1	Natural log of λ_1	Value of $\frac{\Pi}{\log_e \lambda_1}$
4.0	0.3	0.87375	−0.13496	−2.2228
4.0	0.4	0.83520	−0.18009	−2.2212
4.0	0.5	0.79826	−0.22532	−2.2191
4.0	0.6	0.76286	−0.27069	−2.2166
4.0	0.7	0.72890	−0.31622	−2.2137
4.0	0.8	0.68633	−0.36194	−2.2103
10.0	0.3	0.96197	−0.03878	−7.7363
10.0	0.4	0.94957	−0.05175	−7.7298
10.0	0.5	0.93730	−0.06475	−7.7220
10.0	0.6	0.92515	−0.07779	−7.7125
10.0	0.7	0.91312	−0.09089	−7.7018
10.0	0.8	0.90119	−0.10404	−7.6893
20.0	0.3	0.98774	−0.01234	−24.316
20.0	0.4	0.98390	−0.01624	−24.638
20.0	0.5	0.98000	−0.02021	−24.746
20.0	0.6	0.97610	−0.02419	−24.802
20.0	0.7	0.97217	−0.02823	−24.799
20.0	0.8	0.96827	−0.03224	−24.811

$$\|\varepsilon^{(n)}\| = \lambda_1^n \theta_1 \|\mathbf{x}_1\| \tag{9.34}$$

and

$$\|\varepsilon^{(n-1)}\| = \lambda_1^{n-1} \theta_1 \|\mathbf{x}_1\| \tag{9.35}$$

Dividing equation (9.34) by (9.35) generates

$$\frac{\|\varepsilon^{(n)}\|}{\|\varepsilon^{(n-1)}\|} = \lambda_1 \tag{9.36}$$

and

$$\frac{\|\varepsilon^{(n)}\|}{\|\varepsilon^{(0)}\|} = \lambda_1^n = g^{(n)} \tag{9.37}$$

After n cycles, provided $g^{(n)}$ is small enough, the computational open model is said to have attained periodic steady state. During these n cycles, a time $\Theta = n(\Pi + \Pi')$ (or $2n\Pi$ in the symmetric case) elapses. Problems arise because the time Θ to dynamic equilibrium depends upon the severity of the convergence criteria and the fact that Θ is measured as an *integer* number of cycles. These difficulties are overcome to some degree by the introduction of the time constant, κ, in the following way.

Equation (9.37) can be rewritten as

$$n \log_e \lambda_1 = \log_e g^{(n)} \tag{9.38}$$

for chosen values of reduced length, Λ, and reduced period, Π. If the same value of Λ is considered but with a different reduced period, Π^*, and a consequently different dominant eigenvalue, λ_1^*, then, if s cycles to periodic state are required,

$$s \log_e \lambda_1^* = \log_e g^{(n)*} \tag{9.39}$$

It is assumed that $\log_e g^{(n)*} = \log_e g^{(n)}$ (essentially that the same convergence criteria are used in both cases), in which case

$$n \log_e \lambda_1 = s \log_e \lambda_1^* \tag{9.40}$$

In the symmetric case, if the time equilibrium, Θ, is indeed not influenced by Π, then from equation (9.40) is obtained

$$\frac{\Theta \log_e \lambda_1}{2\Pi} = \frac{\Theta \log_e \lambda_1^*}{2\Pi^*} \tag{9.41}$$

or, reciprocating,

$$\frac{\Pi}{\log_e \lambda_1} = \mathit{constant} = \kappa \tag{9.32}$$

Examination of Table 9.2 reveals that $\Pi/(\log_e \lambda_1)$ does decrease slightly as Π increases but κ increases far more markedly with reduced length, Λ.

This development of κ points to the fact that the matrix analysis deals only with the chronological variations of the solid temperature vector $\boldsymbol{T}^{(n)}$ from one cycle to the next and makes no attempt, in what is presented here, to deal with any *spatial* variations in $\boldsymbol{T}^{(n)}$ between cycles in the transient phase. Indeed, the ability to introduce vector norms to yield equation (9.34) from equation (9.22) emphasizes this point.

Following from this, it can be noted that

$$\boldsymbol{T}'^{(n+1)} = M'\boldsymbol{T}'^{(n)} + \Gamma'\boldsymbol{\alpha} \tag{9.42}$$

where $M' = \Gamma'\Gamma$ and where $\boldsymbol{T}'^{(n)}$ is the spatial solid temperature distribution at the end of the cold period in the nth cycle. Following the analysis of equation (9.9), we can deduce that the cycle to cycle variation of $\boldsymbol{T}'^{(n)}$ is determined by the eigenvalues of M'. However, the eigenvalues of M' are equal to the eigenvalues of M where $M = \Gamma\Gamma'$.[†] From this it follows that the variation of $\boldsymbol{T}'^{(n)}$ between cycles in an interval of transient behavior follows a very similar pattern to that of $\boldsymbol{T}^{(n)}$ and that, from both, we yield the same time constant, κ.

The limitation of the matrix method lies in the fact that, as it stands at present, equations (9.9) and (9.42),

$$\boldsymbol{T}^{(n+1)} = M\boldsymbol{T}^{(n)} + \boldsymbol{\alpha} \tag{9.9}$$

$$\boldsymbol{T}'^{(n+1)} = M'\boldsymbol{T}'^{(n)} + \Gamma'\boldsymbol{\alpha} \tag{9.42}$$

do not predict the lag between the cold-side and hot-side responses, εf_1 and εf_2, when these are measured in terms of the *exit* gas temperatures. Such lags only become apparent when the spatial variations of the solid temperatures and thence the spatial variations of the gas temperatures are considered.

9.6 Unifying the Theory of Open and Closed Methods for Solving the Equations that Model Counterflow Regenerators

From the simple equation (9.9) flows a whole series of computational and practical conclusions. For periodic steady state, equation (9.9) becomes

$$\boldsymbol{T}^{(\infty)} = M\boldsymbol{T}^{(\infty)} + \boldsymbol{\alpha} \tag{9.43}$$

[†] If λ is an eigenvalue of $M = \Gamma\Gamma'$, then $\Gamma\Gamma'\boldsymbol{x} = \lambda\boldsymbol{x}$, where $\boldsymbol{x}$ is the corresponding eigenvector. Premultiplying this equation by Γ', we obtain $(\Gamma'\Gamma)\Gamma'\boldsymbol{x} = \lambda\Gamma'\boldsymbol{x}$. From this will be seen that λ is also an eigenvalue of $M' = \Gamma'\Gamma$ but where the corresponding eigenvector is $\Gamma'\boldsymbol{x}$.

It follows that all *closed methods* of regenerator simulation can be represented by

$$\boldsymbol{T}^{(\infty)} = (I - M)^{-1}\boldsymbol{\alpha} \tag{9.44}$$

which, in turn, enables significant economies in the storage and computational requirements for such methods of calculation. In conventional closed methods, for example that of Iliffe [15], it is necessary to solve a set of $2N + 2$ equations to find, simultaneously, the vectors $\boldsymbol{T}^{(\infty)}$ and $\boldsymbol{T}'^{(\infty)}$, where each vector has $N + 1$ elements. Using equation (9.44), only $N + 1$ equations need to be solved in order to find $\boldsymbol{T}^{(\infty)}$. The vector $\boldsymbol{T}'^{(\infty)}$, the temperature distribution at the end of the cold period, can then be found using only a matrix multiplication,

$$\boldsymbol{T}'^{(\infty)} = \Gamma' \boldsymbol{T}^{(\infty)} \tag{9.12}$$

Similar economies of storage were developed by Hill [20] and are described around equations (8.62) and (8.63). Hill's approach is based on algebraic considerations, not on the matrix analysis described in this chapter. The result, nevertheless, is very similar.

In the *open* methods, the model is cycled from an arbitrary solid temperature distribution, $\boldsymbol{T}^{(0)}$. It can be shown readily that $\boldsymbol{T}^{(n)}$ is related to $\boldsymbol{T}^{(0)}$ by equation (9.13). Moreover, this equation throws fresh light on the convergence of an *open* method towards periodic steady state and how a closed method might be developed from an open method. Provided that

$$\lim_{n\to\infty} M^n = 0 \tag{9.14}$$

it has been shown previously that for n large enough, any effect of the initial temperature distribution, $\boldsymbol{T}^{(0)}$, upon $\boldsymbol{T}^{(n)}$ will be lost and the series

$$I + M + M^2 + \cdots + M^{n-1} \to (I - M)^{-1} \quad \text{as} \quad n \to \infty$$

yielding equation (9.44). In this way, an algebraic connection between the closed and open methods of regenerator simulation is provided.

9.7 Overall Performance from the Matrix Method

The overall performance of the regenerator at cyclic equilibrium is given by the thermal ratio η_{REG}. Without loss of generality, we note that for the symmetric case,

$$\eta_{REG} = \frac{\Lambda}{\Pi} \Delta \boldsymbol{T}^{(\infty)} \tag{9.45}$$

where $\Delta \boldsymbol{T}^{(\infty)}$ is the average temperature swing for the packing. $\Delta \boldsymbol{T}^{(\infty)}$ might be denoted using

$$\Delta \boldsymbol{T}^{(\infty)} = \frac{1}{\Lambda}\int_0^{\Lambda}[T^{(\infty)}(\xi) - T'^{(\infty)}(\xi)]\, d\xi \tag{9.46}$$

but here, instead,

$$\Delta \boldsymbol{T}^{(\infty)} = \Phi(\boldsymbol{T}^{(\infty)} - \boldsymbol{T}'^{(\infty)}) = \Phi(I - \Gamma)\, \boldsymbol{T}^{(\infty)} \tag{9.47}$$

is employed, where the Φ denotes an equivalent numerical integration, for example, Simpson's rule, applied to the elements of the vector concerned.

A useful means of testing the correctness of this formulation of Iliffe's method [15] in matrix form has been to compute the value of η_{REG} using equations (9.45) and (9.47) and then to compare our computed results with those in the published literature, for example in the book by Schmidt and Willmott [16]. No errors could be detected in the calculated values of η_{REG}.

9.8 The Effect of a Step Pulse Change in Inlet Gas Temperature

In conclusion of this discussion of the matrix approach to the transient response of regenerators, we point to a connection between periodic steady-state performance and transient behavior following a step pulse in inlet gas temperature. An initial solid temperature profile $\boldsymbol{T}^{(0)} = \boldsymbol{0}$ is considered. For just one cycle, $T_{f,in}^{(1)} = 1$ and cold inlet temperature is left at 0. Thereafter, $T_{f,in}^{(n)} = T_{f,in}'^{(n)} = 0$. This corresponds to a unit step pulse in the hot period of the first cycle. As the regenerator cools down again, the heat from the pulse, temporally stored in the packing, is dispersed at the hot and cold ends of the regenerator until $\boldsymbol{T}^{(\infty)} = \boldsymbol{0}$. It is shown that the manner in which this dispersion of heat takes place is related to the cyclic steady-state performance of the regenerator.

Applying equation (9.9) to the first cycle, we note that

$$\boldsymbol{T}^{(1)} = \boldsymbol{\alpha} \tag{9.48}$$

In the subsequent cycles, since both the "hot" and "cold" inlet temperatures of the gases are zero, then $\boldsymbol{\alpha} = \boldsymbol{0}$. It follows that, for $n = 1, 2, \ldots,$

$$\boldsymbol{T}^{(n)} = M^{n-1}\boldsymbol{\alpha} \tag{9.49}$$

The fraction f_n of the heat dispersed by the cold gas in cycle n is given by

$$f_n = \frac{\Lambda}{\Pi}\Phi(\boldsymbol{T}^{(n-1)} - \boldsymbol{T}'^{(n)}) = \frac{\Lambda}{\Pi}\Phi(I - \Gamma')M^{n-2}\boldsymbol{\alpha} \tag{9.50}$$

The fraction of the heat dispersed by the cold gas over all the transient cycles, until equilibrium is reestablished is given by

$$\sum_{n=2}^{\infty} f_n = \sum_{n=2}^{\infty} \frac{\Lambda}{\Pi}\Phi(I - \Gamma')M^{n-2}\boldsymbol{\alpha} = \frac{\Lambda}{\Pi}\Phi(I - \Gamma')(I - M)^{-1}\boldsymbol{\alpha} \tag{9.51}$$

and therefore

$$\sum_{n=2}^{\infty} f_n = \frac{\Lambda}{\Pi}\Phi(I - \Gamma')\boldsymbol{T}^{(\infty)} = \eta_{REG} \tag{9.52}$$

It can be shown, in a similar manner, that the total fraction of heat dispersed on the hot side is equal to $1 - \eta_{REG}$. The thermal ratio, η_{REG}, is also the chronological average of the cold-period exit dimensionless temperature at equilibrium. It follows that the accumulated fraction of the heat input by a unit step pulse on the hot side which is output in the cold-side gas, is bounded by the maximum and minimum cold-gas exit dimensionless temperatures at cyclic steady state, thereby relating the performance of the regenerator at cyclic equilibrium to its transient performance [21].

9.9 Unsymmetric Balanced Regenerators

Can the results developed above for symmetric regenerators be extended to the unsymmetric case where

$$\Lambda \neq \Lambda' \quad \text{and} \quad \Pi \neq \Pi'?$$

The balanced case, where

$$\frac{\Lambda}{\Pi} = \frac{\Lambda'}{\Pi'} = k$$

and which corresponds to the symmetric case when $k = 1$, is first considered. Here the hot- and cold-side thermal ratios are equal: $\eta_{REG} = \eta'_{REG}$. For periodic steady state, Hausen [22] proposed that the temperature performance of an unsymmetric but balanced regenerator could be replicated by the performance of a symmetric regenerator whose hot- and cold-side descriptive parameters are Λ_H and Π_H where Π_H is the harmonic mean of Π and Π' and where Λ_H is the *weighted* harmonic mean of Λ and Λ', defined by

$$\frac{1}{\Pi_H} = \frac{1}{2}\left(\frac{1}{\Pi} + \frac{1}{\Pi'}\right) \tag{9.53}$$

$$\frac{1}{\Lambda_H} = \frac{1}{\Pi_H}\left(\frac{\Pi}{\Lambda} + \frac{\Pi'}{\Lambda'}\right) \tag{9.54}$$

Iliffe [15] first confirmed the acceptability of this parameterization using Λ_H and Π_H for the computing of the thermal ratio, η_{REG}, at cyclic equilibrium. The lack of symmetry for the balanced case can be specified by a parameter p where

$$\frac{\Lambda}{\Lambda'} = \frac{\Pi}{\Pi'} = p$$

Iliffe examined cases over the ranges $0 < \Pi' \leq 24$ and $0 < \Lambda' \leq 24$ for $p = 2$ and the cases over the ranges $0 < \Pi' \leq 18$ and $0 < \Lambda' \leq 18$ for $p = 3$.

Willmott and Burns [5] verified that the transient behavior of balanced but unsymmetric regenerators can be also parameterized adequately by the use of the harmonic means Λ_H and Π_H defined by equations (9.53) and (9.54). Figure 9.4 illustrates the responses $\varepsilon f_1(\eta_H)$ and $\varepsilon f_2(\eta_H)$ to a step change in hot-period inlet gas temperature for the symmetric regenerator with $\Lambda = \Lambda_H = 24$ and $\Pi = \Pi_H = 8$. It will be observed that εf_1 and εf_2 follow very closely the corresponding responses for the unsymmetric balanced cases tabulated below:

Λ	Λ'	Π	Π'	p	Λ_H	Π_H
36	18	12	6	2	24	8
18	36	6	12	$\frac{1}{2}$	24	8

Note that $\varepsilon f_1(p = 2)$ and $\varepsilon f_1(p = \frac{1}{2})$ lie on opposite sides of the response $\varepsilon f_1(p = 1)$ calculated using $\Lambda_H = 24$ and $\Pi_H = 8$.

The time to equilibrium, Θ_H, therefore, can be computed with a suitably modified version of equation (9.29). This takes the form

$$\Theta_H \approx 0.622\Lambda_H^2 + 4.144\Lambda_H + 6.464 \tag{9.55}$$

for $\Lambda_H \leq 40$ and stopping criterion $\varepsilon = 10^{-4}$. The actual dimensionless time Θ for the unsymmetric balanced regenerator to regain cyclic steady state is given by

$$\Theta = \frac{\Theta_H}{4}(1 + p)\left(1 + \frac{1}{p}\right) \tag{9.56}$$

The approximation (9.55) provides a relative accuracy of 10% or better.

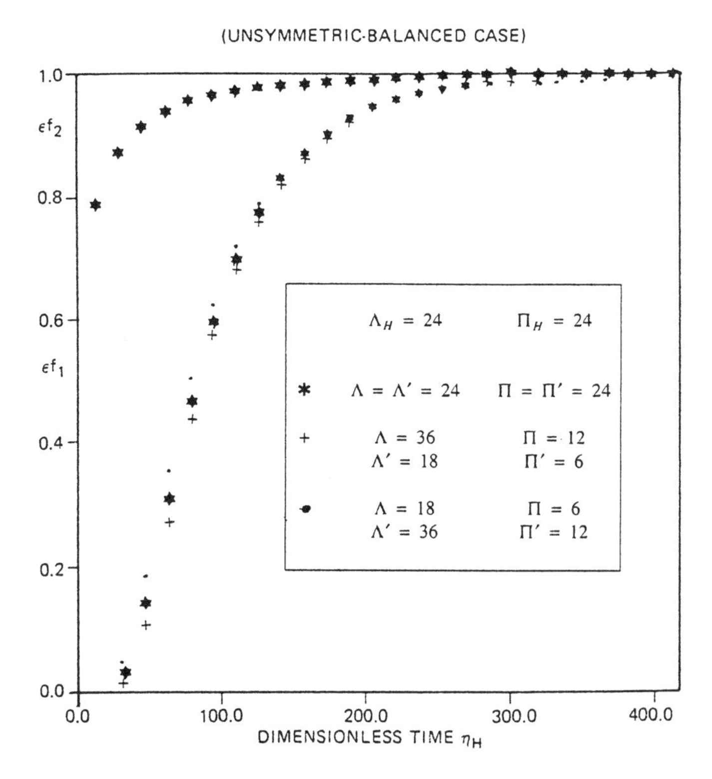

Figure 9.4: The responses εf_1 and εf_2 to step changes in inlet gas temperature. (Schmidt and Willmott [16].)

9.10 Unbalanced Regenerators

In the balanced case, the forces imposing the effects of a step change in the hot period, say, in this case in inlet temperature, might be regarded as being counterbalanced to some extent by opposing forces in the cold period. By way of example, the propagation of thermal energy down the length of the regenerator packing as a consequence of a step increase in inlet gas temperature will be delayed, to a greater or less extent, by the action of the cold gas passing through the same packing in successive cold periods.

Schmidt and Willmott [16] suggest that a regenerator might be regarded as a reservoir of heat from which energy is withdrawn in the cold period and topped up again in the subsequent hot period. If there is a step increase in hot-side inlet gas temperature, the rate at which energy is supplied to the reservoir increases and the level of heat (for example, the average temperature of the regenerator packing) starts to rise. As a consequence, the rate of heat extraction also begins to increase. The rise in level in the reservoir realized in a hot period, over and above the "normal" topping-up of the system at periodic steady state, is counteracted by the drawing-off of heat in the subsequent cold period.

This analogy provides a possible means of understanding the mechanisms at play during a period of transient behavior of an unbalanced regenerator. In this case

$$\frac{\Lambda}{\Pi} \neq \frac{\Lambda'}{\Pi'}$$

and the thermal ratios η_{REG} and η'_{REG} are unequal also. The regenerator is unbalanced, moreover, in the sense that the forces imposing temperature changes in the packing as a result of a step increase in hot-period inlet gas temperature, are not counterbalanced to the same degree, compared with the balanced case, by opposing forces in the cold period. In terms of the reservoir analogy, if the quantity of extra heat supplied in the hot period at the start of a transient phase, is significantly greater or smaller than the quantity of heat withdrawn from the reservoir in the cold period, then the system will move back to cyclic equilibrium more rapidly than if the system were balanced. When a regenerator is unbalanced, therefore, the packing responds temperature-wise more quickly than it would if the system were balanced. It follows that it is necessary to introduce an additional parameter, β, to account for the effect of imbalance upon transient performance:

$$\beta = \frac{\Lambda\,\Pi'}{\Lambda'\,\Pi} \tag{9.57}$$

The use of the weighted harmonic mean reduced length, Λ_H, is extended to the unbalanced case with a view to parameterizing the factors governing transient performance. Displayed in Figure 9.5 is the dependence of the dimensionless time to equilibrium upon Λ_H and the degree of imbalance, β. It is evident that a regenerator exhibits greatest inertia when it is balanced ($\beta = 1$). The time to equilibrium declines significantly for $\beta = \frac{1}{2}$, 2, $\frac{1}{4}$, 4. Indeed for $\beta = 4$, it appears that the effect of dimensionless length, Λ_H, has all but disappeared.

The matrix analysis, as it stands, cannot be extended to the unsymmetric case. Here, the matrix $M = \Gamma\Gamma'$ is not symmetric and there is no assurance, therefore, that all the eigenvalues are real and that the eigenvectors are complete. No presumption can be made, consequently, that a vector $\boldsymbol{T}$ of temperatures can be represented as a linear combination of the eigenvectors. Willmott et al. [1] report, nevertheless, that for the unsymmetric but balanced case where

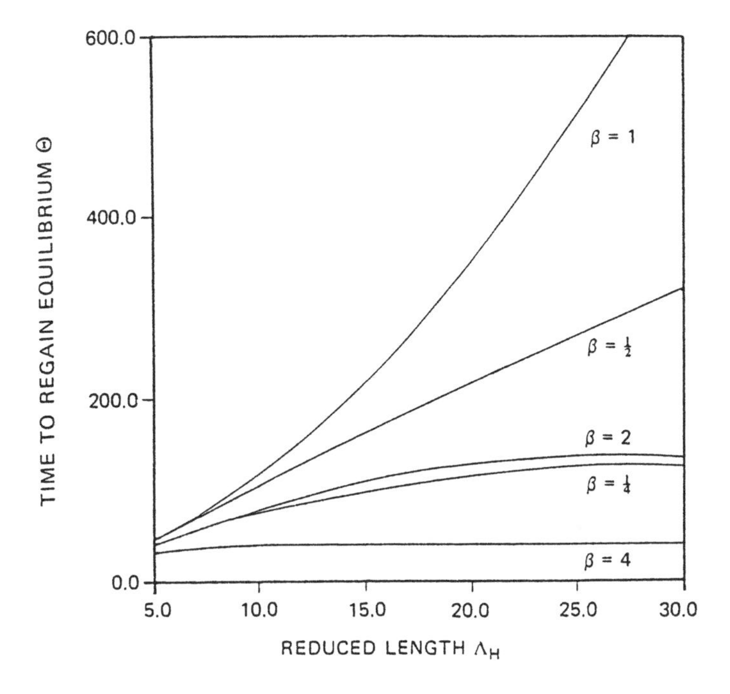

Figure 9.5: Dependence of the time to regain periodic steady state, Θ, upon reduced length, Λ_H, and the measure of imbalance, β, where $\varepsilon = 10^{-4}$. (Schmidt and Willmott [16].)

$$\frac{\Lambda}{\Lambda'} = \frac{\Pi}{\Pi'} = 0.5$$

they found that

$$\frac{\Pi_H}{\log_e \lambda_1} \approx \textit{constant}$$

for varying values for Π_H and a constant value of Λ_H.

9.11 Interpretation of the Relation Between the Transient Performance of a Regenerator and its Dimensionless Parameters

The reduced length, Λ or Λ', is a measure of the size of the regenerator relative to the thermal load imposed upon it in the hot or cold period. On the other hand, the reduced period, Π or Π', is a measure of the period length relative to the heat capacity of the packing and its available heating surface. The reduced period embodies the concept that thin packings of small heat capacity but large heating surface area enforce short periods for efficient operation. On the other hand, packings with a large heat capacity but small heating surface enable regenerators to be operated with longer periods of operation.

The force imposing oscillations in the packing temperature is derived from the alternate washing of the packing surface by the hot and then the cold gas. The same is true for the force causing changes in these oscillations as a consequence of there being a step change in the inlet gas temperature. The "size" of the regenerator can be measured by the product $\bar{\alpha}A$ in either period of operation. The larger the "load," $\dot{m}_f c_p$, the smaller the "relative size" of the regenerator measured as Λ in the hot period, Λ' in the cold. What we might expect is that the smaller the "relative size" of $\dot{m}_f c_p$, that is, relative to $\bar{\alpha}A$, the smaller the force imposing the effects of a step change in regenerator operation, for example, a step change in inlet gas temperature. In other words, as has been observed above, the larger the reduced length, Λ_H, the longer it will take for the packing to respond to the effects of a step change in regenerator operation.

The reduced period does not reflect directly the "relative size" of a regenerator, as does the reduced length. The reduced period, in contrast, represents the heat capacity of the system relative to the "size" of the regenerator measured by $\bar{\alpha}A$ in the hot period and $\bar{\alpha}'A$ in the cold period. A change in operating conditions results in a change in the average temperature of the packing over a complete cycle, that is, a change in the "level of heat" held in the reservoir. The total dimensionless time, Θ, to reestablish cyclic steady state is determined by the size of the reservoir and not by the cycle time. It follows that Θ can be made up of a few long cycles or many short cycles.

In the same way that the actual (as opposed to dimensionless) cycle times are necessarily short for packings with a high surface to heat capacity ratio, so the actual time for a regenerator to attain equilibrium following a step change in operation will be correspondingly small and vice versa.

Example 9.3 We have shown in Example 9.1 that using the convergence criteria with $\varepsilon = 10^{-4}$, the number of cycles required to restore periodic steady state following a step change in inlet gas temperature for the regenerator configuration where $\Lambda_H = 10$ and $\Pi_H = 0.5$ is 110 (extending the analysis to balanced unsymmetric regenerators). If the regenerator is a Cowper stove with a 3 hr cycle time, the actual time to equilibrium will be 330 hr = 13.75 days $\approx$ 2 weeks. On the other hand, with a rotary regenerator spinning at 3 rev/min, that is, a cycle time = 20 s, the actual time to equilibrium will be 2200 s ≈ 37 min. Such differences, frequently, might not be anticipated.

9.12 Step Changes in Gas Flow Rate

The first work on this problem was undertaken by Green [11]. Using computer simulation techniques, he examined the effects of step changes in gas flow rate on a regenerator which has a *symmetric* configuration *prior* to the step change. Green observed that the total time to establish periodic steady state reached a maximum for changes of the order to 10–20% and then was reduced for larger step changes.

The definition of reduced length, Λ, includes the ratio $\bar{\alpha}/\dot{m}_f$. If it is approximated that $\bar{\alpha}$ is linearly proportional to $\dot{m}_f$, it is possible to regard reduced length, Λ, to be independent of flow rate. Step changes in hot period flow rate can then be treated as step changes in the hot-side reduced period, Π alone. This treatment is examined in detail by Burns [23].

Green also used this approximation and his results are shown in Figure 9.6. Willmott and Burns extended their analysis for step changes in inlet gas temperature to step changes in gas flow rate. This was realized by considering the state of the regenerator *after* the step change because the nature of the transient behavior is determined by this final state as will the equilibrium conditions toward which the simulation will converge. It will be shown that this approach can be used to predict the results observed by Green previously.

Whereas the responses $\varepsilon f_1(\tau)$ and $\varepsilon f_2(\tau)$ for step changes in inlet gas temperature are independent of the step size, for the linear model, this is not so in the case of step changes in gas flow rate. The magnitude of the step change in flow rate cannot be neglected. Up to a threshold value in the percentage change made to Π, and implicitly the hot gas flow rate, the time to equilibrium Θ increases with step size. Beyond this threshold, the transient behavior is a function only of the final operating conditions. The threshold increases with reduced length. These features are shown in Figure 9.7 for the case where the final conditions are that the regenerator is symmetric with $\Lambda = \Lambda' = 5$, 10 and 20 for $\Pi = \Pi' = 4$. If step changes beyond the threshold are considered, the transient response is almost the same as for step changes in inlet gas temperature.

The response to changes beyond this threshold can be parameterized in terms of the weighted harmonic mean reduced length, Λ_H, and the imbal-

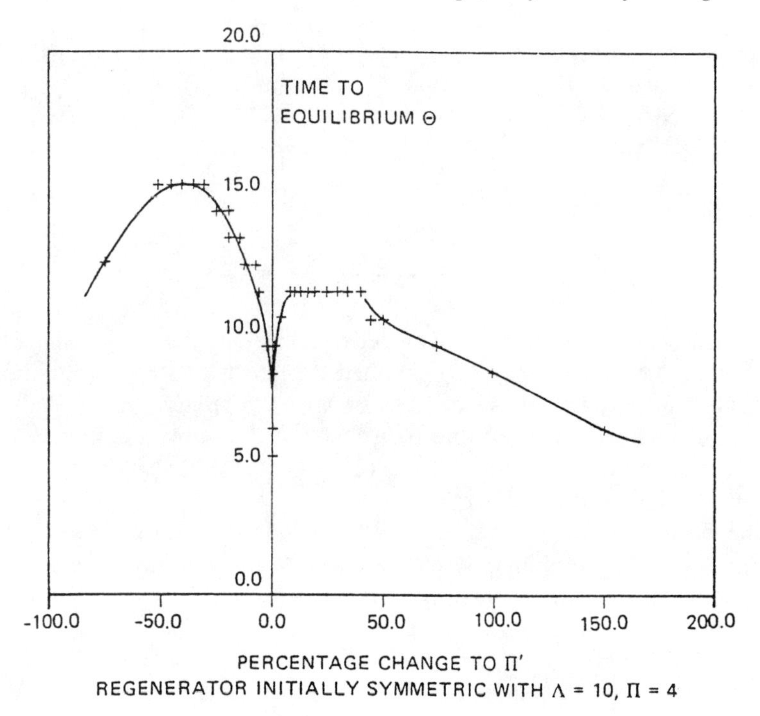

Figure 9.6: Green's representation of the effect of step changes in gas flow rate (symmetric case). (Schmidt and Willmott [16].)

ance factor, β, discussed previously. One additional factor, γ, is required, it turns out, to relate the transient response to step changes in Π below the threshold to the response to changes in Π beyond the threshold. This factor is given by

$$\gamma = \frac{\text{magnitude of step change}}{\text{magnitude of step change at threshold}} \tag{9.58}$$

The parameter $\gamma\,(0 < \gamma < 1)$ has *relatively* little effect on the dimensionless time, Θ, to restore cyclic steady state. Willmott and Burns considered $\varepsilon = 10^{-4}$. The effect is displayed in Figure 9.8 for $2 \leq \Lambda_H \leq 30$ and $\gamma = 1$, 0.5, 0.25. The important feature to note in Figures 9.7 and 9.8 is that the greater the value of the reduced length, Λ_H, the more sensitive the response to the factor γ.

The transient response of a regenerator is parameterized by Λ_H and β alone when $\Delta\Pi$, the step change in hot gas flow rate, is greater than the

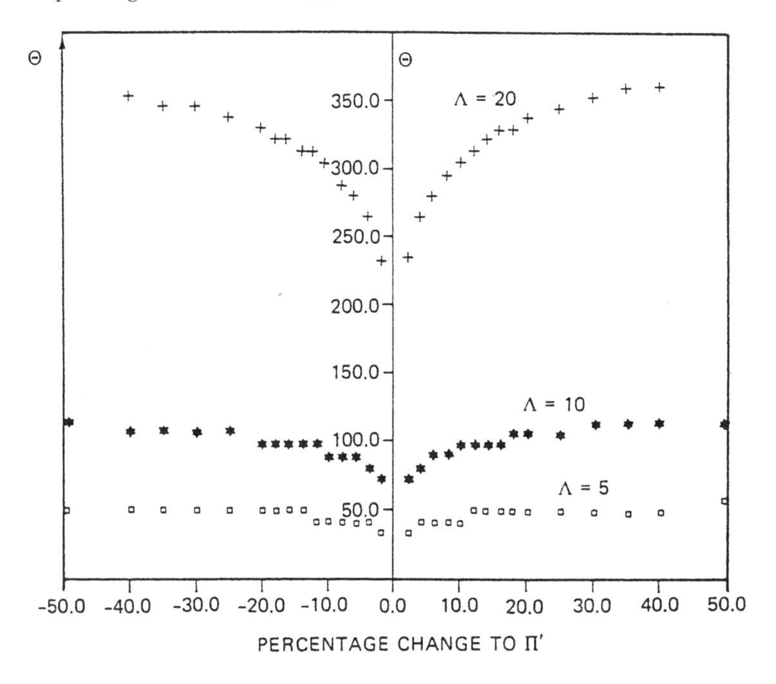

Figure 9.7: Dependence of the time to cyclic periodic state, Θ, for varying degrees of the percentage change to Π with $\varepsilon = 10^{-4}$. (The final state of the regenerator is symmetric with $\Pi = 4$.) (Schmidt and Willmott [16].)

threshold, when $\gamma > 1$. The threshold is usually less than 50%. For $\Lambda = 10$ and $\Pi = 4$, the threshold is 10% and for $\Lambda = 20$ and $\Pi = 4$, it is 35%. The dependence of the response $\varepsilon f_1(\eta)$ upon reduced length, Λ_H, and the imbalance factor, β, for step changes of 50% in flow rate, well beyond the threshold, is shown in Figures 9.9 and 9.10. For such step increases greater than 50%, for the symmetric case, it can be observed that the response $\varepsilon f_1(\eta)$ and the time to reestablish periodic steady state, Θ, are independent of reduced period, Π. The total time to equilibrium can be represented by

$$\Theta = 0.72\Lambda^2 + 4.2\Lambda + 8 \quad \text{for} \quad \Lambda \leq 30 \quad \text{and} \quad \beta = 1 \tag{9.59}$$

Equation (9.55) for step changes in inlet gas temperature and the corresponding equation (9.59) are compared diagrammatically in Figure 9.11. The use of the weighted harmonic mean reduced length, Λ_H, and the imbalance factor, β, for the transient response of regenerators to step changes in inlet gas temperature can be extended to step changes in flow rate. Although small differences between the responses calculated for such unsymmetric

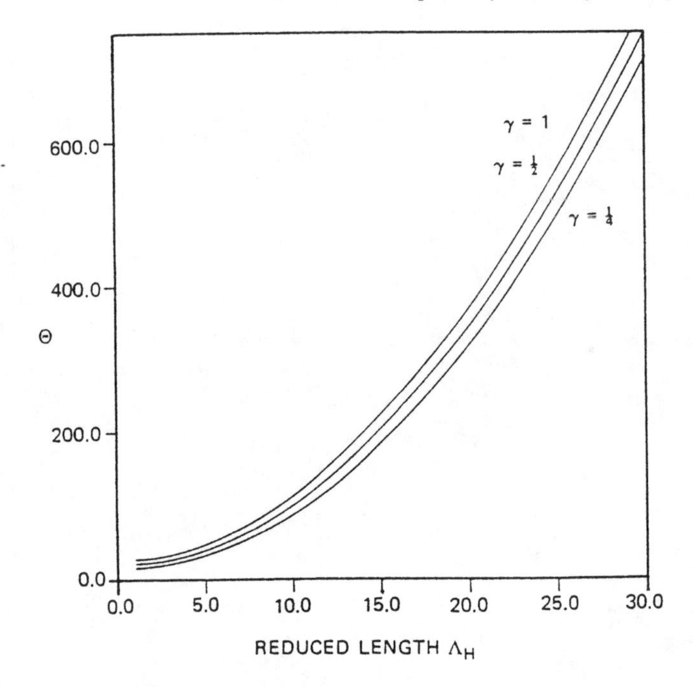

Figure 9.8: Dependence of the time to periodic steady state, Θ, upon reduced length, Λ_H, and γ, the step size parameter. (Schmidt and Willmott [16].)

regenerators and those calculated using the *harmonic means* equivalent symmetric regenerator model, it is doubtful whether such differences could be detected experimentally. In any event, the total time to equilibrium, Θ, computed for $\varepsilon = 10^{-4}$, remains accurately represented by the equivalent symmetric case.

The effect of imbalance, β, is shown in Figure 9.10, where the transient response $\varepsilon f_1(\eta)$ is displayed for reduced length, Λ_H fixed at a value of 10 for $\beta = 1, 2, \frac{1}{2}, 4, \frac{1}{4}$. The same maximal inertia exhibited by a balanced regenerator ($\beta = 1$) is found for step changes both in flow rate and inlet gas temperature.

The initial observations of Green [11], displayed in Figure 9.6, can now be interpreted. Green starts with a regenerator that is symmetric *prior* to a step change. For small changes in flow rate, the regenerator remains approximately symmetric and the time to equilibrium will increase with the relative size of the step change, γ. For larger step increases, the regenerator becomes increasingly unbalanced and the time to reestablish equilibrium, Θ, declines. There is, therefore, a value in the size of the step change

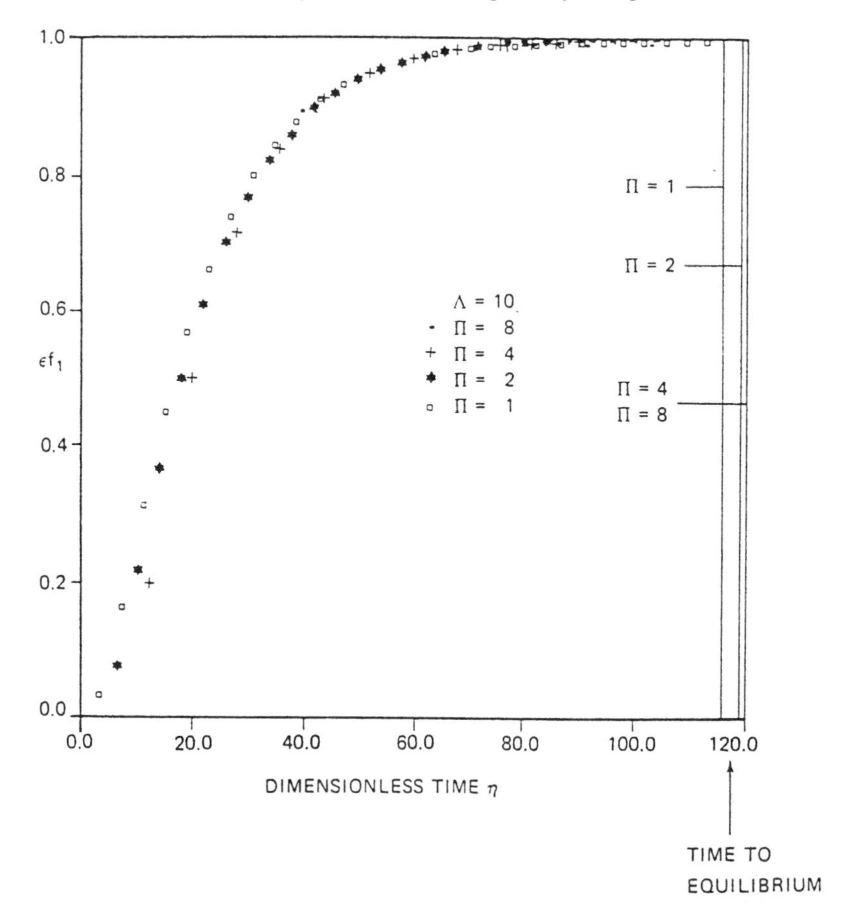

Figure 9.9: Dependence of the response $\varepsilon f_1(\eta)$ upon reduced period, Π, for step changes in Π (symmetric case). (Schmidt and Willmott [16].)

in flow rate for a particular configuration when the time to equilibrium is a maximum. Beyond this maximum, the effect of the imbalance factor, β, becomes increasingly predominant.

9.13 Further Considerations of the Transient Response of a Regenerator

It will have been seen that a limited number of factors, namely reduced length, Λ_H, the imbalance factor, β, and the relative step size for flow rate

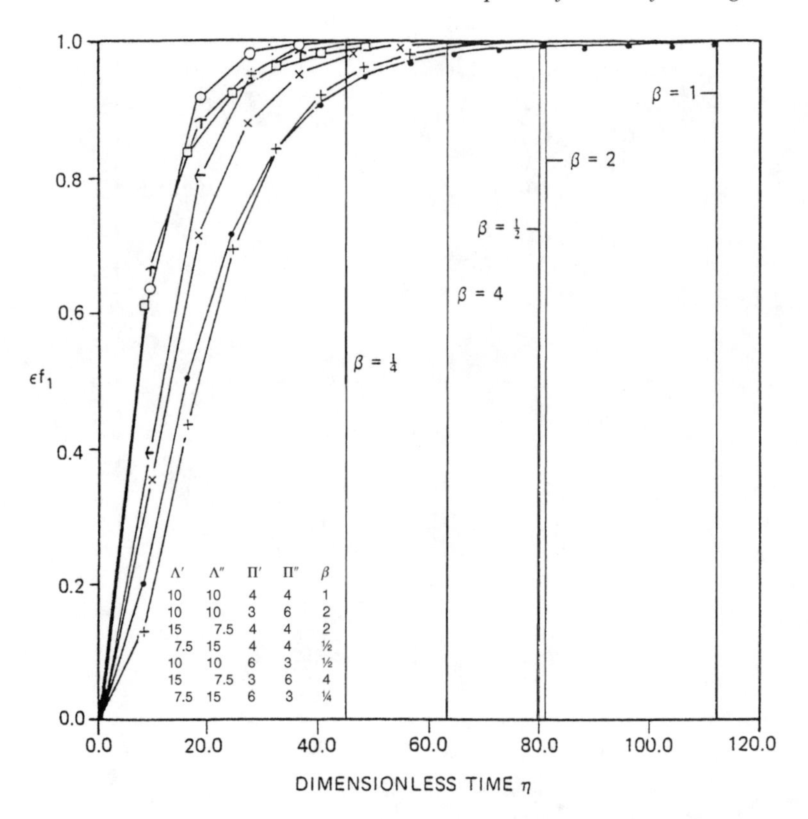

Figure 9.10: Unbalanced regenerator ($\Lambda/\Pi \neq \Lambda'/\Pi'$): variation of εf_1 in response to a step change in Π. (Vertical lines denote time when equilibrium is restored.) (Schmidt and Willmott [16].)

changes, γ, almost fully describe the transient response of a regenerator to step changes both in inlet gas temperature and flow rate. Within this context, Burns [7] found it to be most useful to be able to measure the *rates of convergence* of the responses $\varepsilon f_1(\eta)$ and $\varepsilon f_2(\eta)$. In Figure 9.12 are plotted $\log_e(1 - \varepsilon f_1(\eta))$ and $\log_e(1 - \varepsilon f_2(\eta))$ as a function of dimensionless time, η.

The data points have been joined together by straight lines in order to emphasize the parallel gradients of $\log_e(1 - \varepsilon f_1(\eta))$ and $\log_e(1 - \varepsilon f_2(\eta))$. This corresponds to the observation made previously that the eigenvalues describing the variation of $\boldsymbol{T}^{(n)}$ in transience are the same as those for $\boldsymbol{T}'^{(n)}$ and that the same time constant, κ, applies to $\boldsymbol{T}^{(n)}$ and $\boldsymbol{T}'^{(n)}$.

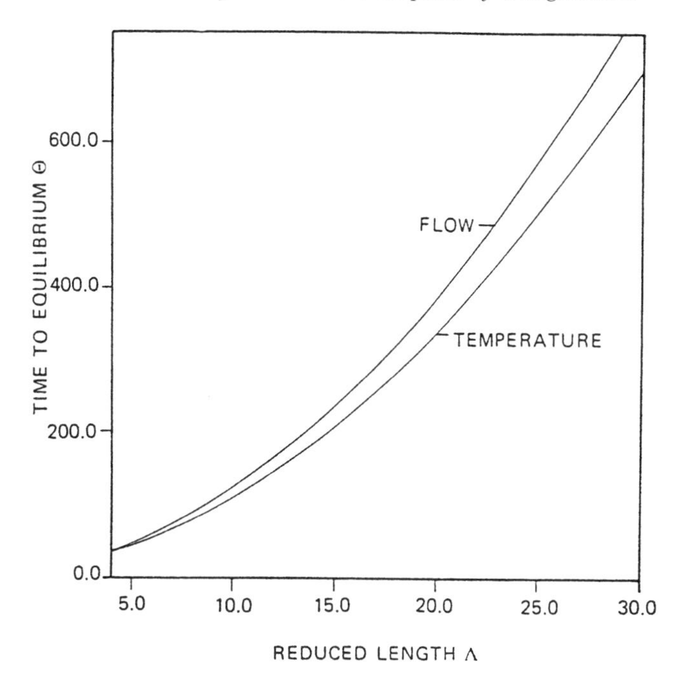

Figure 9.11: Comparison between the times to periodic steady state, Θ, as a function of reduced length, Λ, for step changes in inlet gas temperature and in gas flow rate with $\varepsilon = 10^{-4}$. (Schmidt and Willmott [16].)

Apart from the initial few cycles after the step change, the responses can be represented approximately by the responses εf_1^* and εf_2^* given by

$$\varepsilon f_1^* = 1 - \exp\left(-\frac{\eta_H - C_1}{\eta_c}\right) \tag{9.60}$$

$$\varepsilon f_2^* = 1 - \exp\left(-\frac{\eta_H + C_2}{\eta_c}\right) \tag{9.61}$$

Here η_H is the dimensionless time on the harmonic mean (of Π and Π') scale. Equations (9.60) and (9.61) effectively map the overall asymptotic approach of the regenerative heat exchanger toward a new cyclic steady state following a step change in operating conditions. These equations point toward another time constant, η_c, as a measure of the common rate of convergence of both the responses $\varepsilon f_1(\eta)$ and $\varepsilon f_2(\eta)$ and hence the thermal inertia of the regenerator. The time lag, lag_H, between $\varepsilon f_1(\eta)$ and $\varepsilon f_2(\eta)$ is given by

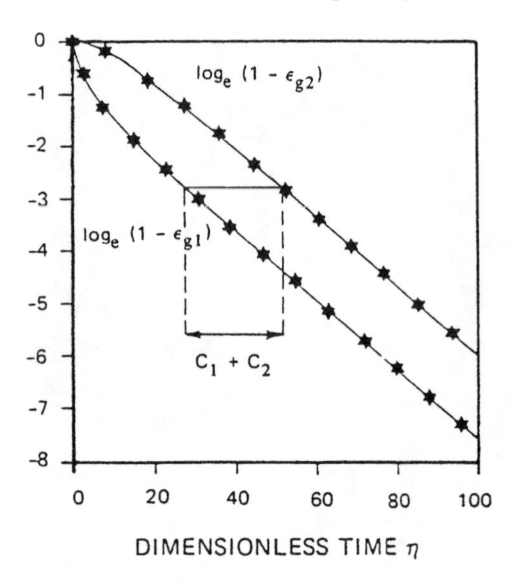

Figure 9.12: The responses $1 - \varepsilon f_1$ and $1 - \varepsilon f_2$ on a $\log_e$ scale, as a function of time η, for step changes in inlet gas temperature. (Schmidt and Willmott [16].)

$$lag_H = C_2 + C_1 \tag{9.62}$$

Willmott and Burns [7] examined the properties of the parameters η_c, C_1, C_2, and lag_H over a whole range of operating parameters for both step changes in flow rate and inlet gas temperature. Their conclusions are summarized below.

1. The responses εf_1 and εf_2 generate the same time constant η_c for any regenerator. This corresponds to the generation of a common time constant, κ, where

$$\kappa = \frac{\Pi}{\log_e \lambda_1}$$

   and where λ_1 is the dominant eigenvalue of both the matrices M and M'.
2. The parameters η_c, C_1, and C_2 and hence the time lag, lag_H, are functions of the weighted harmonic mean reduced length, Λ_H, but exhibit independence of the harmonic mean reduced period, Π_H for $\Lambda_H/\Pi_H > 3$. This is illustrated in Figure 9.13.

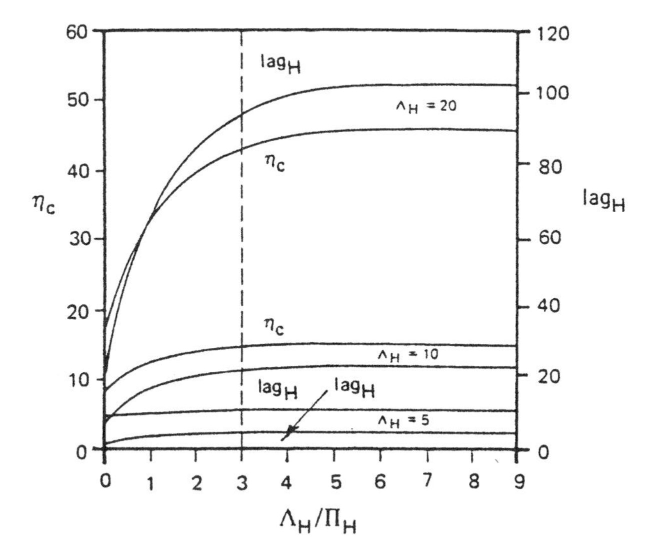

Figure 9.13: The dependence upon the ratio Λ_H/Π_H of reduced length to reduced period, of lag_H and η_c. (Schmidt and Willmott [16].)

3. The time constant, η_c, is determined by the operating parameters Λ_H and β for the *post* step change phase. It is independent of the magnitude of the step change either in gas inlet temperature or flow rate.
4. The constants C_1 and C_2 both increase with reduced length, Λ_H, for step changes in inlet gas temperature. In the case of step changes in flow rate, the situation is complicated: the time lag, lag_H, depends on the magnitude of the step change. For increasingly large step changes in flow rate, however, these changes cross a threshold value when $\gamma > 1$, beyond which the values of C_1 and C_2 and hence the time lag, lag_H, approach the values computed for step changes in inlet gas temperature.
5. The time lag, lag_H, is independent of the size of the step change in inlet gas temperature.

Figure 9.14 depicts the relationship between the time constant, η_c, and the weighted harmonic mean reduced length, Λ_H. Willmott and Burns [7] fitted the quadratic curve (9.61)

$$\eta_c = 0.0922\Lambda_H^2 + 0.489\Lambda_H + 0.928 \tag{9.63}$$

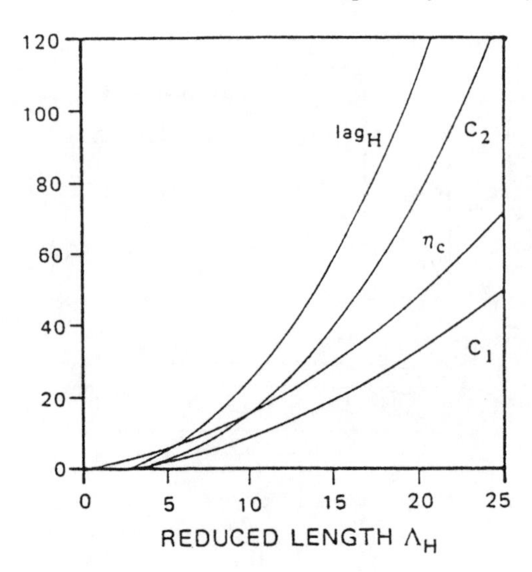

Figure 9.14: The dependence upon Λ_H, the reduced length, of lag_H, η_c, C_1, and C_2. (Schmidt and Willmott [16].)

by the method of least squares. Schmidt and Willmott [16] point to a time constant of $\eta_c = 20$ for a hot-blast stove installation with reduced length $\Lambda_H = 12$.

Previously, it has been mentioned that a regenerator exhibits greatest inertia when it is *balanced*, $\beta = 1$. This property can be exhibited in terms of the variation of the time constant, η_c, as a function of the imbalance factor, β. Figure 9.15 shows this, where it will be observed that the sensitivity of the inertia of a regenerator to the effect of imbalance increases markedly with reduced length, Λ_H.

9.14 The Thermal Inertia Exhibited by Variable Gas Flow Regenerators

The operation of regenerators in *serial* (bypass main) or *staggered parallel* was described in Chapter 8. The means of *control* of such systems was discussed earlier in this chapter. The reader is referred to Figures 8.2 and 8.3 in which the variation of the gas flow rate *within* the cold period of operation is shown, so that an unvarying blast temperature can be realized.

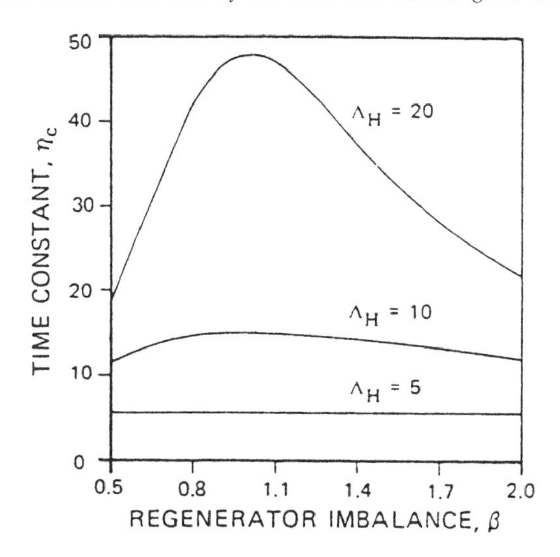

Figure 9.15: The dependence of the time constant η_c upon regenerator imbalance β. (Schmidt and Willmott [16].)

Willmott and Burns [7] made an initial examination of the thermal inertia of such systems. They restricted their considerations to bypass main operation and considered the effect of a step increase in inlet gas temperature or gas flow rate in the hot period upon a hot-blast stove system, initially at periodic steady state.

They modeled the transient phase while equilibrium was being reestablished, the target blast temperature, $t'_{f,B}$, being left unaltered. The thermal load, $\dot{m}'_f c_p(t'_{f,B} - t_{f,in})$, remains unchanged and the result of any step change is forced to become manifest in the hot-side exit gas temperature, $t_{f,x}(\tau)$, and in the way the flow rate through the stove, $\dot{m}'_f(\tau)$, varies with time over successive cycles after the step change equilibrium is restored. A response εw_2 may be defined in terms of the chronological average of $\dot{m}'_f(\tau)$ *within* each cycle of operation. Such a response εw_2, together with the hot-side response εf_1, is shown in Figure 9.16. Such responses, relating as they do to a typical hot blast stove system, are identical to the form extracted for a constant flow regenerator system and the approximations given previously as equations (9.53) and (9.63) may be employed.

Changes in flow rate within the cold period, $\dot{m}'_f(\tau)$, have little influence upon reduced length, Λ', for bypass main operation. It is known that $\Lambda_H/\Pi_H > 3$ for typical Cowper stove configurations and, moreover, variations in $\dot{m}'_f(\tau)$ *within* a cold period of operation have little influence on the

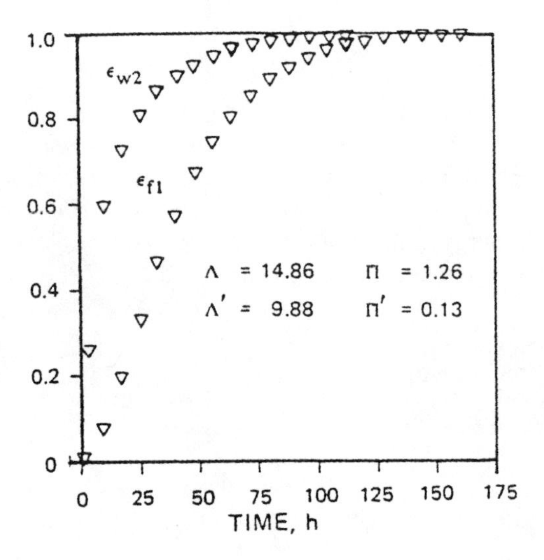

Figure 9.16: The responses εf_1 and εw_2 for a hot-blast stove with a bypass main, for a step change in hot period inlet gas temperature. (Schmidt and Willmott [16].)

cold side reduced period, Π. It follows, as a consequence, that these variations of $\dot{m}'$ have no influence upon the time lag and time constant that parameterize the inertia of such a hot-blast stove system. For transient-response calculations of such variable-flow systems, it is sufficient, therefore, to base the computations upon flow rates and heat transfer coefficients averaged over each period under consideration. It is then possible to use a common basis of the inertia of regenerator systems, namely the reduced length, Λ_H and the imbalance factor, β, for both constant flow and bypass main variable-flow schemes of regenerator operation.

9.15 Concluding Remarks

The steady-state performance of regenerators has been studied widely and has been discussed in earlier chapters. This is not the case for the transient performance. Schmidt and Willmott [16] reviewed the state of the art in this area, mentioning the work of London et al. as well as that of Willmott and Burns, and this has been presented in this chapter. It has been possible here to provide a matrix analysis that underpins this previous work and that is relatively novel. Willmott and Burns point to the need for experimental work in this area and the reader is referred to the work of Cutland [12]. Schmidt and Willmott refer to experimental work of Chao [24] of 1955. They mention that Burns and Willmott found that their theoretical results compared favorably with Chao's experimental values of εf_2, obtained in work for the Ford Motor Company on an experimental rotary regenerator.

Regenerative equipment of one form or another will be embodied in solar and geothermal energy collection and storage systems. Solar systems must be able to accommodate variations in the cloud cover and maintain, simultaneously, a minimal thermal load if at all possible. An understanding of the inertia properties and the transient performance of such regenerators must be important if the full potential of such systems is to be realized.

References

1. A. J. Willmott, D. M. Scott, L. Zhang, "Matrix Formulations of Linear Simulations of the Operation of Thermal Regenerators," *Numer. Heat Transfer Pt B: Fundamentals* **23**(2), 43–66 (1993).
2. R. M. Cima, A. L. London, "The Transient Response of a Two Fluid Counterflow Heat Exchanger – The Gas Turbine Regenerator," *J. Eng. Prog.* **81A**, 1169–1179 (1958).
3. A. L. London, F. R. Biancardi, J. W. Mitchell, "The Transient Response of Gas Turbine Plant Heat Exchangers – Regenerators, Intercoolers, Percoolers and Ducting," *ASME J. Eng. Prog.* **81**, 433–448 (1959).
4. A. L. London, D. F. Sampsell, J. G. McGowan, "The Transient Response of Gas Turbine Plant Heat Exchangers – Additional Solutions for Regenerators of the Periodic Flow and Direct-Transfer Types," *ASME J. Eng. Prog.* **86**, 127–135 (Apr. 1964).
5. A. J. Willmott, A. Burns, "Transient Response of Periodic-Flow Regenerators," *Int. J. Heat Mass Transfer* **20**, 753–761 (June 1977).
6. A. Burns, A. J. Willmott, "Transient Performance of Periodic Flow Regenerators," *Int. J. Heat Mass Transfer* **21**, 623–627 (1978).
7. A. J. Willmott, A. Burns, "Periodic-Flow Regenerators: Parameter Identification for Transient Performance," *Proc. VI Int. Heat Transfer Conf., Toronto* **HX-19**, pp. 297–302 (1978).
8. J. Beets, J. Elshout, "Control Model for a Hot-Blast Stoves System," *Int. Meeting on Iron and Steel (Brussels)* (May 1976).
9. I. Strausz, "Automatic Control of a Hot Stove System at a Blast Furnace, by Use of a Digital Computer," *Q. J. Autom. Control (Belg.)* **1**, 15–20 (Jan. 1970).
10. P. Zuidema, "Non-Stationary Operation of a Staggered Parallel System of Blast Furnace Stoves," *Int. J. Heat Mass Transfer* **15**, 433–442 (1972).
11. D. R. Green, "Some Aspects of Hot Blast Stove Development and Operation," M.Sc. thesis, UMIST, Manchester, UK (1967).

12. N. G. Cutland, "The Unsteady-State Performance of an Experimental Thermal Regenerator," D.Phil. thesis, University of York, UK (Mar. 1984).
13. R. S. Varga, *Matrix Iterative Analysis*, Prentice-Hall, New York (1962).
14. A. J. Willmott, D. M. Scott, L. Zhang, "A Unified Matrix Approach to the Linear Analysis of Thermal Regenerators," *Proc. 3rd United Kingdom Heat Transfer Conference* **1**, pp. 369–376 (Sep. 1992).
15. C. E. Iliffe, "Thermal Analysis of the Contra-flow Regenerative Heat Exchanger," *Proc. Inst. Mech. Eng.* **159**, 363–372 (1948).
16. F. W. Schmidt, A. J. Willmott, *Thermal Energy Storage and Regeneration*, McGraw-Hill, New York (1981).
17. A. J. Willmott, A. Burns, "The Recuperator Analogy for the Transient Performance of Thermal Regenerators," *Int. J. Heat Mass Transfer* **22**, 1107–1115 (1979).
18. A. J. Willmott, "Digital Computer Simulation of a Thermal Regenerator," *Int. J. Heat Mass Transfer* **7**, 1291–1302 (May 1964).
19. P. Henrici, *Elements of Numerical Analysis*, Wiley, New York (1974).
20. A. Hill, A. J. Willmott, "Accurate and Rapid Thermal Regenerator Calculations," *Int. J. Heat Mass Transfer* **32**, 465–476 (1989).
21. L. Zhang, D. M. Scott, "A Unified Approach to Discretisations of the Linear Analysis of Thermal Regenerators," *Int. J. Heat Mass Transfer* **36**(4), 1035–1041 (1993).
22. H. Hausen, "Vervollstandigte Berechnung des Warmeaustausches in Regeneratoren (Improved Calculations for Heat Transfer in Regenerators)," *Z. VDI-Beiheft Verfahrenstech.* **2**, 31–43 (1942) (Iron and Steel Institute translation, June 1943).
23. A. Burns, "Heat Transfer Coefficient Correlations for Thermal Regenerator Calculations: Transient Response," *Int. J. Heat Mass Transfer* **22**, 969–973 (1979).
24. W. W. Chao, "Research and Development of an Experimental Rotary Regenerator for Automotive Gas Turbines," *Proc. Amer. Power Conf.* **17**, pp. 358–374 (1955).

Chapter 10

PARALLEL-FLOW REGENERATORS

10.1 Introduction

It might well seem that a discussion of parallel-flow regenerators, in this short chapter, is an opportunity to explain the superiority of the contraflow operation of these capacitative heat exchangers. Through the packing of such parallel-flow or coflow regenerators, the hot and cold gases pass, alternately, in the *same* direction. It is certainly the case that the counterflow regenerator is far more thermally effective than its parallel-flow counterpart. This can be concluded from the simple observation that the temperature distribution in the packing at the end of a hot period, for example, decreases from the hot-gas entrance to the hot-gas exit. In the subsequent cold period, for counterflow operation, the fluid encounters a steadily increasing temperature in the regenerator packing. In this way, the fluid is increased steadily in temperature.

For parallel-flow operation, however, the cold fluid encounters a high solid temperature and then proceeds to meet solid temperatures that decrease in an attenuating oscillatory fashion. It is thus possible, especially in the early part of such a cold period, for the cold gas to pick up heat from the region of the packing near the entrance of the gas, to exchange no heat with the packing in the middle of the regenerator, and finally to lose heat to the colder solid near the gas exit. When this happens, the gas appears to act as a means of transport of heat from the hot to the cold end of the packing and might be described as a pseudolongitudinal conductivity effect. Such a process continues until, at every position in the regenerator, the solid is at the same temperature or hotter than the gas passing through the channels of the packing. An analogous process may take place in the hot period of parallel-flow regenerator operation.

It is sometimes impossible or inconvenient, nevertheless, for a regenerator to be operated in other than parallel-flow mode. In some industrial applications, the inlet temperature and/or flow rate of the fluid entering the regenerator packing have timewise variations that are repeated in regular fashion. The fluid entering the regenerator need not involve just a single gas stream. During a complete cycle, for example, one gas stream with the higher inlet temperature and thus supplying heat to the packing, might be followed by another gas stream that extracts heat from the same packing.

The initial performance of such a unit is transient in nature and depends to a large extent on the initial temperature condition of the packing as well as the chronological variations of the inlet temperatures of the gases. Under continuous operating conditions, as in contraflow regenerators, the performance becomes cyclic and a periodic steady state is attained. Cyclic equilibrium is said to have been reached when the heat extracted in the packing in the "cold" period is equal to the heat stored in the "hot" period, less the effect of any heat losses from the unit.

Rotary regenerators can, in principle, be operated in parallel-flow mode, but this is unusual. The idealization used here can be applied to both fixed-bed and rotary regenerators.

10.2 Method of Analysis

The linear model described in Chapter 5 and elsewhere in this book is equally applicable to parallel-flow regenerators. The differential equations using dimensionless parameters take the familiar form

$$\text{Fluid:} \quad \frac{\partial T_f}{\partial \xi} = T_s - T_f \tag{10.1}$$

$$\text{Solid:} \quad \frac{\partial T_s}{\partial \eta} = T_f - T_s \tag{10.2}$$

The dimensionless parameters *reduced length*, Λ, Λ', and *reduced period*, Π, Π', prove to be as significant in describing the performance of parallel-flow regenerator operation as they do contraflow. The reversal conditions, however, are different from those that apply to contraflow operation. They take the form

$$T_s(\xi, 0) = T_s'\left(\frac{\Lambda'\xi}{\Lambda}, \Pi'\right) \tag{10.3}$$

$$T_s'(\xi', 0) = T_s\left(\frac{\Lambda\xi'}{\Lambda'}, \Pi\right) \tag{10.4}$$

The same normalized scale, [0, 1], as defined previously as equation (4.5), is used for the solid temperature T_s and the gas temperature T_f. The model assumes that the effect of axial conductivity in the solid can be neglected, although the effect of latitudinal conductivity within the packing can be incorporated within a lumped heat transfer coefficient, $\bar{\alpha}$.

Hausen [1] predicted the performance of parallel-flow regenerators using this model. He examined the periodic steady-state conditions using a closed method, briefly outlined later. Kardas [2] investigated the effect of the thermal conductivity of the packing in a direction perpendicular to gas flow for parallel-flow regenerators. Schmidt and Willmott [3] note, however, that the results obtained by Kardas using the analytical techniques proposed by him do not appear to converge at large values of reduced length, Λ. Kardas restricted his considerations to the case where $P = P'$, that is, where the hot and cold periods of operation are equal.

Kumar [4] proposed an approach to parallel-flow regenerators using a model that embodies both latitudinal *and* axial conductivity within the solid packing. The governing equations for the storage of heat in the packing is given by

$$\frac{1}{\kappa_s}\frac{\partial t_s}{\partial \tau} = \frac{\partial^2 t_s}{\partial x^2} + \frac{\partial^2 t_s}{\partial y^2} \tag{10.5}$$

Here y is the direction of gas flow and the direction x is perpendicular to y. The method with the results that follow are described by Schmidt and Willmott [3] but are not replicated here.

The cyclic steady-state performance of a parallel-flow regenerator can be calculated using an open or closed method, as is the case for counterflow regenerators, as might well be expected. These methods are first discussed in Chapter 5.

10.3 An Open Method for Parallel-Flow Regenerative Heat Exchangers

A typical method is the technique of Willmott [5] and this might well be used. At the start of a period, hot or cold, the solid temperatures are given, usually as those at the end of the previous period. The exception is the start of the hot period at the beginning of the simulation, where it is usually set as

$$T_s(\xi, 0) = \tfrac{1}{2}(T_{f,in} + T'_{f,in}) \quad \text{for} \quad 0 \leq \xi \leq \Lambda \tag{10.6}$$

It is important to note that for parallel-flow regenerators, the solid temperatures are *not* "reversed" at the end of a period in preparation for the start of the next period. This is only required for contraflow regenerator calculations. Instead, the reversal conditions (10.3) and (10.4) are applied.

The gas temperatures are computed from the solid temperatures down the length of the regenerator, together with the inlet gas temperature, using

$$\begin{aligned} T_{f,i,j} &= T_{f,i-1,j} + \frac{\Delta\xi}{2}(T_{s,i,j} - T_{f,i,j} + T_{s,i-1,j} - T_{f,i-1,j}) \\ &= A_1 T_{f,i-1,j} + A_2(T_{s,i,j} + T_{s,i-1,j}) \end{aligned} \tag{10.7}$$

where the subscripts i and $i-1$ denote spatial positions and the subscript j denotes a position in time on the finite difference net and

$$A_1 = \frac{2 - \Delta\xi}{2 + \Delta\xi} \quad \text{and} \quad A_2 = \frac{\Delta\xi}{2 + \Delta\xi} \tag{10.8}$$

The inlet solid temperatures at successive dimensionless time intervals $\Delta\eta$ are computed using

$$\begin{aligned} T_{s,i,j} &= T_{s,i,j-1} + \frac{\Delta\eta}{2}(T_{f,i,j} - T_{s,i,j} + T_{f,i,j-1} - T_{s,i,j-1}) \\ &= B_1 T_{s,i,j-1} + B_2(T_{f,i,j} + T_{f,i,j-1}) \end{aligned} \tag{10.9}$$

where $i = 0$ at the entrance to the regenerator and

$$B_1 = \frac{2 - \Delta\eta}{2 + \Delta\eta} \quad \text{and} \quad B_2 = \frac{\Delta\eta}{2 + \Delta\eta} \tag{10.10}$$

For the general (i, j)th position, with $j > 0$, the solid temperature $T_{s,i,j}$ is calculated using

$$T_{s,i,j} = K_1 T_{s,i,j-1} + K_2 T_{f,i,j-1} + K_3 T_{s,i-1,j} + K_4 T_{f,i-1,j} \tag{10.11}$$

where

$$K_1 = \frac{B_1}{X}, \quad K_2 = \frac{B_2}{X}, \quad K_3 = \frac{A_2 B_2}{X}, \quad K_4 = \frac{A_2 B_2}{X}, \quad \text{and}$$
$$X = 1 - A_2 B_2$$

Having computed $T_{s,i,j}$, the corresponding gas temperature, $T_{f,i,j}$ is calculated using equation (10.7).

Schmidt and Willmott [3] suggested that closed methods require considerably less computation time than is required for open methods for parallel flow regenerators. This remains true but is no longer an issue with the considerably more powerful personal computers now available compared with 20 years ago!

10.4 Closed Methods for Parallel-Flow Regenerators

Provided the regenerator is not "long," say for cases where $\Lambda < 25$ and $\Lambda/\Pi \leq 10$, it might well be possible to use Iliffe's method [6] as described in Chapter 5. The temperature distribution at the end of the hot period is given by

$$F(\xi) = T_s(\xi, \Pi) = 1 - e^{-\Pi}[1 - F(\xi)] + \int_0^{\xi} K(\xi - \varepsilon)[1 - F(\varepsilon)]\, d\varepsilon \tag{10.12}$$

and at the end of the cold period by

$$F'(\xi') = T_s'(\xi', \Pi') = e^{-\Pi'} F'(\xi') + \int_0^{\xi'} K'(\xi' - \varepsilon) F'(\varepsilon)\, d\varepsilon \tag{10.13}$$

where the functions K and K' are defined by equations (5.15) and (5.16). The reversal conditions for parallel-flow regenerators can be applied directly within these integral equations to yield

$$F'(\Lambda'\xi/\Lambda) = 1 - e^{-\Pi}[1 - F(\xi)] + \int_0^{\xi} K(\xi - \varepsilon)[1 - F(\varepsilon)]\, d\varepsilon \tag{10.14}$$

$$F(\Lambda\xi'/\Lambda') = e^{-\Pi'} F'(\xi') + \int_0^{\xi'} K'(\xi' - \varepsilon) F'(\varepsilon)\, d\varepsilon \tag{10.15}$$

The integrals in equations (10.14) and (10.15) are replaced by Simpson's rule approximations and these equations are reduced to a set of simultaneous linear algebraic equations in $\{F_k \mid k = 0, 1, 2, \ldots, m\}$ and $\{F_k' \mid k = 0, 1, 2, \ldots, m\}$ where $m\Delta\xi = \Lambda, m\Delta\xi' = \Lambda'$ and where F_k and the F_k' are calculated at equally spaced positions, $\Delta\xi$ and $\Delta\xi'$ apart, respectively.

10.5 Symmetric Regenerators: Reversal Conditions

In Chapter 5, it was shown that for the *symmetric* case, where $\Lambda = \Lambda'$ and $\Pi = \Pi'$, for counterflow regenerators, a symmetry exists between the solid temperature distributions at the ends of the hot and cold periods. Figure 10.1 reveals an equivalent symmetry for parallel-flow operation. The two temperature profiles on the [0, 1] scale are therefore related as follows:

$$T_s'(\xi, \Pi') = 1.0 - T_s(\xi, \Pi) \tag{10.16a}$$

or

$$F'(\xi) = 1.0 - F(\xi) \tag{10.16b}$$

This simplifies considerably any discussion of closed methods applied to symmetric parallel flow regenerators. Substituting equation (10.16b) into equation (10.13) gives

$$F'(\xi) + e^{-\Pi}F'(\xi) + \int_0^{\xi} K(\xi - \varepsilon)F'(\varepsilon)\,d\varepsilon = 1 \tag{10.17}$$

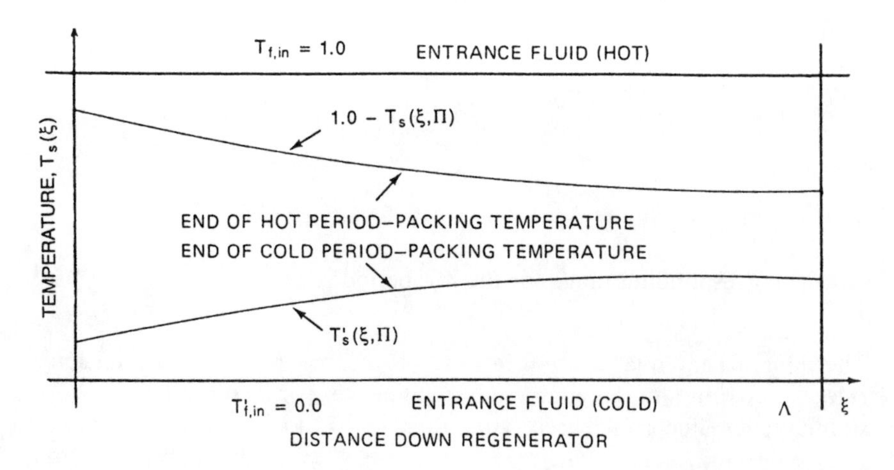

Figure 10.1: Final solid temperature distributions – parallel-flow operation of regenerators. (Schmidt and Willmott [3].)

10.6 Other Closed Methods for Parallel-Flow Regenerators

Schmidt and Willmott [3] chose to describe the heat pole method within the context of parallel-flow regenerators, if only to mention Hausen's earlier work in this field. They pointed out that for infinitely narrow poles, Hausen's method could be regarded as reducing to the integral equation (10.17) for the symmetric case.

The robust methods described in Chapter 7 for counterflow regenerators, however, are equally applicable to the parallel-flow scheme. The fast Galerkin method, for example, using Legendre polynomials can be applied. Equation (7.44) for contraflow takes the form

$$\int_0^\Lambda \sum_{j=0}^{n} \left[\alpha_j \left(Q_j(\xi) + e^{-\Pi} Q_j(\xi) + \int_0^\xi K(\xi - \varepsilon) Q_j(\varepsilon)\, d\varepsilon \right) - 1 \right] Q_i(\xi)\, d\xi = 0 \tag{10.18}$$

Advantage can still be taken of the orthogonality of the $Q_j(\xi)$ functions on the $[-1, 1]$ scale as well as the economies available by solving

$$F(\xi) - \mathrm{f}_1(\mathrm{f}_2(F(\xi))) = 0$$

with $F'(\xi')$ being located by using

$$F'(\xi') = \mathrm{f}_2(F(\xi))$$

as discussed in Chapter 7 for the nonsymmetric case.

Hausen [1, 7] points to a possible analytical *closed* solution to the differential equations for the symmetric case where $\Lambda = \Lambda'$ and $\Pi = \Pi'$. Hausen uses different but similar dimensionless parameters. He considers that the cold period covers the range

$$-\Pi \le \eta \le 0$$

and the corresponding range for the hot period is

$$0 \le \eta \le +\Pi$$

The same gas enters the packing at $\xi = 0$ in square wave fashion at temperature $t_f = t_{f,in}$ in the hot period and at $t_f' = -t_{f,in}$ in the cold, that is, at the dimensionless temperature $T_{f,in} = +1$ in the hot period, at $T_{f,in}' = -1$ in the following period. Hausen established, using a closed method, a series solution for periodic steady state of the form

$$T_f(\xi, \eta) = -\frac{4}{\pi}\left(\phi_1 + \frac{1}{3}\phi_3 + \frac{1}{5}\phi_5 + \cdots\right) \tag{10.19}$$

$$T_s(\xi, \eta) = -\frac{4}{\pi}\left(\mu_1 + \frac{1}{3}\mu_3 + \frac{1}{5}\mu_5 + \cdots\right) \tag{10.20}$$

where thc functions $\{\phi_j, \mu_j \mid j = 1, 3, 5, \ldots\}$ are eigenfunctions and the coefficients $\{1/j \mid j = 1, 3, 5, \ldots\}$ were determined by Hausen using a Fourier method.

The functions are offered by Hausen in the form

$$\phi_j(\xi, \eta) = \exp\left(-\frac{j^2\pi^2}{\Pi^2 + j^2\pi^2}\xi\right)\sin\left(\frac{j\pi}{\Pi}\eta - \frac{\Pi j\pi}{\Pi^2 + j^2\pi^2}\xi\right) \tag{10.21}$$

$$\mu_j(\xi, \eta) = \left(\frac{\Pi^2}{\Pi^2 + j^2\pi^2}\right)^{1/2} \exp\left(-\frac{j^2\pi^2}{\Pi^2 + j^2\pi^2}\xi\right) \times \sin\left(\frac{j\pi}{\Pi}\eta - \frac{\Pi j\pi}{\Pi^2 + j^2\pi^2}\xi - \arctan\frac{j\pi}{\Pi}\right) \tag{10.22}$$

where $\phi_j(\xi, \eta) = \mu_j(\xi, \eta) = 0$ for $j = 0$.

10.7 Parallel-Flow Regenerator Performance

Hausen [1] published a graph displaying the thermal ratio, η_{REG} (thermal effectiveness), as a function of reduced length Λ for different values of reduced period Π for the symmetric case. This is reproduced here as Figure 10.2. Also displayed are the corresponding curves for $\Pi = 0$ and $\Pi = \Lambda$ for counterflow operation.

One key difference between parallel and contraflow performance is that

$$\text{Counterflow:} \quad \lim_{\Pi \to 0} \eta_{REG}(\Lambda) = \frac{\Lambda}{\Lambda + 2}$$

whereas

$$\text{Parallel flow:} \quad \lim_{\Pi \to 0} \eta_{REG}(\Lambda) = \tfrac{1}{2} \quad \text{for} \quad \Lambda > 5$$

so that, for $\Pi = 0$, η_{REG} steadily increases and approaches the value of 1.0 as $\Lambda \to \infty$ for counterflow, whereas η_{REG} quickly settles down to the value of 0.5 for $\Lambda > 4$ for parallel flow.

For larger values of reduced period, Π, the effect, or the lack of the effect of pseudoaxial conductivity in parallel flow, as discussed earlier in this chapter, makes the situation quite complex. For Λ large enough,

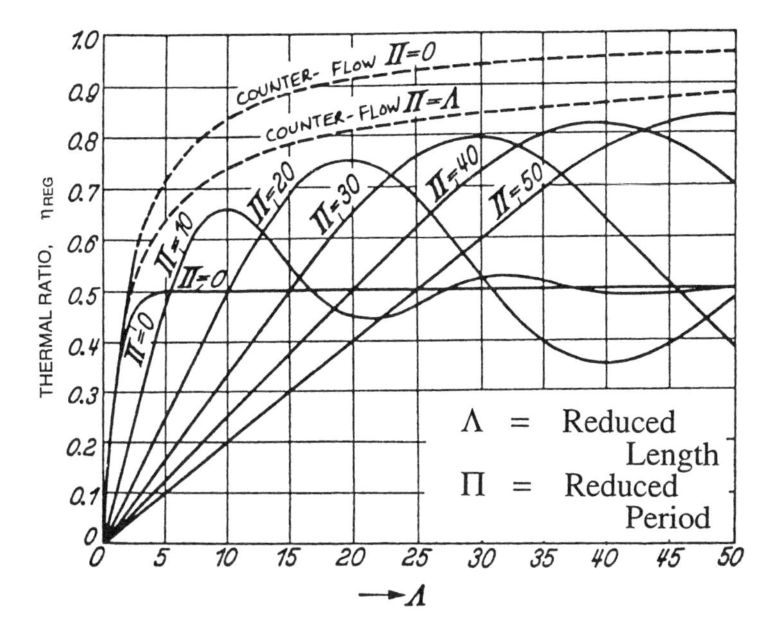

Figure 10.2: Thermal ratio in relation to reduced length and reduced period for symmetric parallel-flow regenerators. (Schmidt and Willmott [3].)

$\eta_{REG} \to 0.5$ for parallel flow, whereas $\eta_{REG} \to 1.0$ for counterflow. For $\Pi = 10$, for example, η_{REG} varies in a damped oscillatory manner, moving above 0.5 and then below 0.5 as Λ increases until it settles down to $\eta_{REG} = 0.5$. There is no equivalent damped oscillatory variation of η_{REG} for contraflow operation.

Schmidt and Willmott [3] computed the values of the thermal ratio, η_{REG}, for $\Lambda = \Pi$ over the range $2.5 \leq \Lambda \leq 15$ for parallel and counterflow operation, and these results are shown in Table 10.1. They also computed the thermal effectiveness, η_{REG}, for parallel flow for $2.5 \leq \Lambda \leq 25$ for $\Pi = 1.0$, 2.5, 5.0, 7.5, 10.0, and 15.0. They employed the method of Nahavandi and Weinstein [8] applied to parallel-flow regenerators. This method is discussed earlier with relation to contraflow regenerators. These computed values are displayed in Table 10.2.

The periodic form of the temperature distribution in the solid, calculated by Hausen, is illustrated by Figure 10.3 where the solid temperature spatial variation at the start/end of the cold/hot period is presented for different reduced periods, Π, for the symmetric case. The chronological variation of the solid temperature spatial distribution for $\Pi = \Pi' = \pi$ is given in Figure 10.4, and for the gas in Figure 10.5. For $\Lambda > 5$ in this case, $T_f(\xi, \eta) \to T_s(\xi, \eta) \to \frac{1}{2}$.

Table 10.1: Comparison in thermal effectiveness, η_{REG}, between parallel-flow and counterflow regenerators

$\Lambda = \Pi$	Parallel flow η_{REG}	Counterflow η_{REG}
2.5	0.462	0.512
5.0	0.562	0.638
7.5	0.618	0.700
10.0	0.658	0.738
12.5	0.689	0.764
15.0	0.713	0.784

10.8 Concluding Remarks

The superiority of counterflow operation is indicated by the observation that η_{REG}(counterflow) > η_{REG}(parallel flow) for any value of reduced length, Λ, for $\Lambda > \Pi$. A counterflow regenerator can realize an effectiveness of 83% for $\Pi < 3$, whereas a parallel-flow regenerator four times as large, $\Lambda = 40$, must be used to realize the same effectiveness. This performance by a parallel-flow regenerator is very sensitive to the duration of the cycle; the effectiveness collapses if the cycle time is halved in this case. Again, there is no equivalent to this in counterflow operation.

Table 10.2: The thermal effectiveness, η_{REG}, of symmetric, parallel-flow regenerators

Reduced	Thermal effectiveness, η_{REG}					
length, Λ	$\Pi = 1.0$	$\Pi = 2.5$	$\Pi = 5.0$	$\Pi = 7.5$	$\Pi = 10.0$	$\Pi = 15$
2.5	0.461	0.462	0.407	0.137	0.247	0.167
5.0	0.499	0.514	0.562	0.542	0.467	0.331
7.5	0.5	0.504	0.547	0.618	0.613	0.486
10.0	0.5	0.499	0.505	0.583	0.658	0.613
12.5	0.5	0.5	0.491	0.516	0.620	0.692
15.0	0.5	0.5	0.495	0.474	0.543	0.713
17.5	0.5	0.5	0.5	0.470	0.476	0.681
20.0	0.5	0.5	0.501	—	—	0.613
22.5	0.5	0.5	0.5	0.502	0.447	0.531
25.0	0.5	0.5	0.5	—	—	0.458

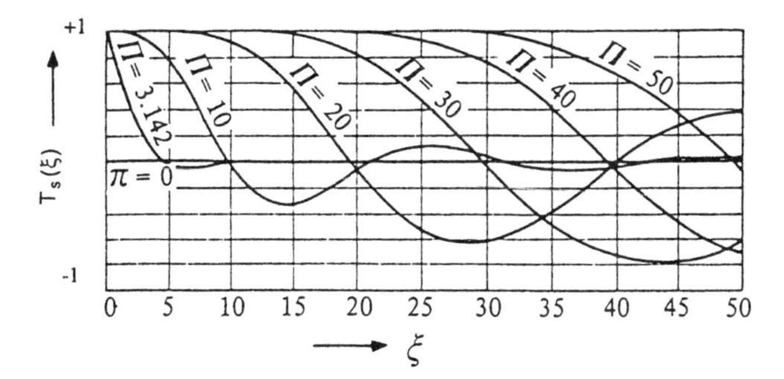

Figure 10.3: Spatial variation of the temperature of the packing of a symmetric parallel-flow regenerator at the beginning of the cold period at periodic steady state. (Hausen [7].)

It is clear, therefore, that it is advantageous to operate a regenerator in counterflow mode if this is at all feasible, which seems, in practice, to be the case. Where parallel flow is preferred from constructional or operational considerations, great care must be exercised in matching the flow rates and regenerator physical size, on the one hand, and cycle time, on the other, if thermal effectiveness is to be maximized. It is just not the case for parallel-flow operation of regenerators that a simple reduction of cycle time, if possible, will improve regenerator performance as is the case for counterflow operation. Quite the reverse can occur in some circumstances for parallel flow!

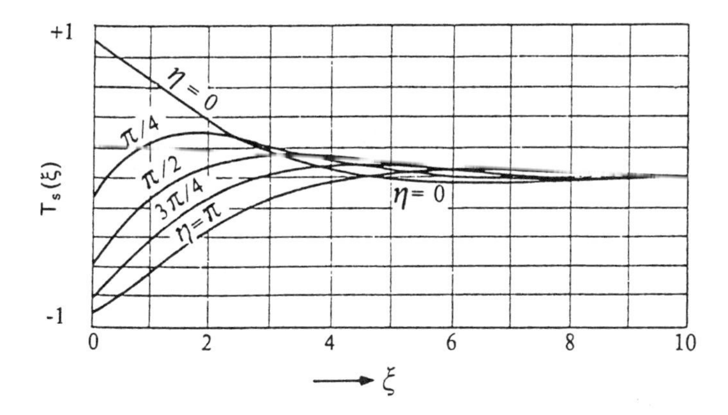

Figure 10.4: Spatial variation of the temperature of the packing of a symmetric parallel-flow regenerator during the cold period at periodic steady state for $\Pi = \pi$. (Hausen [7].)

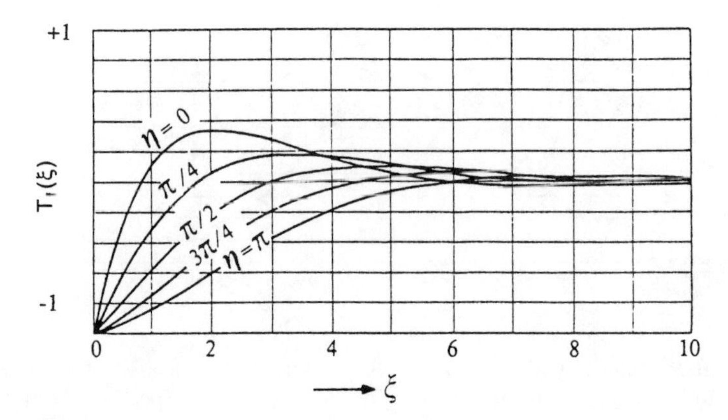

Figure 10.5: Spatial variation of the temperature of the gas of a symmetric parallel-flow regenerator during the cold period at periodic steady state for $\Pi = \pi$. (Hausen [7].)

References

1. H. Hausen, *Warmeubertragung in Gegenstrom, Gleichstrom und Kreuzstrom*, Springer-Verlag, Berlin (1950).
2. A. Kardas, "On a Problem in the Theory of the Unidirectional Regenerator," *Int. J. Heat Mass Transfer* **19**, 567 (1966).
3. F. W. Schmidt, A. J. Willmott, *Thermal Energy Storage and Regeneration*, McGraw-Hill, New York (1981).
4. M. Kumar, "Periodic Response of a Parallel Flow, Solid Sensible Heat Thermal Storage Unit," M.S. thesis, Pennsylvania State University, USA (1978).
5. A. J. Willmott, "Digital Computer Simulation of a Thermal Regenerator," *Int. J. Heat Mass Transfer* **7**, 1291–1302 (May 1964).
6. C. E. Iliffe, "Thermal Analysis of the Contra-flow Regenerative Heat Exchanger," *Proc. Inst. Mech. Eng.* **159**, 363–372 (1948).
7. H. Hausen, *Heat Transfer in Counterflow, Parallel Flow and Crossflow*, English translation (1983) edited by A. J. Willmott, McGraw-Hill, New York (originally published 1976).
8. A. N. Nahavandi, A. S. Weinstein, "A Solution to the Periodic-Flow Regenerative Heat Exchanger Problem," *Appl. Sci. Res.* **10**, 335–348 (1961).

INDEX